Springer

Berlin
Heidelberg
New York
Barcelona
Budapest
Hong Kong
London
Milan
Paris
Santa Clara
Singapore
Tokyo

Environmental Science

Series Editors: R. Allan U. Förstner W. Salomons

W.Calmano, U.Förstner (Eds.)

Sediments and Toxic Substances

Environmental Effects and Ecotoxity

With 106 Figures and 50 tables

 Springer

Volume Editors

PROF. DR. W. CALMANO

Hamburg-Harburg University of Technology
Section of Environmental
Science and Technology
Eissendorfer Straße 40
21073 Hamburg
Germany

PROF. DR. U. FÖRSTNER

Hamburg-Harburg University of Technology
Section of Environmental
Science and Technology
Eissendorfer Str. 40
21073 Hamburg
Germany

Cataloging-in-Publication Data applied for

Die Deutsche Bibliothek - CIP-Einheitsaufnahme

Sediments and toxic substances : environmental effects and ecotoxity ; with 50 tables / W. Calmano :
U. Förstner (ed.). - Berlin ; Heidelberg ; New York ; Barcelona ; Budapest ; Hong Kong ; London ;
Milan ; Paris ; Santa Clara ; Singapur ; Tokyo : Springer, 1996

(Environmental science)

NE: Calmano, Wolfgang [Hrsg.]

ISBN-13: 978-3-642-79892-4 e-ISBN-13: 978-3-642-79890-0
DOI: 10.1007/978-3-642-79890-0

© Springer-Verlag Berlin Heidelberg 1996
Softcover reprint of the hardcover 1st edition 1996

Media conversion: Anne Marie de Grosbois
Typesetting: Camera ready by authors/editors
SPIN: 10497704 32/3136 – 5 4 3 2 1 0 Printed on acid free paper

Preface

In modern sediment research on contaminants five aspects are discussed which, in an overlapping succession, also reflect development of knowledge on particle-associated pollutants during the past twenty-five years: (1) identification of sources and their distribution; (2) evaluation of solid/solution relations; (3) study of transfer mechanisms to biological systems; (4) assessment of environment impact; and (5) selection and further development of remedial measures, in particular, of dredged materials. Scientific research and practical development are still expanding in all these individual aspects.

Similar to other waste materials, management of contaminated sediments requires a holistic approach. This means that assessment of biogeochemical reactions, interfacial processes and transfer mechanisms as well as the prognosis of long-term borderline conditions, in particular of capacity-controlling properties, should be an integrated part of the wider management scheme, i.e., the analytical and experimental parameters should always be related to potential remediation options for a specific sediment problem.

The underlying coordinated project, which was funded by the German Federal Ministry for Science and Technology (now the Federal Ministry for Education, Science, Research and Technology) provided excellent opportunities for multidisciplinary effort, bringing together biologists, chemists, engineers, geologists and other researchers. During its active phase, the group attracted much interest nationally and internationally. The group members highly appreciate the manifold contacts and invitations during the past five years.

As a result of this encouragement and in view of the widely increased interest in the present subject, it was decided that the outcome of our joint project, together with additional contributions to important aspects, should be published in an English version. In particular we would like to thank Dr. Jochen Gruber and Prof. Wim Salomons for their review of the articles as well as for invaluable suggestions to the content and structure of the present book.

Hamburg, November 1995 Wolfgang Calmano and Ulrich Förstner

Table of Contents

List of Contributors

Abbt-Braun, Dr. Gudrun
Engler-Bunte-Institute
Department of Waterchemistry
University of Karlsruhe
Richard-Willstätter-Allee 5
D-76131 Karlsruhe
Germany

Adamczyk, Dr. Z.
Institute of Catalysis and
Surface Chemistry
al. Niezapominajek
PL-30-239 Krakow
Poland

Ahlf, Dr. Wolfgang
Hamburg-Harburg University
of Technology
Section of Environmental
Science and Technology
Eissendorfer Str. 40
D-21073 Hamburg
Germany

Belouschek, Prof. Dr. P.
Institute of Environmental
Research Schlieben e.V.
Im Teelbruch 61
D-45219 Essen
Germany

Calmano, Prof. Dr. Wolfgang
Hamburg-Harburg University
of Technology
Section of Environmental
Science and Technology
Eissendorfer Str. 40
D-21073 Hamburg
Germany

Ebinghaus, Dr. Ralf
GKSS Forschungszentrum
Geesthacht GmbH
Institut für Physikalische und
Chemische Analytik
Max-Planck-Str. 1
D-21502 Geesthacht
Germany

Flemming, Dr. Hans-Curt
Lehrstuhl für Wassergüte und
Abfallwirtschaft der Technische
Universität München
Am Coulombwall
D-85748 Garching
Germany

Förstner, Prof. Dr. Ulrich
Hamburg-Harburg University
of Technology
Section of Environmental
Science and Technology
Eissendorfer Str. 40
D-21073 Hamburg
Germany

Frimmel, Prof. Dr. Fritz H.
Engler-Bunte-Institute
Department of Waterchemistry
University of Karlsruhe
Richard-Willstätter-Allee 5
D-76131 Karlsruhe
Germany

Gorny, Dr. B.
 Institute of Environmental
 Research Schlieben e.V.
 Im Teelbruch 61
 D-45219 Essen
 Germany

Gramm, Helmuth
 GKSS Research Centre
 Institute of Chemistry
 Max-Planck-Str.
 D-21502 Geesthacht
 Germany

Gruber, Dr. Joachim
 University of Technology
 Hamburg-Harburg
 Section of Environmental
 Science and Technology
 Eissendorfer Str. 40
 D-21073 Hamburg
 Germany

Hansen, Prof. Dr. Peter-Diedrich
 Berlin University of
 Technology, FB 7
 Institute for Ecological
 Research and Technology
 Department of Ecotoxicology
 Keplerstr. 4–6
 D-10589 Berlin
 Germany

Hof, K.
 Institute of Environmental
 Research Schlieben e.V.
 Im Teelbruch 61
 D-45219 Essen
 Germany

Jantzen, Dr. Eckard
 GKSS Research Centre
 Institute of Physical and
 Analytical Chemistry
 Max-Planck-Str. 1
 D-21502 Geesthacht
 Germany

Knauth, Prof. Dr. Hans-Dieter
 GKSS Forschungszentrum
 Institut für Gewässerphysik
 Max-Planck-Str.
 D-21502 Geesthacht
 Germany

Kuballa, Jürgen
 GKSS Research Centre
 Institute of Physical and
 Analytical Chemistry
 Max-Planck-Str. 1
 D-21502 Geesthacht
 Germany

Li, Pinglin
 Fujian Provincial Institute for
 Science of Environmental
 Protection
 Hualin Road
 Fuzhou, Fujian
 P.R. China

Marshall, Prof. Dr. Kevin C.
 24 Calypso Ave.
 Mosman NSW 2088
 Australia

Neitzel, Dr. V.
 Ruhrverband, Essen
 Kronprinzenstr. 37
 D-45128 Essen
 Germany

Nowicki, S.
 Institute of Environmental
 Research Schlieben e.V.
 Im Teelbruch 61
 D-45219 Essen
 Germany

Petersen, Dr. Wilhelm
 GKSS Forschungszentrum
 Institut für Gewässerphysik
 Max-Planck-Str.
 D-21502 Geesthacht
 Germany

Reichardt, Dr. Wolfgang
 IRRI-International Rice
 Research Institute
 Soil and Water Science
 Division
 Research Centre Los Banos
 Laguna,
 P.O. Box 933
 1099 Manila
 Philippines

Reiners, Claudia
 GKSS Research Centre
 Institute of Chemistry
 Max-Planck-Str.
 21502 Geesthacht
 Germany

Schmitt, Jürgen
 Institut für Siedlungswasserbau
 Wassergüte und
 Abfallwirtschaft
 Universität Stuttgart
 Bandtäle 1
 D-70569 Stuttgart
 Germany

Schroeder, Dr. Friedhelm
 GKSS Forschungszentrum
 Institut für Gewässerphysik
 Max-Planck-Str.
 D-21502 Geesthacht
 Germany

Skiba, Dr. U.
 Institute of Environmental
 Research Schlieben e.V.
 Im Teelbruch 61
 D-45219 Essen
 Germany

Wallmann, Dr. Klaus
 GEOMAR
 Abteilung Marine
 Umweltgeologie
 Wischhofstr. 1–3
 D-24148 Kiel
 Germany

Wilken, Dr. Rolf-Dieter
 GKSS Research Centre
 Institute of Physical and
 Analytical Chemistry
 Max-Planck-Str. 1
 D-21502 Geesthacht
 Germany

1 Sediment Quality Assessment: Chemical and Biological Approaches

W. Calmano, W. Ahlf and U. Förstner

1.1 Introduction

Sediments are repositories for physical and biological debris and sinks for a wide variety of chemicals. The concern associated with the chemicals sorbed to sediments is that many commercial species and food-chain organisms spend a major portion of their life-cycle living in or on aquatic sediments. This provides a pathway for these chemicals to be consumed by higher aquatic life and wildlife, including avian species as well as humans. Direct transfer of chemicals from sediments to organisms is now considered to be a major route of exposure for many species. These issues are focusing attention on sediment contamination and highlight the fact that sediments are an important resource (Adams et al. 1992).

Because sediments are both carriers and potential sources of contaminants in aquatic systems, these materials may also affect groundwater quality and agricultural products when disposed on land. Contaminants are not necessarily fixed permanently in the sediments but may be remobilized when physico-chemical conditions change. Bioavailability of toxic chemicals and food-chain transfer may be strongly affected by such processes and by the type of chemical binding on the sediment particles. If the concentrations in the water phase are low, sediment bound contaminants may have greater impact on aquatic organisms than dissolved concentrations. The basic premise that has been used to protect water quality has been to restrict contaminants from occurring in water at concentrations above the known chronic or toxic levels. This approach should also applied for sediments, but it has to include the long-term behaviour of the contaminants which is determined by the physico-chemical and biochemical reactions in the system.

New objectives regarding the improvement of water quality as well as problems with the resuspension and land deposition of dredged materials require

standardized guidelines for sediment quality. Reasons for the development of sediment quality criteria in the last years have been (Di Toro et al. 1991; Förstner et al. 1994) that:

- In contrast to the strong temporal variability in the water phase, sediments integrate contaminant concentrations over time, although spatial variability in sediments is considerably greater than in the water column.
- Long-term perspectives in water management need "integrated strategies", in which sediment-associated pollutants have to be considered as well.
- Sediment criteria could provide a simple and cost-effective means of screening sediment measurements to identify areas of concern and could provide regulators with information in a short period of time.
- Sediment criteria could be readily incorporated into existing regulatory operations because a unique numerical sediment-specific criterion can be established for any chemical and compared to field measurements to assess the likelihood of significant adverse effects.
- Management plans have increasingly been based on the "assimilative capacity" of a distinct receiving system, and this requires knowledge of the properties of sedimentary components as the major sinks for toxic chemicals.
- There is the wide spectrum of problems with dredged sediments. Permissions for dredging activities and deposition of dredged material have to be based on standardized sampling protocols and test procedures.

First efforts have been undertaken by the US Environmental Protection Agency (US-EPA) to develop standard procedures and criteria for the assessment of environmental impact of sediment-associated pollutants. Initial review papers provided by consultants, JRB-Associates (1984) and Battelle (1985) indicated five approaches for closer consideration. Further discussions led to the differentiation of biological and chemical-numerical approaches (e.g. Chapman et al. 1987). Since this time, considerable technical progress has been made in assessment methods. This chapter is based on a recently published excellent report by Adams et al. (1992) who reviewed the current methods available to assess sediment quality. Table 1.1 gives an overview of the various chemical and biological methods.

With respect to chemical methods there is no immediate indication on biologi-

Table 1.1. Methods for assessing sediment quality

Chemical Methods	Integrated Chemical and Biological Methods	Biological Methods
Background approach	Equilibrium partitioning	Bulk sediment toxicity
Acid volatile sulphides	Apparent effects threshold	
Substrate composition	Interstitial water toxicity	Freshwater benthic
Long-term mobility	Spiked sediment toxicity	approach
	Tissue residue approach	
	Sediment quality triad	

cal effects. Their major advantages lie in their easy application and amendment to modelling approaches. Biological methods integrate sediment characteristics and pollutant loads but they do not generally indicate the cause of effects.

1.2 Chemical Methods

1.2.1 Background Approach

This method compares the actual data with sites comprising natural or insignificant pollutant concentrations. Particularly useful are samples from deeper layers of the sediment sequence at a given site, for example from drill holes, since this material is derived from the same catchment area and usually similar in its substrate composition. Nonetheless, standardization with respect to grain size distribution is indispensable.

In Germany a first quantitative measure of metal pollution in aquatic sediments has been proposed by Müller (1979) which is called the *Index of Geoaccumulation* (I_{geo}), i.e. the enrichment on geological substrates:

$$I_{geo} = \log2\ Cn/1.5 \cdot Bn \tag{1}$$

The I_{geo} compares the measured concentration of the elements in the fine grained sediment fraction (C_n) with geochemical background values (B_n) in fossil clay and silt sediments. The Index of Geoaccumulation consists of 7 grades, whereby the highest grade (6) reflects 100-fold enrichment above background values. Table 1.2 presents examples for the River Rhine catchment. Also shown is a

Table 1.2. Comparison of IAWR water quality indices (based on biochemical data) and Index of Geoaccumulation (I_{geo}; Müller 1979) of trace metals in sediments of River Rhine

IAWR Index	IAWR Water Quality (Pollution Intensity)	I_{geo}[a]	I_{geo} Class[a]	Metal Examples	
				Upper Rhine	Lower Rhine
4	very strong pollution	> 5	6		Cd
3-4	strong to very strong	> 4–5	5		
3	strongly polluted	> 3–4	4		Pb, Zn
2–3	moderate to strong	> 2–3	3	Cd, Pb	Hg
2	moderately polluted	> 1–2	2	Zn, Hg	Cu
1–2	unpolluted to moderate	> 0–1	1	Cu	Cr, Co
1	unpolluted	< 0	0	Cr, Co	

[a] Sediment Accumulation Index $I_{geo} = \log_2 C_n/1.5 \cdot B_n$

Table 1.3. Index of Geoaccumulation (I_{geo}; after Müller 1979) for sediments from five rivers in Germany

	Neckar R.	Main R.	Rhine R.	Elbe R.	Weser R.
Cu	0	1	2	2	0
Pb	1	2	2	2	2
Zn	1	2	3	4	2
Cd	3	2	6	4	4
Hg	0	1	4	6	1
Average	1.0	1.6	3.4	3.3	1.8

$I_{geo} = \log_2 C_n/1.5 \cdot B_n$

comparison of these sediment indices with the water quality classifications of the International Association of Waterworks in the Rhine catchment (IAWR).

Highest rates of accumulation – at different levels for the Upper and Lower Rhine sections – have been found for cadmium, lead, zinc and mercury, whereas copper, chromium and cobalt are less concentrated in these sediments. No further consideration is given to the ecological relevance of the values.

In Chapter 1, for the comparison of five major river systems in Germany, the I_{geo} has been used. Of these river system examples, the sediment sample from Lower Rhine River (collected from the deeper part of the sediment pile) exhibits highest overall metal accumulation, followed by the sample from the Elbe River. In both examples mercury and cadmium are typically influencing the average factor of enrichment The sediment samples from the rivers Weser, Main and Neckar are by far less contaminated, however, there is still a considerable enrichment of cadmium in sediments of both the Weser and Neckar Rivers.

The major deficiency of this method is that it does not take into account the biological effects of the various pollutants. This situation may partly be overcome by introducing a factor for the relative toxicity. Such a factor can point more clearly to the critical compounds and may therefore stimulate setting priorities for control and rehabilitation measures. In Table 1.4 a factor of toxicity has been used which is derived from the Microtox Test System. In the case of the extreme mer-

Table 1.4. Factor of enrichment x factor of toxicity (standardized Mikrotox values after Walker 1988)

	Neckar R.	Main R.	Rhine R.	Elbe R.	Weser R.
Cu (x 5)	7	10	17	22	6
Pb (x 1)	2	4	4	6	4
Zn (x 5)	10	15	35	70	30
Cd (x 10)	62	53	500	340	360
Hg (x 35)	46	98	805	2520	81
Total	127	180	1381	2958	481

Table 1.5. Elution index for selected river sediments as determined from exchangeable proportions (1 N ammonium acetate). Calculated relative to background data from Elbe River sediments, these values are multiplied by a factor 100

	Neckar R.	Main R.	Rhine R.	Elbe R.	Weser R.
Cu	0.2	-	1	1	-
Pb	1	1	2	1	1
Zn	7	9	28	36	9
Cd	30	30	230	30	-
Total oxic	38	40	261	68	10
(Anoxic	0.5	0.3	8	> 4	4)

cury pollution of the Elbe River sediments, reduction of a specific source – emissions from chlorine alkali plants – can contribute to the improvement of the overall sediment quality. With available technology and reasonable costs, the concentration of mercury in the Elbe River could be reduced to such an extent that the sediment quality would be at least comparable to the sample from the Weser River.

A further example refers to the elution properties of sediment samples. In Table 1.5 an *elution index* is based on the metal concentrations which are extractable with 1 N ammonium acetate at pH 7. This fraction should be mobilized under more saline conditions, e.g. in the estuarine mixing zone. Comparison of the release rates from oxic and anoxic sediments clearly indicates that oxidation of sediments affects a very significant increase of the metal mobility. This effect is particularly important for cadmium.

At the river mouths along the southern coast of the North Sea approximately 20 million m^3 of sediments have to be dredged from Rhine/Meuse (Rotterdam harbour) and ~10 million m^3 from the rivers Scheldt (Antwerp), Weser (Bremerhaven) and Elbe (Hamburg). The total quantity of sediment to be dredged in The Netherlands, Belgium and Germany amounts to about 12 times the total suspended matter supply from the Rhine. The disposal possibilities for these enormous quantities are severely limited because of the pollutants present in the dredged material.

In Germany the critical area, for which "orientation values" for priority pollutants have been designed, is the Elbe River estuary. During the last 10 years the harbour authorities used – in addition to land disposal – the method of "sludge harrowing" for resuspension of the sediments deposited in harbour basins and shipping channels. During ebb stream the suspended particles should be exported to the North Sea. This effect can be seen from recent data (Förstner 1990). In the upper reaches of the Elbe estuary mercury, cadmium and zinc are enriched by factors of 10 to 100 compared to natural background concentrations. Meanwhile, however, there is a significant increase of mercury and cadmium in the outer estuary close to the North Sea.

Table 1.6. Proposal for reference values of pollutant concentrations on the basis of a ten-fold enrichment (exception: Cd and Hg: five-fold enrichment) compared to background values of Elbe River sediments (fraction < 63 μm). Concentration values given in mg/kg

	Reference Rotterdam Harbour		Background Concentration		Proposal for LAWA
	Target value	Standard value	Elbe R. sediments		Reference values [a]
Hg	0.3	1	0.2	– 0.4	1
Cd	0.8	4	0.3	– 0.5	2
As	25	40	3	– 5	40
Pb	50	125	25	– 30	250
Cu	25	70	20	– 30	250
Zn	180	750	90	– 110	1000
PCB [b]	1	10			0.01
PAH [b]	50	500			0.5

[a] Förstner (1990)
[b] for each individual PCB or PAH

We have proposed relatively simple reference values for decisions on active re-suspension. A ten-fold enrichment was considered as an undisputed significant pollution effect, and sediments above this limit should not be resuspended. For cadmium and mercury, both elements from the blacklist, a five-fold increase in the sediment was considered as a critical value. Our proposal (Förstner 1990) is similar to the Dutch sediment draft (Van Veen and Stortelder 1988) designed for the management of Rotterdam harbour sludge, subsequent to the filling of the artificial peninsula by Rotterdam (see Table 1.6).

1.2.2 Acid Volatile Sulphides (Metals)

The total metal concentrations in sediments are not good estimations of the "free" or bioavailable fraction. Different sediments can vary in toxicity by a factor of 10 or more for the same total metal concentration. To use toxicity estimates based on chemical measurements, there needs to be a way to estimate the bioavailable fraction of the total amount present. A number of approaches to determine metal bioavailability associated with sediments have been reviewed or tried, including carbon normalization and sorption of metals in oxic freshwater sediments to particulate carbon and the oxides of iron and manganese (Jenne et al. 1986; Jenne 1987; Tessier et al. 1985)

Recently the dominant role of the sediment sulphides in controlling metal bioavailability has been demonstrated (Ankley et al. 1991; Carlson 1991; Di Toro et al. 1990; Di Toro et al. 1992). Sulphides are common in many freshwater and marine sediments and are the predominant form of sulphur in anaerobic sediments. The ability of sulphides to complex with metal ions is well known. This accounts

for the lack of toxicity when even high metal concentrations are present in some sediments. The solubilities of metal sulphide complexes are well below the toxic threshold. It has been shown that the solid-phase sediment sulphides that are soluble in cold acid, termed *acid volatile sulphides* (AVS), are the key factor for controlling the toxicity of cadmium and nickel and potentially several other heavy metals (Di Toro et al. 1990; Di Toro et al. 1992). No toxicity is observed from these metals when bound to sediments and when, on a molar basis, the concentration of AVS is greater than the sum of the molar concentrations of the sulphide-binding metals. When the ratio of the AVS to metal concentration drops below 1, toxicity begins to appear.

The advantages of this method are the same as those of the equilibrium partitioning approach for non-ionic organics (see Sect. 1.3.1) because AVS can be used to normalize sediment metal concentrations in the same way that sediment organic carbon is used to normalize non-ionic chemicals. If this normalization would be valid the analytical efforts could be reduced considerably. However, there is increasing evidence that metal availability cannot generally be deduced by this parameter. The normalization of metal concentrations by AVS would be invalid if the sediment AVS content is very low. This could occur in fully oxidized sediments. Most sediments have at least a small oxic zone near the sediment–water interface. The importance of this zone, relative to AVS normalization, has not been evaluated (Adams et al. 1992). Aquatic plants have adapted to anoxic sediments and by oxidizing the rhizosphere they make both essential elements and toxic metals more bioavailable. Sooner or later resuspension of sediments, particularly in rivers, will induce oxygenation, and by this process remobilization of metals will occur.

1.2.3 Substrate Composition (Metals)[1]

Criteria for the prognosis of the middle- and long-term behaviour of metals should include the abilities of sediment matrices for producing acidity and for neutralizing such acid constituents. Important acid-producing oxidation reactions leading to a decrease of pH in low buffered sediments are listed in Table 1.7.

Experimental approaches for calculating acid producing capacity (APC) and acid consuming capacity (ACC) for sulphidic mining residues have been summarized by Ferguson and Erickson (1988). A test described by Sobek et al. (1978) involves the analysis of total pyritic sulphur. Potential acidity is then subtracted from neutralizing potential which can be obtained by adding a known amount of

[1] This text (Sects. 1.2.3 and 1.2.4) is a modified extract from a section on implications on sediment quality criteria assessment of the contribution "Non-Linear Release of Metals from Aquatic Sediments" to the book of W. Salomons and W.M. Stigliani *Biogeodynamics of Pollutants in Soils and Sediments* in this series.

Table 1.7. Important acid-producing oxidation reactions in aquatic systems (Calmano et al. 1993)

Elements		Reactions	f[a]
Inorganic	S	$H_2S + 2 O_2 = SO_4^{2-} + 2 H^+$	2
	S	$S^- + 3/2 O_2 + H_2O = SO_4^{2-} + 2 H^+$	2
	S, Fe	$FeS + 9/4 O_2 + 3/2 H_2O = FeOOH + SO_4^{2-} + 2 H^+$	2
	S, Fe	$FeS_2 + 15/4 O_2 + 5/2 H_2O = FeOOH + 2 SO_4^{2-} + 4 H^+$	4
	Fe	$Fe^{2+} + 1/4 O_2 + 5/2 H_2O = Fe(OH)_3 + 2 H^+$	2
	N	$NH_4^+ + 2 O_2 = NO_3^- + H_2O + 2 H^+$	2
	N	$NO_x + 1/4(5-2x)O_2 + 1/2 H_2O = NO_3^- + H^+$	1
Organic	N	$R-NH_2 + 2 O_2 = R-OH + NO_3^- + H^+$	1
	S	$R-SH + H_2O + 2 O_2 = R-OH + SO_4^{2-} + 2 H^+$	2

[a] acid formation coefficient

HCl, heating the sample and titrating with standardized NaOH to pH 7. Bruynesteyn and Hackl (1984) calculated APC from total sulphur analysis; here, APC was then subtracted from ACC, obtained by titration with standardized sulphuric acid to pH 3.5. The APC-relationships of sediments are more complex than that in sulphidic ores because the APC from organic matter must be considered; time-scale plays a major role in sediments.

The acid neutralization capacity of an aquatic system is composed of the acid-consuming capacity of the solids ACC(s) and that of the dissolved phase ACC(aq):

$$ACC = ACC(s) + ACC(aq) \qquad (2)$$

The ACC(aq) of the dissolved phase – in aquatic systems it is always much smaller than ACC(s) – may be described by the following Eq. (3):

$$ACC(aq) = [HCO_3^-] + 2 [CO_3^{2-}] + 2 [S^{2-}] + [HS^-] + [NH_3] - [H^+] \qquad (3)$$

A similar rule shall apply to the ACC(s) of the solids at pH > 5:

$$ACC(s) = 2[CaO] + 2[MgO] + 2[Na_2O] + 2[K_2O] + 2[MnO] + 2[FeO]$$
$$- 2[SO_3] - 2[P_2O_5] - HCl] \qquad (4)$$

From the two parameters ACC and APC an effective acid producing capacity (APC_{eff}) can be calculated. APC_{eff} of a sediment suspension may be defined as:

$$APC_{eff} = V/W \; ([H^+]_e - [H^+]_o) \qquad\qquad (5)$$

where APC_{eff} = effective acid producing capacity
 V = suspension volume
 W = solid mass
 $[H^+]_{o,e}$ = $[H^+]$ before and after oxidation of the suspension.

For example, for an anoxic Elbe River sediment with an original $pH_o = 7.03$ and a $pH_e = 3.29$ after an oxidation time of 35 days, an $APC_{eff} = 5.12$ mmol kg^{-1} could be calculated.

The most efficient fixation process within anoxic sediments for trace metals is the formation of free sulphide during anaerobic organic matter degradation and reduction of sulphate. Therefore, study of heavy metal associations with sulphide and carbonate formations in anoxic sediments provides insight into early diagenetic processes (Berner 1981, Morse and Mackenzie 1990).

Whereas the ability of the sediment to produce free sulphide is determined by the sulphate reduction rates, the ability to remove all produced free HS$^-$ is given by the reactive metal – predominantly reducible Fe^{3+} concentrations – available to form sulphide minerals (*Available Sulphide Capacity* [ASC]; Williamson and Bella 1980). Simultaneous application of standard sequential leaching techniques (BCR-Version; Ure et al. 1993) on critical trace elements and matrix components can be used for geochemical characterization of anoxic, sulphide-bearing sediments in relation to the potential mobility of critical trace metals (Kersten and Förstner 1991):

Step 1	Exchangeable ions	0.5 g sediment – 5 h	Ca^{2+} and Fe^{2+}
	Carbonates	– 20 ml of acetic acid 0.11 mol l^{-1}	from carbonates and phosphates
Step 2	Reducible metals	16 h – 20 ml NH$_2$OH·HCl 0.1 mol l^{-1}, HNO$_3$ pH 2	Fe^{3+} from reactive oxides
Step 3	Sulphidic and organic bound metals	10 ml hydrogen peroxide 8.8 mol l^{-1}(twice) - extr. 50 ml NH$_4$Ac/HOAc pH 5	S from easily oxidizable sulphides (FeS)

For determining the acid producing capacity (*maximum APC*) in anoxic sediments, both the FeS pool (*actual APC*) and the maximum ferrous sulphide (worst case: pyrite)-producing capacity upon disposal has to be taken into consideration. The latter is given by the *sulphide binding capacity* (ASC, see preceding), which may be determined by using the Fe-concentrations in step 2 of the sequential leaching results (see Table 1.8).

The stoichiometry of the oxidizable S- and Fe-fractions of step 3 indicates that the ferrous sulphide extracted in this step was in the FeS form. As shown in Table 1.8, the sum of both the ASC and the *actual APC* gives the *maximum APC* for the samples. The ACC value for these samples was more simply determined from

Table 1.8. Balance between APC and ACC values for an anoxic sediment sample from Hamburg Harbour (Kersten and Förstner 1991)

Compound	Function	Value	Parameter (mmol/kg)	
Ca in step 1	base potential	(x 0.5 =)	ACC	90
Fe in step 2	ASC	(335)		
S in step 3	sulphide sulphur	(+ 85 =)	max. APC	420
Balance			ACC - APC=	-330

the Ca-concentration released from reactive carbonates by the Na-acetate solution of step 1, which has to be multiplied by a factor of 0.5 to account for the stoichiometric ratio between pyrite and calcite within the redox reaction:

$$FeS_2 + 2\ CaCO_3 + 15/4\ O_2 + 1/2\ H_2O = FeOOH + 2\ SO_4^{2-} + 2\ Ca^{2+} + 2\ CO_2$$

The negative balance between the ACC and APC indicates, that the mud sample from Hamburg harbour has a significant acidification potential.

1.2.4 Assessing Long-term Mobility (Metals)

With regard to prediction of long-term effects of sediment-bound metals, chemical extraction procedures are of limited value because they usually involve neither reaction-mechanistic nor kinetic considerations. This lack can be avoided, e.g. by an experimental approach, originally used by Patrick et al. (1973) and Herms and Brümmer (1978), where sediments can be treated in a circulation system under controlled intensification of significant release parameters such as pH-value, redox-potential and temperature. Our experimental design (Schoer and Förstner 1987) includes an ion-exchange system for extracting and analysing the released metals at an adequate frequency, and compares sequential extraction results before and after treatment of the sample in the circulation apparatus. Individual metal species are released at different time intervals. Taking into account both element contents released during the 10-week-experiments (equivalent to several thousand years of solid/water interaction) and those extrapolated from extraction "pools", concentrations can be calculated for different scenarios.

While these extrapolations have been made from pH 5 conditions, titration curves from investigations on a wide spectrum of metal-bearing waste materials (Obermann and Cremer 1992) suggest, that pH 4 may be more appropriate for long-term predictions of potential metal releases from contaminated sediments. In this pragmatic approach, which is outlined in Fig. 1.1, the pH is automatically adjusted to 4 during the time period of 24 h. Apart from the release rates of metals,

which can be determined from samples taken at different time intervals, the sum curve (diagram of discrete acid additions would be valuable!) of acid consumption provides information on the potential changes of the matrix composition during acidification and the availability of buffer capacity at different time-scales.

The acid consumption curve in Fig. 1.1 reflects slow long-term metal release from sediments. Because calcite dissolution was fast, the acid consumption in the first stage increased drastically within a short period of time. Cation release and alumosilicate dissolution were dominant factors consuming acid in the later stage. The reactions can be treated as:

$$SO=Me + 2\,H^+ = SO=H_2 + Me^{2+}, \text{ and} \tag{6}$$

$$Al_2O_3 + 6\,H^+ = 2\,Al^{3+} + 3\,H_2O \tag{7}$$

where SO= and Me are reaction groups in solid phase and metal ion, respectively. The reactions of hydrogen ion with sorbed metal and alumosilicate are delayed due to the complex sediment structure and matrix. For example, special penetration of hydrogen ion is required for reacting with cations on clay minerals coated with organic matter or biofilms. Rates of reactions can be estimated by measurements of metal concentrations in solution.

Regarding the potential release of metals from contaminated sediments, changes of redox and pH conditions are of prime importance. In practice, therefore, characterization of sediment substrates with respect to their buffer capacity is

Fig. 1.1. Patterns of a suspension titration (Obermann and Cremer 1992). The initial phase, during which the target-pH-Value is not attained, is followed by a 50-fold longer main phase with stable pH-conditions

an indispensable step for the prognosis of middle- and long-term processes of metal mobilization.

(1) Description of Basic Processes. Experimental studies in the multi-chamber device and reactor cell provided information on the metal transfer between different sedimentary matrices and on interrelations between mobilization and readsorption mechanisms (Calmano et al. 1988) Further studies should focus on the role of iron sulphides in the large-scale changes in metal species, i.e. solid/solution interactions and bioavailability of trace metals in aquatic systems. The combination of matrix and trace element data in sediments, extracted simultaneously with different reagents, can provide additional information, which can be used to establish models on chemical interactions occurring under characteristic borderline conditions, with respect to sulphide, chloride and complexing agents (Wallmann et al. 1993).

Reliability of long-term prognoses on the impact of metal-bearing sediments is restricted due to the dominance of non-linear and delayed processes in redox-controlled systems. While measurements of both actual acid producing and consuming capacities can be undertaken with available methods (Sect. 1.2.3), a possible "split" or translocation of their inventories between solid and dissolved phases renders such parameters difficult to be predicted in a long-term perspective. It seems, that calculation of maximum acid producing capacity (APC) involving "available sulphide capacity", i.e. reducible Fe(III)-concentration, gives a relative safe basis for such considerations, because iron generally is less mobile than some sulphur species. From data on long-term acid producing potential of anoxic sediments, appropriate management techniques for dredged materials can be derived.

(2) Definition of Parameters. The pragmatic pH_{stat}-approach for evaluating the behaviour of solid waste subsequent to landfilling involves the "spot test" of lowering the pH value and – in addition – provides data on the buffer capacity of the solid matrix. Direct assessment of the acid producing capacity can be performed by ventilation with air or oxygen; our data, however, have shown that these experiments should be designed for relative long time periods, i.e. for several weeks duration.

Although the three parameters forming the framework of future criteria for assessing long term effects from metal bearing sediments acid producing potential, acid consuming capacity and metal concentration – are closely interrelated, regulatory activities will focus on the concentration term, mainly because of its priority function as effect parameter in biological systems. In this respect, however, the long-term prognostic approach is different from the assessment of the short-term impact by metal-contaminated sediments, which is preferentially done by applying bioassays on sediment pore water and elutriates. Long-term prediction, i.e. extrapolations of experimental or field data to a hypothetical situation, necessarily implies some type of numerical method. From the available approaches, the group of elution techniques generally seems to be best suited for this

purpose; within this group, long-term predictive capacity is predominantly reflected by the more rigid extractants, in particular acid solutions.

Guideline values based on acid extraction data have been proposed for soil and solid waste materials. The most common pH level of about 4 was primarily aimed to simulate conditions of acid precipitation; however, as has been shown from sediment and solid waste data, internal processes such as oxidation of sulphides can by far be more effective in the lowering of pH.

(3) Criteria for Dredging and Up-Land Deposition of Sediments. In dredged material management, two different target areas for combined matrix/metal criteria can be distinguished: (1) sediment resuspension and (2) dredged material disposal. With respect to "resuspension of aquatic sediments", which involves more short-term effects than the disposal of dredged material, special emphasis should be posed on the factor "available metal species". Within certain categories of acid producing and consuming capacities, guideline values for individual metals should be based on elutriate data, preferentially at pH 4, for better comparison with other solid matrices (e.g. Swiss Ordinance for Waste Materials, which includes two categories of limit values: *Inert-/Construction Materials* and *Residue Material Deposits*, the latter mainly comprising pretreated products of municipal solid waste incineration; Anonymous 1990).

Environmental impact of sediment deposits is influenced by the internal chemical conditions rather than by the concentration and extractability of metals. Priority, therefore, should be given to the optimization of long-term chemical stability ("geochemical engineering").

Above discussions have shown that potential mobility, bioavailability and toxicity of metals in sediments strongly depend on redox conditions. From earlier chemical extraction procedures, as proposed by Tessier et al. (1979) to more recent SEM/AVS ratio determinations (Di Toro et al. 1990, 1992), metal speciation is a major characteristic describing its potential mobility in sedimentary environments. The inherent link between the parameters to describe "metal speciation" is the characterization of chemical stability and bioavailability of metal compounds in sediments under certain redox conditions.

In a sulphidic anoxic environment, even of the sediment is heavily polluted by metals, organisms are considered to be still safe due to the strong fixation of metal ions by S^{2-} or HS^- as source of acid volatile sulphide. Such heavily polluted sediment, however, can behave as a time bomb which is triggered by only one factor – redox increase to a critical point, i.e. by exposure to oxygen-rich overlying water or directly to air. Once this situation occurs (a possible pathway also is oxygen transfer via plant roots), toxic metals in the sediments will be released to the water phase or transformed into more bioavailable species (Calmano et al. 1992).

At the moment, research on long-term effects of redox variations on metal behaviour in sediments is mostly based on thermodynamic considerations. Future research should emphasize studies on the kinetics of metal species transformations, hydrogen ion production and metal release as affected by changing redox

conditions. Additional important aspects involve the bridging of the gaps between numerical criteria approaches, as reflected by matrix composition and metal mobility and biological approaches. Assessment for bioavailability of metal species is still based on results of water phase and elutriate, mostly using acute toxicity data (Ankley et al. 1991; Ahlf et al. 1992) Initial data on solid phase bioassays suggest good perspectives for future applications in sediment quality assessment. It may well be, that for such systems, which are much less disturbed than artificial sediment elutriates, relationships between matrix conditions – as reflected, e.g. by redox indices – and metal species bioavailability may be found, which may serve as a more solid basis for the interpretation of results from bioassays, eventually with respect to chronic toxicity.

1.3 Integrated Chemical and Biological Methods

1.3.1 Equilibrium Partitioning Approach

This approach is based on the theory that contaminants sorbed to sediments achieve thermodynamic equilibrium over appropriate periods of time between sediment and sediment pore water concentrations. At equilibrium, the mass of the chemical present in either phase can be estimated by measuring the mass present in the other phase. For various organic chemicals the sediment pore water concentration and the bulk sediment concentration are related by the carbon normalized sediment–water partition coefficients.

The US-EPA is currently proposing the equilibrium partitioning approach for deriving sediment quality criteria for non-ionic organic chemicals (Adams et al. 1992). This approach has been chosen because it addresses the two principal technical issues that must be resolved: the varying bioavailability of chemicals in sediments and the choice of the appropriate biological-*effects concentration* (concentrations of chemicals above which biological effects are expected). It has been found that if different sediments are compared, there is essentially no relationship between sediment chemical concentrations on a dry weight basis and biological effects. However, if the chemical concentrations in the pore water of the sediment are used (for chemicals that are not highly hydrophobic) or the sediment chemical concentrations on an organic carbon basis, then the biological effects occur at similar concentrations – within a factor of two – for the different sediments. In addition, the effects concentrations are the same as, or can be predicted from, the effects concentrations determined in water-only exposures (see excellent review by Di Toro et al. 1991).

Because the methodology presumes that partitioning of the chemicals between sediment and pore water is at equilibrium, as shown in Fig. 1.2, it is assumed that the organisms receive an equivalent exposure from a water-only exposure or from

Fig. 1.2. Organism exposure routes via sediments and pore water (after Di Toro et al. 1991)

any equilibrated phase; either from pore water via respiration, or from sediment carbon via ingestion, or from a mixture of the routes. Thus the pathway of exposure is not significant.

Sediment quality criteria for non-ionic organic chemicals are based on the chemical concentration in sediment organic carbon. For highly hydrophobic chemicals this criteria is necessary since the pore water concentrations give poor estimates of the chemical activities for those compounds. The pore water concentration is the sum of the free chemical concentration, which is bioavailable and represents the chemical activity, and the concentration of chemical complexed to dissolved organic carbon which is not bioavailable. Using the chemical concentration in sediment organic carbon eliminates this ambiguity.

Sediment quality criteria (SQC) also require that a chemical concentration be chosen that is sufficiently protective of benthic organisms. The *final chronic value* (FCV) from the US-EPA water quality criteria is proposed. An analysis of the data compiled in the water quality criteria documents, demonstrates that benthic species, defined as either epibenthic or infaunal species, have a similar sensivity to water column species. This is the case if the most sensitive species are compared and if all species are compared. The results of benthic colonization experiments also support the use of the FCV.

Equilibrium partitioning cannot remove all the variation in the experimentally "observed sediment-effects concentration" and the concentration predicted from the "water-only exposures". A variation of approximately a factor of 2 or 3 remains. Hence, it is recognized that a quantification of this uncertainty should accompany the sediment quality criteria.

The calculation procedure for establishing SQC is as follows (Di Toro et al. 1991). If FCV is the final chronic water quality criterion (WQC) for a given chemical, then the SQC ($\mu g/kg$ sediment) are computed using the partition coefficient K_p (l/kg sediment) between sediment and pore water:

$$SQC = K_p \cdot FCV \tag{8}$$

The partitioning of a non-ionic organic chemical among soil and sediment particles is reasonably well understood, and a standard model exists for describing the process. The hydrophobicity of the chemical is quantified by using the octanol/water partition coefficient, K_{ow}. The sorption capacity of the sediment is determined by the mass fraction of organic carbon for the sediment, f_{oc}. For sediments with $f_{oc} > 0.2\%$ by weight, the organic carbon appears to be the predominant phase for chemical sorption. The partition coefficient, K_p (the ratio of chemical concentration in the sediment, C_s, to chemical concentration in pore water, C_d) is given by

$$K_p = \frac{C_s}{C_d} = f_{oc}K_{oc} \tag{9}$$

where K_{oc} is the partition coefficient for sediment organic carbon.

The other environmental variable that has a dramatic effect on partitioning appears to be the particle concentration in the suspension in which K_p is measured. There is a considerable controversy regarding the mechanism responsible for the particle concentration effect and a number of explanations have been offered. However, all the interpretations yield the same result for sediment–pore water partitioning, namely that $K_{oc} \approx K_{ow}$ for sediments.

Using Eqs. (8) and (9), a SQC is calculated from

$$SQC = f_{oc}\, K_{oc}\, FCV \tag{10}$$

This equation is linear in the organic carbon fraction, f_{oc}. As a consequence, the relationship can be expressed as

$$\frac{SQC}{f_{oc}} = K_{oc}FCV \tag{11}$$

If we define

$$SQC_{oc} = \frac{SQC}{f_{oc}} \tag{12}$$

as the organic carbon-normalized SQC concentration (μg chemical/kg organic carbon), then

$$SQC_{oc} = K_{oc}FCV \tag{13}$$

This leads to the following important conclusion: for a specific chemical having a specific K_{oc}, the organic carbon-normalized sediment concentration, SQC_{oc}, is independent of sediment properties.

Therefore, it is expected (Di Toro et al. 1991) that toxicity of non-ionic organic chemicals in sediments can be predicted from the "water-only effects concentra-

tion" and the K_{oc} of the chemical. In this regard the concept of sediment toxic units is useful. These are computed as the ratio of the organic carbon-normalized sediment concentrations (C_s/f_{oc}) and the predicted sediment LC_{50} (the concentration causing 50% mortality among the test organisms) using K_{oc} and the water-only LC_{50}. That is,

$$\text{predicted sediment toxic unit} = \frac{C_s/f_{oc}}{K_{oc}(\text{water - only } LC_{50})} \tag{14}$$

The principal advantages of the *equilibrium partitioning approach* are, that it allows for the derivation of chemical-specific sediment values from WQC and accounts for differences in the bioavailability of non-ionic chemicals sorbed to sediments from different sites. Major limitations of the method are that there are relatively few WQC available, the method is only for non-ionic chemicals and limited field validation of the method has occurred. Also limitations to be considered are: the effects of sample preparation, e.g. the drying procedures and the separation techniques, such as filtration or centrifugation (Calmano 1979); the strong effects of grain size composition (Spies 1989); and the before-mentioned influence of particle concentration in the aquatic system which is even more important if the sorption/desorption kinetics are too slow to reach equilibrium in a given time of interactions. In particular, these effects have been observed for heavy metals, PCBs and chlorophenols. Therefore, the *equilibrium partitioning approach* should not be taken into consideration for those chemicals at present.

1.3.2 Apparent Effects Threshold Approach

This method uses empirical field and laboratory data to identify concentrations of chemicals above which biological effects are always expected. In a given data set, the apparent effects threshold (AET) for a particular chemical is the sediment chemical concentration above which biologically adverse effects are observed relative to an appropriate reference sediment. A wide variety of organisms and biological tests can be used to obtain the effects data (data on the adverse effects). These may include benthic infaunal field surveys and bulk sediment bioassays with various organisms and endpoints. Adams et al. (1992) summarized the procedure for calculation of the AET:

• Chemical and biological effects data on numerous field sediment samples have to be obtained. This may be done by analytical measurements of the sediment chemical concentrations, benthic infaunal measurements and bulk sediment bioassays.
• Impacted and non-impacted sediment sites are identified. This must be done for each sediment sample if the biological effects are statistically different from a "clean" reference site.

• The AET has to be determined for each given chemical. This is done by using the paired data sets for all sediment sites in a given area (e.g. Weser estuary or Elbe estuary). From these paired data, unimpacted sites (no observed biological effects) are selected and used to determine the highest chemical concentration present in the sediment samples not causing biological effects for each chemical of interest. This value is reported as the AET for a given chemical based on a specific biological response (e.g. infaunal abundance).

An AET based on a different endpoint (growth or survival in a laboratory bioassay with a field sample) can also be calculated. By calculating several AET values for different biological endpoints, a range of values can be determined and the data evaluated for consistency, accuracy and sensivity.

The AET approach is probably most useful as a screening tool to provide a consistent basis on which to evaluate sediment contamination within a given region and to define which sediment samples or sites have the greatest potential to adversely impact aquatic life or – conversely – the lowest probability of ecological impact. A major limitation to this approach is its inability to directly assess which contaminant causes which biological effect. Empirical data are used to derive the apparent effects threshold and it is not possible to separate individual chemical effects when multiple chemicals are present in the same sample. Also the interpretation is not always clear since it depends on the applied effect indicators. In Germany the AET-method cannot be used for freshwater sediments since there are not enough data available from such environments. Data from United States cannot be used because they are mainly related to marine sediments.

1.3.3 Interstitial Water Toxicity Approach

The composition of interstitial water is the most sensitive indicator of the types and extent of reactions that take place between contaminants on the sediment particles and the aqueous phase which contacts them. Particularly for fine grained material the large specific surface area related to the small volume of its entrapped interstitial water ensures that minor reactions with the solid phases will be indicated by major changes in the composition of the aqueous phase.

The interstitial water approach is a procedure for measuring the toxicity of sediment-bound chemicals to aquatic organisms by exposing the organisms to pore water extracted from sediments. Pore water toxicity tests can be performed with a variety of test organisms, both benthic and water column species. The use of pore water to assess sediment quality is based on the assumption that the pore water is in equilibrium with the surrounding sediment and that the water phase provides a direct and important route of exposure for organisms that live in sediments. This method also assumes that the "soluble" or "free" uncomplexed fraction of any chemical in the pore water is the fraction most responsible for observed sediment toxicity (Di Toro et al. 1991). Recently it has been expanded to incorporate toxic-

ity identification evaluation procedures developed for effluent toxicity tests (Mount and Anderson-Carnahan 1988 a,b; Mount 1988). This addition allows the identification of the contaminant responsible for the sediment toxicity.

The interstitial water toxicity test involves 3 steps:

- Isolation of the pore water from the sediment samples
- Conducting of toxicity tests with pore water
- Application of toxicity identification evaluation procedures to pore water fractions (Adams et al. 1992)

The amount of pore water available is often the most limiting factor regarding the selection of test species and duration of the test. Pore water often has to be diluted to obtain a realistic dose-and-response relationship but different dilution methods may affect the toxicity (Giesy et al. 1990). In addition, there remains the unbridgeable conflict that biological tests need oxygen-rich water and contaminated sediments are in most cases anoxic. Oxidation of the pore water before or during the test procedure leads to a precipitation of iron- and manganese oxihydrates. A concurrent coprecipitation of metals means that metal toxicity may be underestimated in pore water tests (Schubauer-Berigan and Ankley 1991). Precipitation of iron also removes portions of dissolved natural organic substances from the solution (Sholkovitz 1976). Therefore, it is possible that the concentration of organic pollutants may be affected by such processes. The significance of this finding needs to be further evaluated.

To simulate resuspension effects of sediments which especially occur at dredging operations and high-water events in rivers, similar approaches have been applied with water extracts of sediments. These elutriate methods go back to a procedure which has been developed in the United States in the mid-seventies and is still used today with minor modifications (US-EPA and US-ACE 1977). The use of test organisms is possible under the same aspects as in the interstitial water approach. Although that method has not furthered the level of ecotoxicological knowledge, it has major importance in practice by allowing an indicative assessment of highly contaminated sediments by the use of simple methods (Ahlf and Wild-Metzko 1992).

1.3.4 Spiked Sediment Toxicity

In this approach the toxicity of a specific chemical to one or more benthic organisms is measured by the addition of the chemical to the test sediments in different doses. The method uses laboratory data to:

- Establish cause-and-effect relationships in order to predict concentrations of specific chemicals that would be harmful to resident biota under field conditions (Lamberson and Swartz 1989). Numerous combinations of sediment types and chemicals either individually or in mixtures can be tested.

- Carry out risk assessment and risk management of substances and preparations which are expected to reach sediments. For example, highly hydrophobic substances (having a lack of affinity for water) may be assessed as being low risk to pelagic fauna, but could, however, exert significant toxic effects in sediment.
- Set environmental quality standards (EQS) for substances known to be found in sediments. If EQS are to be set on the basis of chemical concentrations, "safe" levels must be determined in part, by means of sediment toxicity tests (SETAC 1993).

The advantages of this method are that it relates toxicity to a specific chemical and dose and could be used to develop chemical-specific assessment values. It provides a means to test a range of organisms and sensitive life stages, and it most closely approximates the approach for generating water quality criteria.

Disadvantages are that it does not test "real-world" contaminated sediments or in-place organisms. The ageing of the sediments often results in much slower sediment desorption rates than is experienced from unaged laboratory spiked samples (Adams et al. 1992). Also, the above-mentioned oxidation and precipitation reactions may affect binding properties of chemicals and sorption/desorption equilibria.

1.3.5 Tissue Residue Approach

This approach relies on an ability to establish a maximum concentration of a chemical in sediment that will result in an acceptable concentration in aquatic organisms or consumers of aquatic biota like humans, birds or mammals. Van der Kooij et al. (1991) summarized such an approach, recently proposed in The Netherlands to use the equilibrium partitioning method to derive a coherent set of quality criteria for aquatic systems. Two sorts of data can be used:

- *Toxic effect data.* By means of ecotoxicological extrapolation methods critical concentrations in water are derived from direct aquatic effect data. These critical concentrations are then "translated" into critical concentrations in solids by means of solids–water partition coefficients.

- *Product standards.* For many chemicals, maximally allowable concentrations – generally to protect man – in e.g. fish, are available. These critical concentrations in organisms are "translated" into critical concentrations in water by means of bioconcentration factors. These, in turn, are "translated" into critical concentrations in solids using solids–water partition coefficients.

With both methods quality criteria can be calculated. This procedure yields two criteria for one chemical, the lowest of which is to be taken as decisive. As mentioned above (see Sect. 1.3.1) concentrations in water and solids are related through a partition coefficient:

$$K_p = \frac{C_s}{C_w} \qquad (15)$$

where: K_p = solids–water partition coefficient
$\quad\quad\quad C_s$ = concentration in the solid phase
$\quad\quad\quad C_w$ = concentration in the water phase

In the case of equilibrium, values for K_p can be derived theoretically on the basis of properties of the chemical and characteristics of the solids. Alternatively, values for K_p can be derived from field measurements of the concentration in water before and after filtration:

$$K_p = \frac{C_{tot} - C_w}{SM \cdot C_w} \qquad (16)$$

where: C_{tot} = total concentration in water, before filtration
$\quad\quad\quad C_w$ = concentration in the water phase, after filtration
$\quad\quad\quad SM$ = concentration of suspended matter

It can be inferred from field measurements that concentrations of micropollutants in sediments differ systematically from the concentrations in the suspended solids in the water column above the sediment (Aiking and Bruggeman 1987). For metals the concentration ratio, suspended matter:sediment, is approximately 1.5; for organic chemicals this ratio is approximately 2.0. These values are valid for Dutch sediments and suspended matter, and may differ in other countries. If the ratio of the concentration in suspended matter and water is taken as a measure for K_p, which is implicitly done when K_p is empirically derived according to Eq. 16, the concentration in sediments C_{sed} is to be calculated as follows:

$$C_{sed} = \frac{C_{susp}}{r} = \frac{C_s}{r} = \frac{K_p \cdot C_w}{r} \qquad (17)$$

where: C_{sed} = concentration in sediment
$\quad\quad\quad C_{susp}$ = concentration in suspended matter
$\quad\quad\quad r$ = empirical concentration ratio suspended matter:sediment
$\quad\quad\quad\quad\quad$ (metals: $r = 1.5$; organic chemicals: $r = 2.0$)

To calculate the total (dissolved + particulate) concentration of a chemical, if a critical concentration C_w for water is given, Eq. 16 is used again. After rearrangement Eq. 16 becomes:

$$C_{tot} = C_w \cdot (1 + K_p \cdot SM) \qquad (18)$$

In a similar way as for the relation between concentrations in water and suspended matter, water and biota are – by definition – related to each other through a bioconcentration factor:

$$BCF = \frac{C_{org}}{C_w} \tag{19}$$

where: $BCF =$ bioconcentration factor
$C_{org} =$ content of chemical in organisms
$C_w =$ concentration of the chemical in the water phase

Products standards for the consumption of fish, mussels, etc. have been established for a number of chemicals. If for these chemicals BCFs are also known, Eq. 19 can be used to derive a critical concentration in water C_w from critical concentrations in organisms C_{org}. For this purpose Eq. 19 is rewritten into:

$$C_w = \frac{C_{org}}{BCF} \tag{20}$$

Substituting Eq. 20 into Eq. 17, the equation for the concentration in sediment becomes:

$$C_{sed} = \frac{C_{org} \cdot K_p}{r \cdot BCF} \tag{21}$$

For organic chemicals K_p values are derived from the octanol–water partition coefficient of the compound and the organic carbon content of the solid material using the referencing procedure formulated by Karickhoff et al. (1979). Bioconcentration factors are usually expressed on a fat basis. As both the fat-based bioconcentration factor and the solids–water partition coefficient are proportional to the octanol–water partition coefficient, it is theoretically expected that the ratio of BCF_{fat} and K_{oc} has the same value for all compounds. Field observations confirm this and it appears that this ratio takes a value of approximately 2 (Bruggeman et al. 1985; Bruggeman 1987). Taking the value, 5%, as a typical fat fraction for fish, this leads to:

$$BCF = 0.05 \cdot BCF_{fat} \approx 0.05 \cdot K_{ow} \tag{22}$$

if: $K_{oc} = 0.6 \cdot K_{ow}$, and
$BCF_{fat} \approx 2 \cdot K_{oc} \approx 2 \cdot 0.6 \cdot K_{ow} \approx K_{ow}$.

Although this relation between K_{ow} and BCF holds for many organic chemicals (Mackay 1982), it should be noted that deviations from this relation may be observed, e.g. for more polar chemicals, very hydrophobic compounds and also for chemicals which are biotransformed. For metals K_p values show a great variability and depend on many physico-chemical factors, e.g. salinity, pH, DOC, etc. as previously discussed (Sect. 1.3.1).

1.3.6 Sediment Quality Triad

This approach involves a combination of biological laboratory and field data in comparison with chemical analyses which compensates the limitations of the individual methods. It consists of three components (Chapman 1986):

- Identification and quantification of inorganic and organic contaminants in the sediment with the lowest limit of detection possible
- Measurement and quantification of toxicity based on bulk sediment laboratory toxicity tests. Bioassays are selected to span the widest possible range of organism type, life cycle, exposure route, and feeding type.
- Evaluation of in situ biological effects or alteration. Benthic infaunal community structure and function data are often used because the organisms are relatively immobile and site specific. Other endpoints such as benthic fish histopathological data have also been used.

The three data components are compared on a quantitative basis and are normalized to "clean" reference site values by converting them to ratio-to-reference (RTR) values.

The principal advantages of the triad approach are that it can be used for any sediment type; it does not require a priori assumptions concerning the specific mechanism of interaction between organisms and contaminants; and it integrates laboratory, field, biological, and chemical data. Major limitations are, that statistical approaches have not been fully developed for assessing triad data, the method

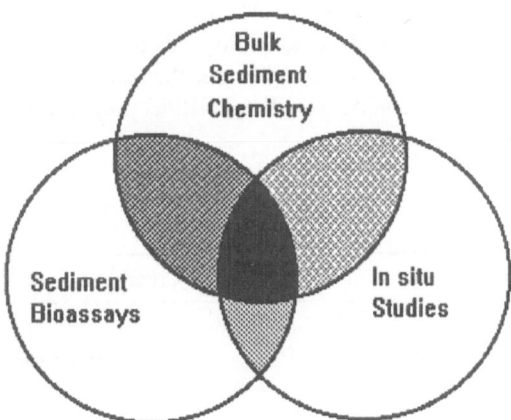

Fig. 1.3. Conceptual model of the sediment quality triad, which combines data from chemistry, bioassay and in situ studies. Areas where the three facets of the triad show the greatest overlap (in terms of either positive or negative results) provide the strongest data for determining numerical sediment criteria

does not provide direct cause-and-effect data when multiple chemicals are present and it does not eliminate the possibility that the presence of unmeasured contaminants may be responsible for the observed toxic effects (Adams et al. 1992). But although the method is not standardized at present, it may nonetheless lead to general sediment quality criteria, however, at relatively high costs.

In Europe, two integrated procedures have been proposed during recent years. The one by Van de Guchte (1991) is directly based on the sediment triad and it is aimed for the site-specific assessment of highly contaminated sediments – "hotspots" – in The Netherlands. The triad parameters include:

- Chemical analyses of 49 substances, e.g. metals, PCB, PAH, chloro-pesticides
- Biological analyses of field parameters for site-specific assessment of bioavailability of characteristic pollutants and their adverse effects – on benthic organisms (Investigations on accumulation processes, e.g. on molluscs, will be included)
- Toxicological analyses using bioassays (*Chironomus riparius* sediment toxicity test and *Daphnia magna* pore water/elutriate test)

Integration of the data from the three components of the triad reveals an effect-oriented assessment of sediment quality. Using two criteria for each parameter (see Table 1.9) a first classification into three classes can be realized for each of the three components, based upon the most sensitive parameter in each basic or extended parameter set. Ranking of sites and priority setting can be realized by using original data within each class.

This concept has been applied on Rhine sediments. At a given chemically based sediment quality, two contamination levels can be derived from the studied parameters: "Safe Level" and "Warning Level". If the three groups of measured parameters (chemical, field-biological, ecotoxicological) congruently deviate from the reference values, then the "Warning Level" is reached and there will be a high priority for further action. This could mean that a distinct proportion of contamin-

Table 1.9. Some indicative criteria in the triad assessment of contaminated Rhine sediments (Van de Guchte 1993)

Parameter	Criterion 1	Criterion 2
Chemicals		
– Metals, PAH, PCB, OCP	SQC-safe level [a]	SQC-warning value [a]
Field Surveys		
– Chironomids	10% deformities	20% deformities
– Maximum density	1500 individuals/m^2	500 individuals/m^2
Bioassays		
– *Daphnia* reproduction	10% effect	50% effect

[a] SQC = Sediment Quality Criteria

ated sediments cannot be resuspended or that special precautions have to be applied during dredging operations.

The second integrated procedure for the assessment of sediment-bound pollutants is based on the concept of Ahlf et al. (1991) and relates exclusively to toxicological data. It is mainly based on microbial toxicity tests using bacteria and algae. An overall biological assessment scheme includes:

- Field description of benthic communities
- Benthos bioassay on total sediment
- Pore water bioassay (bacteria, Chromotest) and elutriate bioassay (algae, bacteria, Chromotest)

In addition, tests can be performed on fractions of the sediment which have been extracted or treated with a cosolvent. For example, a non-polar surfactant has been applied which is commercially used to solubilize hydrocarbons from contaminated soil. On the other hand, the concept may be modified according to users requests in that only parts of this structure may be needed for a site specific problem.

In the Hamburg harbour area a sediment study was carried out on the basis of 50 sediment cores (Ahlf and Wild-Metzko 1992). A portion of the sediment was extracted with a non-polar surfactant. Factor analysis has been performed for 64 samples with chemical characterization (see Table 1.10). The bulk concentration of cadmium, copper, mercury, hydrocarbons, extractable organic halogens (EOX) and Lindane were selected as relevant variables.

The first factor joins the main contaminants and indicates the same contamination source and/or the same sedimentary behaviour. The second factor suggests a relationship of Lindane to bacterial toxicity. The third factor represents the high phytotoxicity of mercury and its assessment in the algae assay. Finally, separation

Table 1.10. Representation of the four main factor loadings resulting from a factor analysis of 64 sediment samples (Ahlf and Wild-Metzko 1992)

	Factor 1	Factor 2	Factor 3	Factor 4
Algae assay			0.91	
Bacteria assay		0.89		
Bacteria/detergent		0.86		
Cadmium	0.90			
Copper	0.91			
Mercury	0.67		0.51	
Hydrocarbons				0.94
EOX	0.86			
Lindane		0.54		

of the hydrocarbons in factor four indicates that the distribution of hydrocarbons is affected by distinct inputs.

An integrated concept of sediment toxicity evaluation which is presently designed for surface waters in Germany will consist of the two components "bulk sediment" and "elutriate". The bulk sediment bioassay, which also includes pore water, involves analysis of effects on micro-organisms, plants, benthos and on genotoxicity. New developments refer to solid phase bioassays using micro-organisms. Plants are important indicators of pollutants in dredged material disposed on land. Elutriate tests comprise standardized procedures on bacteria, algae, daphnia and DNA. Both compartments can be compared and further interpretation on extrapolation from laboratory data to nature is mainly derived from the description of microflora. Its major advantage lies in the fact that, it reflects complex ecological responses in the sediment below a cm^2 scale. A new solid phase bioassay is presently developed on the basis of studies on artificial sediment in comparison with elutriates and the bacteria were more sensitive in contact with the spiked sediments.

1.4 Tiered Sediment Assessment Strategy

A research program on contaminated sediments may require a conceptual framework supporting the current approaches for assessing data on chemical exposure and biological effects on organisms that is collected in a stepwise or tiered manner. For example, an analysis of the benthic community is ecologically necessary but very expensive and can be nonsensical if the sediment is anoxic. A tiered sediment assessment strategy allows for periodic decisions to stop the analysis if:

- Adequate safety is demonstrated
- Toxicity is well characterized
- Significant questions remain – to collect more data

Such tiered integrated biological and chemical sediment assessment methods have been offered as management strategies for the treatment of contaminated sediments and decision making frameworks for the disposal of dredged material because these methods lay particular stress on the cost factor and offer a clearly arranged scheme (Adams at al. 1992; IJC 1988; US-EPA 1988). According to a procedure proposed by Reynoldson and Day (1993) we want to introduce a tiered strategy for a *general sediment assessment* (Fig. 1.4) which takes some critical aspects into a stronger consideration.

Tier I

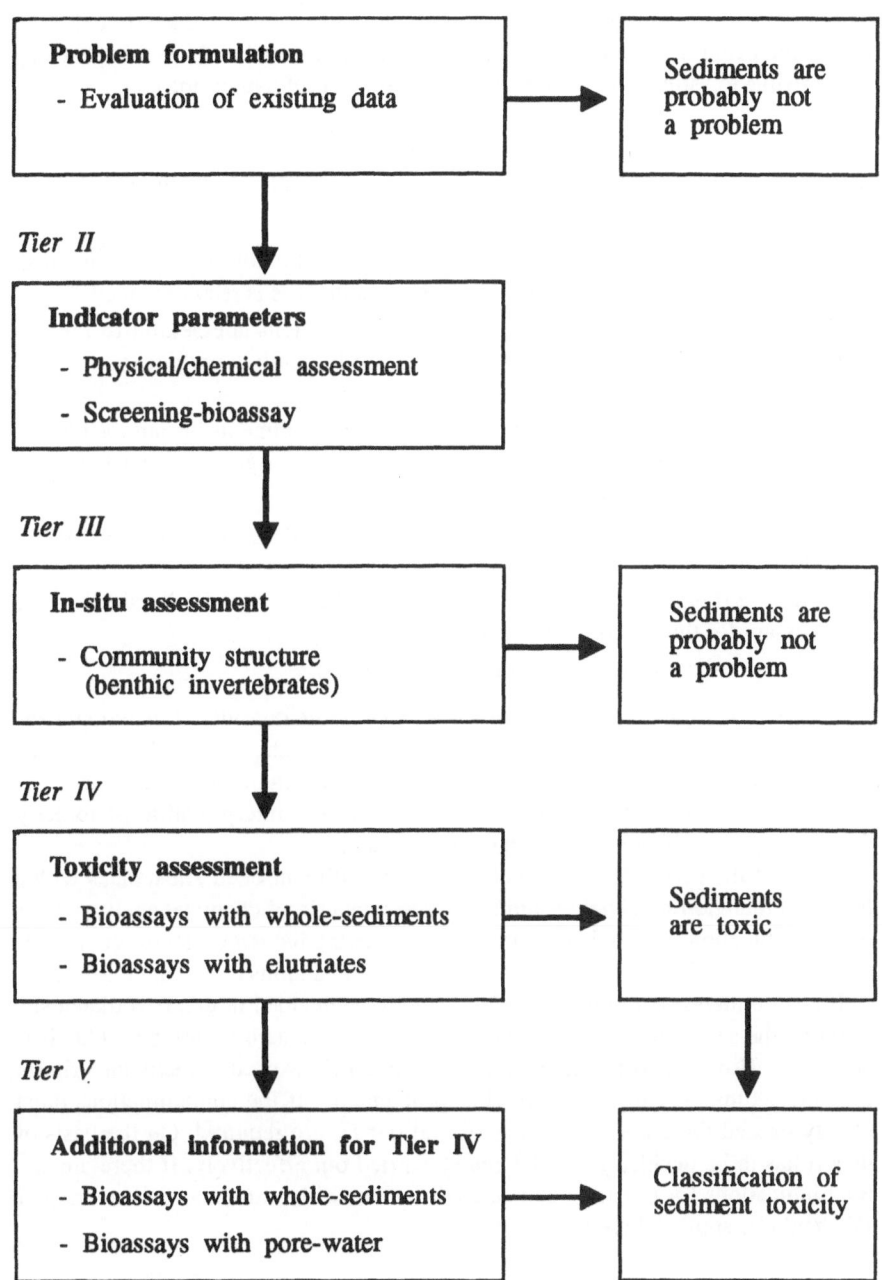

Tier II

Tier III

Tier IV

Tier V

Fig. 1.4. A tiered sediment assessment strategy

Tier I. We recommend in the first step an evaluation of existing data. It is essentially that not only data from chemical monitoring programs are used, but also details about ecological inventories and biological studies. Already in this first step it is possible to estimate if sediments are probably a problem or not.

Tier II. If there is suspicion of a contamination with hazard chemicals, tier II has to be applied. A minimum amount of data is required which is essentially to characterize the hazard potential in the sediment:

- *Grain size distribution* – a minimum size analysis of the fine grain fraction ($< 6 3 \ \mu$m) and additionally of five medium grain sizes is recommended.
- *Wet and dry weight* – determination of water content is essential to relate a toxicity value to a sediment concentration. Additionally a rough estimation of the porosity is possible.
- *pH value* – is an essential parameter in freshwater sediments because acidification of the system shows direct effects on aquatic organisms and increases availability of heavy metals.
- *Redox potential* – strongly anoxic sediments are scarcely colonized by invertebrates.
- *Organic carbon content* – affects the bioavailability of organic chemicals and is necessary for the interpretation of sediment toxicity.
- *Ammonia* – shows acute toxicity for test organisms.

The role of ammonia in sediments is more and more considered. Ammonia enrichment in greater amounts may occur in eutrophic aquatic systems by the microbial catabolism of nitrogen compounds under anoxic conditions. The presence of ammonia in sediments can extremely complicate the interpretation of toxicity data, e.g. by coincident effects with other substances.

In tier II the presence of toxic chemicals in sediment or of substances with a high accumulation potential is studied. A basic data set of chemical analyses with indicator function is sufficient. A microbial screening bioassay should provide security for the chemical assessment and give first indications of sediment toxicity.

If contaminations are undetectable, it cannot be decided in this first diagnostic step that the sediments are unproblematic. In any case, a verification by tier III is necessary. However, if the contaminations obviously exceed the existing criteria then the sediments can be classified as problematic. If the contaminations don't clearly exceed the criteria, then tier III and tier IV are required. On the basis of data it has to be decided if tier III can be carried out effectively. If there are distinct indications for a strictly anoxic environment with ammonia enrichment, tier IV should be applied directly.

Tier III. The third step provides measurements in the zone of impact to confirm the significance of the chemicals to benthic life. From an ecological point of view a detailed analysis of the benthic community structure is most important. Until to-

day, only ecological determinations of the species diversity of macro-invertebrates (sometimes down to the species level) have been fully developed for a sediment assessment. But this strategy shows weaknesses, in that, the analyses are expensive and depend on season and locality. Nevertheless, with adequate experience it is possible at this step to identify a sediment as probably unproblematic. Generally, it must be stated that assessments by simplified biological methods (using key organisms) need to be further evaluated. To confirm that impairments in the benthic community are caused by a sediment toxicity, tier IV has to be carried out.

Tier IV. In this step sediment toxicity is established with bioassays. Recommended is a whole sediment test with *Chironomids* as test organisms. Additionally, an elutriate test with *Daphnia* should complete the result. The elutriate test is proposed because it is standardized and shows a strong indicator function for polluted sediments.

As long as there exist no criteria for bioassays with the whole sediment, the test result requires a statistically significant difference (95% probability) in comparison to control studies. A lacking toxicity proof needs in any case an extension of ecotoxicological measurements to confirm the absence of a chemical. If toxic effects are confirmed then the extent of the zone of impact must be determined in additional tests. If the zone of impact is small, it may be decided to stop further investigations. From the individual data a preliminary classification can be undertaken but it may also be that additional toxicity studies will be more selective.

Tier V. In this step the hitherto not-applied test procedures of the recommended bioassays combination should be carried out. The classification of toxicity should be discriminatory enough to permit ranking of the relative toxicity of many samples. Additional bioassays of different sensivities and exposure pathways could ensure a comparing classification of sediment toxicity (Ross et al. 1992). The classification of toxicity can be made on different discriminatory levels. An impairment in only one of the bioassays can be sufficient to fail the classification "unproblematically". Such a simple biological sediment assessment method for marine sediments based on effect criteria recently entered into legislative rules in Washington (Ginn and Pastorok 1992).

1.5 Conclusions

Sediment quality assessment is considerably more complex than water quality assessment due to the many site-specific parameters that need to be considered which are not a factor for water. These factors include bioavailability, sorption kinetics, sediment characteristics (buffer capacity and acid-producing capacity), sediment deposition and erosion (redox potential) and temporal and spatial differ-

ences. The methodologies developed to date do not adequately deal with the complex nature of sediments.

Against the background of increasing growth in testing methods and data, SETAC-Europe took on a workshop, held at Renesse (The Netherlands) and the initiative to enhance harmonization by means of a guidance document for sediment toxicity testing and bioassay procedures (SETAC 1993). The objective of this workshop was to harness practical experience with sediment testing methods from academic, regulatory and industrial centres of expertise, to consolidate the state of the art into a set of guidelines and to identify areas of priority for further research.

For most purposes, assessments of the biological effects, which may be exerted by contaminants in sediments, are most appropriately made by exposing sediment dwellers to whole sediment, although some insights can be obtained by measuring the toxicity of pore water or elutriates to pelagic fauna or flora. It was generally agreed, however, that only substances which are likely to enter and persist in sediments need to be assessed in sediment tests.

There are no indications that sediment dwellers are inherently any more susceptible to contaminants at the biochemical and physiological levels than pelagic organisms. Sediment tests are thus essentially a means of measuring the different exposure which sediment dwellers may experience as a result of the processes of adsorption, physical contact and oral uptake, which do not occur to such a great extent in the overlying water column.

Sediment toxicity tests are still in their infancy. The way in which sediments are sampled and stored for bioassays or spiked for toxicity tests can lead to unacceptable variability in results. It is probable that some of the variability in toxicity tests could be eliminated by the use of artificial sediments, although useful tests can be conducted with natural sediments providing that the range of acceptability for certain parameters such as the particle size spectrum and the organic carbon content is tightly defined. There is also a need to develop statistical methods that can be used for analyzing dose-response relationships in sediment bioassays which usually include the complex mixture of substances present in contaminated sediments.

The most useful sediment toxicity tests and bioassays are those which measure the chronic toxicity resulting from long-term exposure because some sedimentary contaminants can be persistent. However, partly due to the problems associated with feeding sediment dwellers in a natural way and partly due to our lack of understanding the optimum living conditions for these organisms, there are very few sediment tests which are currently able to measure chronic effects.

As with all ecotoxicological data the most difficult problem in sediment toxicity assessment is the reliable comparison of data obtained under different sedimentary conditions and the confident extrapolation of laboratory-derived sediment test results to the field situation. Some progress has been made for neutral organic compounds by normalizing for the organic carbon content of the test sediment. Normalizing for acid volatile sulphide (AVS) may have eqivalent utility for

certain metals. However, there is a need to develop similar normalizing proce-dures for the many other confounding factors in sediment test (as discussed in Sect. 1.2.4) to enable all test results to be reliably compared. Extrapolation to field situations is even more difficult at present because we have little information about the impact of sedimentary contaminants under natural conditions. Until we have a much larger data base of information on pollution effects in experimental mesocosm sediments and are able to relate bioassay responses in a more structured way to known impacts on wild communities of benthic organisms, we will not be able to derive specific applications or safety factors to be used when extrapolating sediment toxicity test results.

Management of contaminated sediments requires a holistic approach. In addi-tion to the common predictive techniques for estimating contaminant behaviour, the interactive nature of various parameters has to be recognized. This means that particular emphasis should be placed on the evaluation of the two data sets: "driv-ing force/geochemical gradient" parameters (mostly redox sensitive components which may be critical for the acid-producing capacity) and "capacity controlling properties" (with a prevalence on carbonate content). Assessment of these parameters and their long-term borderline conditions, as well as the potential links between the two groups "driving force/capacity controlling properties" and of individual parameters should be an integrated part of an extended management scheme, i.e. the combined chemical and biological parameters should always be related to potential remediation options for a specific sediment problem.

Finally we want to summarize the most important research recommendations stated by the SETAC workshop in Renesse, The Netherlands (SETAC 1993):

- There is an urgent need for a decision making framework to enable specific rec-ommendations to be made on when to use whole sediment bioassay as opposed to water phase bioassays.

- Further comparative studies (i.e. whole sediment versus pore waters and elutri-ates) are required before the general use of pore water or elutriates as surrogates can be recommended.

- There is a need for standardization of sediment pore water extraction and elutri-ation methods because these contribute substantially to the variability between test results in these systems.

- Information is needed on the maximum "no-effect levels" of organic solvents in sediments which may be used during sediment spiking procedures.

- The relationship between hydrophobicity and the time required to reach equilib-rium partitioning needs to be investigated for a range of sediment properties and contaminant types.

- There is a need to develop an acceptable artificial sediment, in order to mini-mize the confounding factors caused by sediment variability in sediment toxic-ity tests.

- If natural sediments are to be used, the degree to which the confounding factors contribute to variability needs to be investigated so that recommended limits can be set.

- There is a need to develop a statistical method for the analysis of dose-response relationships of dilution of complex unknown mixtures of pollutants in sediments.

- There is a general need for new test designs which use small, short-lived animals that grow and reproduce in laboratory sediment systems and do not require additional feeding.

- Reliable long-term, sublethal endpoints with a much wider range of species are required for sediment toxicity tests and bioassays.

- Reliable normalization procedures are needed for chemicals other than neutral organics.

- There is a need for effects data on chemicals in sediments obtained under both laboratory and field conditions which will permit the development of sediment-specific application factors.

- There is a need to generate data from multispecies sediment toxicity assessments in the field in order to provide validation of predictions/extrapolations made from laboratory-based sediment bioassays.

References

Adams WJ, Kimerle RA, Barnett JW jr (1992) Sediment quality and aquatic life assessment. Environ Sci Technol 26(10):1865–1875

Ahlf W, Dahm M, Förstner U, Wild-Metzko S (1991) Biologisches Bewertungskonzept für Sedimente. Vom Wasser 76:215–223

Ahlf W, Gunkel J, Liß W, Neumann-Hensel H, Rönnpagel K, Förstner U (1992) Mikrobielle Biotests mit Sedimenten. Schriftenreihe Wasser-, Boden- und Lufthygiene, Gustav Fischer Verlag, Stuttgart, 89:427–435

Ahlf W, Wild-Metzko S (1992) Bioassay responses to sediment elutriates and multivariate data analysis for hazard assessment of sediment-bound chemicals. Hydrobiologia 235/236:415–418

Aiking H, Bruggeman WA (1987) Project micropollutants in sediments. Interpretation of measurements on suspended matter 1983–1984. Report of the Institute for Inland Water Management and Waste Water Treatment, Lelystad

Ankley GT et al. (1991) Environ Toxic Chem 10:1299–1307

Anonymous (1990) Technische Verordnung über Abfälle (TVA) Der Schweizerische Bundesrat (Swiss Federal Parliament), SR 814.015, Dec 10, 1990, Bern, Switzerland

Battelle Washington Operations (1985) Sediment quality criteria development workshop. Nov 28–30, 1984. Washington, 25 pp

Berner RA (1981) A new geochemical classification of sedimentary environments. J Sediment Petrol 51:359–365

Bruggeman WA, Kerkhoff MAT, Wegman RCC (1985) Organic micropollutants in sediments: distribution and bioaccumulation. In: Van den Bergen VWJ, Kerkhoff MAT, Wegman RCC (eds) Onderwaterbodems, Rol en Lot, Royal Netherlands Chemical Society, Environ Chem Section, Rotterdam, pp 57–71

Bruggeman WA (1987) Chemical and biological approaches to standardization. Technical Committee for Soil Protection, Symp on Sediment Quality, Ede, 10 Dec 1986, VTCB M86, Leidschendam

Bruynestein A, Hackl RP (1984) Evaluation of acid production potential of mining waste materials. Miner Environ 4:5–8

Calmano W, Ahlf W, Förstner U (1988) Study of metal sorption/desorption processes on competing sediment components with a multi-chamber device. Environ Geol Water Sci 11:77–84

Calmano W (1979) Untersuchungen über das Verhalten von Spurenelementen an Rhein- und Mainschwebstoffen mit Hilfe radioanalytischer Methoden. PhD thesis, TH-Darmstadt

Calmano W, Hong J, Förstner U (1992) Einfluß von pH-Wert und Redoxpotential auf die Bindung und Mobilisierung von Schwermetallen in kontaminierten Sedimenten. Vom Wasser 78:245–257

Calmano W, Hong J, Förstner U (1994) Binding and mobilization of heavy metals in contaminated sediments affected by pH and redox potential. In: Christensen E et al. (Eds) Proc 1st Intern Spec Conf on Contaminated Aquatic Sediments: Historical Records, Environmental Impact, and Remediation. Water Sci Technol (in press)

Carlson AR (1991) Environ Toxic Chem 10:1309–1319

Chapman PM (1986) Sediment quality criteria from a sediment quality triad: An example. Environ Toxicol Chem 5:957–964

Chapman PM, Barrick RC, Neff JM, Swartz RC (1987) Four independent approaches to developing sediment quality criteria yield similar values for model contaminants. Environ Toxicol Chem 6:723–725

Di Toro DM et al. (1990) Toxicity of cadmium in sediments: The role of acid volatile sulfide. Environ Toxicol Chem 9:1487–1502

Di Toro DM, Zarba CS, Hansen DJ, Swartz RC, Cowan CE, Pavlou SP, Allen HE, Thomas NA, Paquin PR (1991) Technical basis for establishing sediment quality criteria for nonionic organic chemicals using equlibrium partitioning. Environ Toxicol Chem 10:1541–1583

Di Toro DM et al. (1992) Environ Sci Technol 26:96–101

Ferguson KD, Erickson PM (1988) Pre-mine prediction of acid mine drainage. In: W Salomons, U Förstner (eds) Environmental Management of Solid Waste – Dredged Material and Mine Tailings. Springer Berlin, pp 24–43

Förstner U (1990) Bewertung sedimentbezogener Maßnahmen in Ästuar- und Küstengewässern der Bundesrepublik Deutschland. Wasser und Boden 8/90:508–512

Förstner U, Ahlf W, Calmano W (1994) Sediment quality objectives and criteria development in Germany. Wat Sci Technol (in press)

Giesy JP, Rosiu CJ, Graney RL, Henry MG (1990) Benthic invertebrate bioassays with toxic sediment and pore water. Environ Toxicol Chem 9:233–248

Ginn TC, Pastorok RA (1992) Assessment and management of contaminated sediments in Puged Sound. In: GA Burton Jr (ed) Sediment Toxicity Assessment. Lewis Publ, Boca Rotan, FL, 371–401

Herms U, Brümmer G (1978) Löslichkeit von Schwermetallen in Siedlungsabfällen und Böden in Abhängigkeit von pH-Wert, Redoxbedingungen und Stoffbestand. Mitt Dt Bodenk Ges 27:23–43

IJC-International Joint Commission (1988) Procedures for the assessment of contaminated sediments problems in the Great Lakes. Sediment Subcommittee and its Assessment Work Group, Windsor, Ontario

Jenne EA et al. (1986) An activity-based model for developing sediment criteria for metals: A new approach. Chemicals in the Environment, Int Conf, Lisbon, Portugal

Jenne EA (1987) Sediment quality criteria for metals: IV, Surface comlexation and acidity constants for modelling cadmium and zinc adsorption onto iron oxides. Prep. for US-EPA, Office of Water Regulations and Standards, Criteria and Standards Div, Washington DC

JRB Associates, McLean/Virginia (1984) Bachground and review document on the development of sediment criteria. Sept 30, EPA Contract No 68-01-6388, 32 pp

Karickhoff SW, Brown DS, Scott TA (1979) Sorption of hydrophobic pollutants on natural sediments. Wat Res 13:241–248

Kersten M, Förstner U (1991) Geochemical characterization of the potential trace metal mobility in cohesive sediment. Geo-Marine Letts 11:184–187

Lamberson JO, Swartz RC (1989) Sediment classification methods. Draft Final Report, Watershed Protection Division, US-EPA, Washington, DC, Chapter 3

Mackay D (1982) Correlation of bioconcentration factors. Environ Sci Technol 16:274–278

Morse JW, Mackenzie FT (1990) Geochemistry of Sedimentary Carbonates. Elsevier Publ Co, New York

Mount DI, Anderson-Carnahan L (1988 a) Methods for aquatic toxicity identification evaluations: Phase I Toxicity Identification Procedures. US-EPA, Duluth, MN, EPA 600-3-88/034

Mount DI, Anderson-Carnahan L (1988 b) Methods for aquatic toxicity identification evaluations: Phase II Toxicity Characterization Procedures. US-EPA, Duluth, MN, EPA 600-3-88/035

Mount DI (1988) Methods for aquatic toxicity identification evaluations: Phase III Toxicity Confirmation Procedures. US-EPA, Duluth, MN, EPA 600-3-88/036

Müller G (1979) Schwermetalle in den Sedimenten des Rheins – Veränderungen seit 1971. Umschau in Wissenschaft und Technik 79:778–783

Obermann P, Cremer S (1992) Mobilisierung von Schwermetallen in Porenwässern von belasteten Böden und Deponien: Entwicklung eines aussagekräftigen Elutionsverfahrens. Landesamt für Wasser und Abfall Nordrhein-Westfalen, Düsseldorf/Germany, Vol 6, 127 p

Patrick WH, Williams BG, Moraghan JT (1973) A simple system for controlling redox potential and pH in soil suspensions. Soil Sci Soc Amer Proc 37:331–332

Reynoldson TB, Day KE (1993) Freshwater sediments. In: P Calow (ed) Handbook of Ecotoxicology, Blackwell Scientific Publications, Oxford, 83–100

Ross PE, Burton Jr GA, Crecelius EA, Filkins JC, Giesy Jr JP, Ingersoll CG, Landrum PF, Mac MJ, Murphy TJ, Rathbun JE, Smith VE, Tatem HE, Taylor RW (1992) Assessment of sediment contamination at Great Lakes areas of concern: the ARCS program Toxicity-Chemistry Work Group strategy. J Aquatic Ecosystem Health 1:193–200

Schoer J, Förstner U (1987) Abschätzung der Langzeitbelastung von Grundwasser durch die Ablagerung metallhaltiger Feststoffe. Vom Wasser 69:23–32

Schubauer-Berigan MK, Ankley GT (1991) The contribution of ammonia, metals and non-polar organic compounds to the toxicity of sediment interstitial water from an Illinois river tributary. Environ Toxicol Chem 10(2):201–207

SETAC (1993) Guidance document on sediment toxicity assessment for freshwater and marine environments. Workshop on Sediment Toxicity Assessment, Slot Moermond Congrescentrum, Renesse, The Netherlands, 8–10 Nov 1993

Sholkovitz ER (1976) Flocculation of dissolved organic and inorganic matter during the mixing of river water and sea water. Geochim Cosmochim Acta 40:831–845

Sobek AA, Schuller WA, Freeman JR, Smith RM (1978) Field and laboratory methods applicable to overburden and mine soils. US Environmental Protection Agency Report EPA-600/2-78-054

Spies RB (1989) Sediment bioassays, chemical contaminants, and benthic ecology: New insights or just muddy water? Mar Environ Res 27:73–75

Tessier A, Campbell PGC, Bisson M (1979) Sequential extraction procedure for the speciation of particulate trace metals. Anal Chem 51:844–851

Tessier A et al. (1985) Geochim Cosmochim Acta 49:183–194

US-EPA and US-ACE (1977) Environmental Protection Agency and US Army Corps of Engineers: Ecological evaluation of proposed discharge of dredged material into ocean waters. Environmental Effects Laboratory, Experimental Waterways Experiment Station, Vicksburg, MS

US-EPA (1988) Toxic sediments: approaches to management. EPA-68/01-7002 Epa Science Policy Integration Branch, Washington, DC, 90 p

Van de Guchte K (1991) Macroinvertebrates and quality assessment of Rhine sediments. In: JAW de Wit, MA van der Gaag, K van de Guchte, CJ van Reeuwen, J Koeman (eds) The effect of micropollutants on components of the Rhine ecosystem. EHR-Publication No 35-1991, Lelystad, The Netherlands, pp 39–58

Van de Guchte K (1993) The sediment quality triad: an integrated approach to assess contaminated sediments. Proc Int Conf on River Water Quality – Ecological Assessment and Control, Dec 16–18, 1991, Bruxelle

Van der Kooij LA, Van de Meent D, Van Leeuwen CJ, Bruggeman WA (1991) Deriving quality criteria for water and sediment from the results of aquatic toxicity tests and product standards: application of the equilibrium partitioning method. Wat Res 25(6):697–705

Van Veen HJ, Stortelder PBM (1988) Erforschung belasteteter Sedimente in den Niederanden. In: Wolf KH, Van der Brink WJ, Colon FJ (Hrsg) Altlastensanierung '88. Kluwer Academic Publisher, Dordrecht, pp 1297–1309

Walker JD (1988) Effects on microorganisms. J Water Pollut Control Fed 60:1106–1121

Wallmann K et al. (1993) Intern J Environ Anal Chem 51:187

Williamson KJ, Bella DA (1980) Estuarine sediments: Successional model. J Environ Engin Div ASCE 106:695–710

2 The Influence of Diagenetic Processes on the Exchange of Trace Contaminants at the Sediment–Water Interface

W. Petersen, K. Wallmann, P. Li, F. Schroeder and H.-D. Knauth

2.1 Introduction

In heavily contaminated rivers like the Elbe, sediments are highly polluted and may act as a long-term source of trace elements for the bottom water. A flux from the sediment to the bottom water is caused by diffusion processes from the much higher concentration of contaminants in the interstitial water. This higher concentration is due to mineralization processes as well as by direct degradation of organic substances at sediment surface and the release of the trace elements bound on it. On the other hand, in the oxic boundary layer the trace elements can be fixed both by possible formation of new organic matter and by adsorption and co-precipitation on freshly formed hydrous ferric oxide (Petersen et al. 1993). Again these hydrous ferric oxides can be dissolved under anoxic condition by reduction of Fe(III) to Fe(II) and thus causing a release of trace elements to the water phase.

To investigate the fluxes at the sediment–water interface it is possible to do in situ measurements using benthic flux–chambers (Sundby et al. 1986, Westerlund et al. 1986, Schroeder et al. 1990). These investigations allow the quantification of fluxes of nutrients and provide some information on the more important processes. However, the experiments are costly and it is impossible to control all determining parameters. Thus, the processes that occur are often difficult to interpret. Although laboratory experiments on flooded sediment cores cannot simulate natural conditions precisely they allow much better control and reproducibility of the variables such as redox, pH, salinity and oxygen. Furthermore, it is possible to acquire more information on the changes of the characteristics of the sediments by investigating depth profiles in parallel incubated cores at different times. The measurements also provide information about early diagenetic processes in the sediment. From the measured concentration gradients in the porewater the diffusive controlled fluxes can be estimated. Recently the early diagenetic processes and the mobility and the

binding forms of trace elements in sediment suspensions under different redox conditions were investigated by Wallmann (1992). He found that under oxic conditions Cd, Cu and Ni were released. Under post-oxic conditions ($\sum H_2S \leq 1$ $\mu mol/l$) Co, As and also Ni were mobilized, whereas Zn, Cd, Cu and Pb remained fixed due to sulphide precipitation. Subsequently after anaerobic incubation over a longer time, sulphide arose at higher concentrations so that all trace elements were controlled by the solubility of their sulphide minerals.

The aims of this work are (1) to look into the evolution of early diagenetic processes in the different zones of a artificial homogenized and oxidized sediment, (2) to investigate the different conditions that determine the processes with regard to a tidal river, (3) to examine the influence of these diagenetic processes on the mobility of contaminants and (4) to estimate the rate of mobilization or fixation and the amounts of contaminants released to, or removed from, the overlying water.

2.2 Materials and Methods

The investigations were carried out either on single cores (plexiglas tubes, diameter 10 cm) with sediment depths of 10–20 cm and an overlying water column of approximately 8 cm or in a laboratory simulation apparatus (LABOSI) which was specially designed to investigate the fluxes of nutrients and the mobility of trace elements under well-defined conditions by controlling environmental parameters like pH, salinity, redox potential and oxygen concentration. A schematic view of the LABOSI apparatus is shown in Fig. 2.1.

The LABOSI apparatus contains six sediment cores (diameter = 19 cm) resulting in a total sediment area of 1700 cm^2 over that water (water volume: 26 litres) is streamed with a well defined velocity. The cores and the water are thermostated at 20°C. The water circuit is coupled via an efficient gas exchange device to a closed gas loop (1.9 litres). By regulation of the partial oxygen pressure in the gas loop, a stable oxygen concentration in the water phase can be achieved between 0.05 mg/l and saturation and the oxygen consumption can be continuously observed via the the consumed volume of oxygen. The pH in the system is controlled by a pH-meter and a pH-stat (Metrohm Impulsomat 616 and burette 632) that doses an acid (HCl) or a base (NaOH) into the system. The changes of the alkalinity in the water phase can be calculated directly from the consumed acid or base. The signals of all sensors (oxygen concentrations for gas and water phase, pH, redox potential, flow rate and turbidity) are continuously recorded. The pressure inside the closed system will be automatically kept at 1.06 bar by a motor driven piston, regardless of sampling in the water or gas phase. A detailed description can be found by Schroeder et al. (1992). At fixed intervals water samples were automatically taken using a peristaltic pump and analysed on-line for trace elements with a computer controlled polarographic system (Metrohm processor VA646 coupled with a DOS compatible computer). By differential pulse polarography (DPP) Fe^{2+} and Mn^{2+}

Fig. 2.1. Scheme of the laboratory simulation apparatus (LABOSI)

were determined. The concentration of $\sum H_2S$ was measured with differential pulse cathodic stripping voltammetry (DPCSV). The free and labile fraction of the heavy metals (Cu, Pb, Cd and Zn) was determined with differential pulse anodic stripping voltammetry (DPASV).

Water samples for nutrient analysis were taken with a cooled autosampler and measured off-line with ion chromatography (NO_3, NO_2 and Cl) and photometry (NH_4, PO_4). Carbon dioxide in the water phase was determined by an ion selective electrode (samples acidified to pH = 5). The total concentration of trace elements (As, Co, Ni, Zn, Cr, Cd, Cu and Pb) was also off-line measured by atomic absorption spectrometry (AAS).

Sediments were collected in March 1991 from the location *Mühlenberger Loch*, a tidal mud-flat 5 km downstream of Hamburg characterized by low water velocities and high sedimentation rates (Petersen et al. 1990). For the incubation experiment we used 30 litres of surface sediment (0–7 cm depth) which was suspended in 100 litres of Elbe water from the same sample site. The suspension was homogenized with a stirring-device and vented with air. After 15 h the solid phase was separated by centrifugation and filled in six tubes of a 19 cm diameter for the LABOSI apparatus and in four tubes of a 10 cm diameter for single core experiments. River Elbe water (filtered through a 0.2 μm membrane filter) was added to both the LABOSI experiment and the single core experiment (water volume \approx 1.2 l). In parallel, the overlying water of the single cores and the water of the LABOSI experiment were bubbled with air. In the LABOSI apparatus the water was streamed across the sediment, whereas in the single core experiment the water was only stirred. All

incubation experiments were performed in the dark at 20°C. After 5 days (core I1), 12 days (core I2), 19 days (core I3) and 33 days (core I4) each time, one of the single cores was segmented in 0.2- to 3-cm intervals down to a depth of 10 cm in a glove box under nitrogen atmosphere. The porewater of the slices was squeezed out with a porewater press. For nutrient analysis the interstitial water was frozen to 15°C and for heavy metal analysis the water samples were acidified with HNO_3. The samples of the solid phase were stored in a gas-tight glass vessel filled with nitrogen at -15°C for determining the iron species and the acid volatile sulphide (AVS). These samples were extracted with 1 N HCl for 24 h at room temperature and the extracted ferric and ferrous iron was determined with DC and AC polarography. During the first 1 h of extraction, AVS was determined from H_2S trapped in zinc acetate. Details are described by Wallmann et al. (1993). The fraction ≤ 20 μm was obtained by sieving. The total concentration of trace elements in the solid phase was determined after decomposition in concentrated HNO_3 with AAS at 80°C. The cores of the LABOSI experiment were not analysed. After 8 days in both the single core and LABOSI experiments the pH in the water phase was adjusted to pH = 7.5 by dosages of 0.1N HCl. In the LABOSI experiment this adjustment was performed automatically whereas the four single cores were adjusted periodically by the addition of HCl.

2.3 Results and Discussion

The different phases of the experiment and the corresponding concentrations in the overlying water of the LABOSI experiment (oxygen, pH, nitrate and ammonium) are shown in Figs. 2.2 and 2.3.

In phase 1, the LABOSI apparatus was open and the flooded water was still vented with air. The pH increased soon after the starting of the experiment to values of pH = 8.3 due to CO_2 volatilization. After regulation of the pH with HCl, the pH was kept constant at 7.5. After 26 days the apparatus was closed and the pH was no longer regulated. The oxygen was reduced down to 0.1 mg/l by flushing the gas–loop with nitrogen and then kept constant for the next 123 h (phase 2). The pH decreased to pH = 6.8. In phase 3 after a total experimental period of 813 h, the oxygen had been completely removed and for 182 h the water phase was anoxic. In the last phase (phase 4) the oxygen concentration was increased to 0.2 mg/l and then to saturation (8 mg/l). The pH in the anaerobic phase (phase 3) increased to pH = 7, whereas in the following oxic phase the pH decreased again down to 6.8 due to the release of protons by oxidation processes.

In the oxic phase during the first 70 h (phase 1) the measured concentrations of nutrients (NH_4, NO_3) showed a strong decrease. After that time only NO_3 slowly further decreased from 200 to 120 μmol/l, whereas NH_4 had relatively stable values of 11 μmol/l. This indicates that at that time, the nitrate formed by nitrification of ammonium, was nearly compensated by the diffusive flux of nitrate into the sediment. After 400 h there was a leakage in the water circuit and 20% of the water

Fig. 2.2. Evolution of oxygen and pH in the different phases of the LABOSI experiment

Fig. 2.3. Evolution of nitrate and ammonium in the different phases of the LABOSI experiment

Fig. 2.4. Cadmium and zinc release in the LABOSI experiment in the oxic phases 1 and 2 and the anoxic phase 3

was replaced by Elbe water with higher ammonium and nitrate concentrations. In phase 2 (O_2 = 0.1 mg/l) nitrification that resulted in an increase of NH_4 in the water phase due to diffusion from the sediment no longer occurred, whereas nitrate decreased to a level of 14 μmol/l and was completely depleted in the anoxic phase (phase 3).

2.3.1 Behaviour of Trace Elements in the Oxic Phases

The behaviour of selected trace elements in the LABOSI experiment for the two oxic phases is shown in the Figs. 2.4 to 2.6. Figure 2.4 depicts the mobilization of the trace elements zinc and cadmium after adjustment of the pH. The cadmium concentrations increased from 0.2 nmol/l to 13 nmol/l (net flux = 110 nmol $m^{-2}d^{-1}$) and the zinc concentrations from 500 to 1900 nmol/l (net flux = 28 μmol $m^{-2}d^{-1}$). This can be attributed to the mineralization of organic substance as well as to the oxidation of sulphidic species that, however, cannot be observed in the concentrations of sulphate due to the relatively high background levels (1.5 mmol/l) of sulphate.

Both elements show a decrease after 400 h due to the dilution by the added new water as described in Sect. 2.3. The decrease continued over 620 h, after which both metal concentrations increased again. This may be attributed to the higher organic content (algae) of the new water which first bound the trace metals and after degradation of this organic matter released the trace elements again. In

Fig. 2.5. Copper and nickel release in the LABOSI experiment in the phases 1–2

phase 2 at low oxygen concentration, first cadmium and then also zinc decreased, revealing a precipitation of the corresponding sulphide minerals at very low sulphide concentrations which were not detectable (\leq10 nmol/l).

A similar behaviour was observed for copper and nickel (Fig. 2.5). Following the pH regulation, both elements were released (net flux: Cu = 1.2 μmol m^{-2}d^{-1}, Ni = 1.7 μmol m^{-2}d^{-1}). After dilution with fresh water, rich in organic matter, the concentrations reflected a similar behaviour as with Cd and Zn. At low oxygen concentrations in phase 2, only copper was already partly removed from the bottom water due to precipitation of sulphidic minerals. This was not the case for nickel because the nickel sulphide is much more soluble.

Since the ratio between the bottom water volume (V_w) and the covered sediment area (A_s) is comparable for the experiment with single cores and the LABOSI experiment (single cores: $V_w/A_s \approx 130$, LABOSI: $V_w/A_s \approx 150$) the rates of trace element releases should also be comparable. Figure 2.6 depicts this comparison for the release of nickel in both experiments within the first 400 hours. The concentrations of the single cores were calculated from the average of three incubated cores. As the Fig. 2.5 shows, the results agree well within an estimated experimental error of approximately 20%. However, due to the smaller sediment area the single cores showed a higher variability.

Fig. 2.6. Comparison of the evolution of the nickel concentration between the incubation in the LABOSI experiment and the incubated single cores (average of three parallel incubated cores)

2.3.2 Behaviour of Trace Elements in the Anoxic Phase

In the following anoxic phase (phase 3 after 810 h) the evolution of trace element concentrations was determined by the release of sulphide and reductive formed Fe(II) and Mn(II). Thus, cadmium and zinc were nearly completely removed due to sulphide precipitation as depicted in Fig. 2.4.

The evolution of the sulphide concentrations together with the observed phosphate release is shown in Fig. 2.7 for phases 2 and 3. Soon after oxygen concentration reached zero, phosphate was mobilized and after 115 h detectable concentrations of $\sum H_2S$ (\geq 10 nmol) arose.

After complete depletion of other electron acceptors (e.g. NO_3 and NO_2) the onset of iron release, as well as of arsenic and chromium, could also be observed as depicted in Fig. 2.8.

The evolution of the concentrations in the overlying water of manganese and cobalt is shown in Fig. 2.9. Manganese was already released into the bottom water at low oxygen concentrations (O_2 = 0.1 mg/l)) due to diffusion of Mn(II) through the oxic layer (net flux = 1.2 mmol m^{-2}d^{-1}). Mn(II) already increased in the low oxic phase in contrast to Fe(II) because it was not reoxidized as fast as Fe(II) and the thickness of the oxic layer becomes thinner. However, the release of manganese is further accelerated in the anoxic phase with a net flux of 5.3 μmol m^{-2}d^{-1}, which is in the same range as the net flux of Fe(II) (net flux = 7.8 μmol m^{-2}d^{-1}). In addition

Fig. 2.7. Sulphide and phosphate concentrations in the overlying water in phases 2 and 3

Fig. 2.8. The evolution of iron, arsenic and chromium in the water phase (phases 2–4)

Fig. 2.9. The evolution of manganese and cobalt in the water phase (phase 2–4)

to the accelerated release of manganese, the cobalt concentration increased.

The migration of phosphate as well as of arsenic from the sediment to the water phase under anoxic conditions can be attributed to iron mobilization since they are associated to iron compounds (Sundby et al. 1986, Schwedhelm et al. 1988) whereas Co should be more associated with manganese (Heggie et al. 1984). In the case of chromium this element is itself redox sensitive and will be reduced to Cr(III) (Emerson et al. 1979).

In the following reoxidation phase, both phosphate and the trace elements were removed from the overlying bottom water as fast as iron decreased, as a result of adsorption and/or co-precipitation on the freshly formed hydrous ferric oxides.

In summary, the measurements show that many trace elements which are not so strongly associated to the organic matter only escape to the bottom water if the sediment–water interface becomes anoxic even though the porewater profiles show a strong concentration gradient which should result in a diffusive flux to the bottom. To investigate the barrier that stops the diffusive flux of trace contaminants we examined the trace element profiles of the solid phase of cores which were segmented after different periods of time. At the start of the experiment the sediments were completely homogeneous. After 5 days, the concentration profiles of the porewater and the solid phase of the first segmented core showed the formation of typical zones which have been determined by early diagenetic processes. The porewater profiles showed high iron and also arsenic concentrations at depths \geq 5 mm, whereas in the upper millimeters no or much less iron or arsenic was observed (Fig. 2.10 A).

The single core segments were extracted with 1 N HCl to investigate the "re-

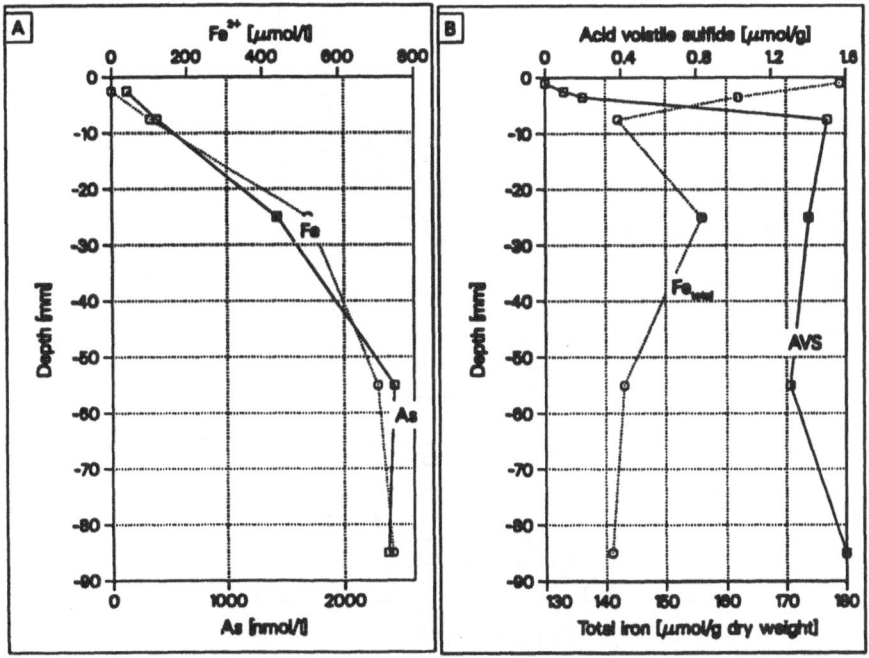

Fig. 2.10. A Concentrations of Fe and As in the interstitial water in core I3 (incubation time of 19 days). B Concentration profiles of the extractable "reactive" iron and acid volatile sulphides (AVS) of core I2 (incubation time of 12 days) after extraction with 1 N HCl

active" extractable iron from the solid. As can be seen in Fig. 2.10 B the total extractable "reactive" iron (Fe(II) and Fe(III)) is enriched in the upper zones of the sediment and shows much higher values than the extracted iron in the homogenous sediment at the start of the experiment (127 μmol/g). Conversely, the acid volatile sulphides (AVS) are enriched in the deeper anoxic zones of the sediment (Fig. 2.10 B).

The oxygen and pH profiles that were measured with microelectrodes with much higher vertical resolution showed a penetration depth of oxygen of 2 mm. At this depth the pH profile also first reached constant values of pH = 7.5. The measured pH and oxygen profiles are depicted in Fig. 2.11. According to Revsbech and Joergensen (1986) the oxygen demand of the core can be calculated from the diffusive boundary layer of these profiles. A flux of 14 mmol m^{-2} d^{-1} was estimated from these profiles.

The enrichment of iron can be attributed to the oxidation of Fe(II) formed in the anoxic zones that was transported by diffusion to the oxic layer of the sediment and

Fig. 2.11. Oxygen and pH profiles of core I3 (incubation time of 19 days)

there reoxidized.

Parallel to the enrichment of iron, an enrichment of the trace element arsenic was detectable. These measured concentrations were compared with the calculated fluxes to the uppermost layer. The diffusive flux from deeper depths was calculated by Fick's first law from the gradients of the concentration in the interstitial water. Diffusion coefficients were taken from Li and Gregory (1974) and the porosity was determined from the water content. The fluxes to the overlying water were calculated from the changes of the concentration in the overlying water. The calculations were compared with measured concentrations of the elements at the solid phase of the sediments in this layer after certain periods of time.

Figure 2.12 shows the calculated enrichment of arsenic in the oxic layer (depth = 0–5 mm) of the parallel incubated sediment cores which were segmented after 5, 12 and 19 days of incubation.

The diagram shows that the measured concentrations (enrichment = 40%) were even higher than the estimated concentrations (enrichment = 17%). It is likely that other contaminants also will be enriched in the oxic layer of the sediment. However, due to the higher backgrounds in the sediment and lower concentration gradients in

Fig. 2.12. Estimated and measured concentrations of arsenic in the solid phase (fraction $\leq 20 \ \mu m$) in the uppermost oxic sediment layer after different period of time of parallel incubated cores

the interstitial water due to the analytical errors this cannot be detected.

In summary, under oxic conditions the experiments show that from freshly deposited material, cadmium, zinc and copper are released to the bottom water as a result of the mineralization of organic substances. The diffusive flux of manganese was enhanced under low oxygen concentrations (0.1 mg/l) due to the reduction of the thickness of the oxic layer. The diffusion of reductively formed Fe(II) from deeper layers to the oxic layer leads to a formation of iron oxides which are enriched at the boundary between oxic and anoxic layer. These freshly formed ferrous iron oxides may act as a natural barrier for the diffusive flux of trace contaminants into the overlying water as could be shown for arsenic. Under anoxic conditions the reductive dissolution of hydrous ferric oxides leads to a strong release of the oxide bound trace elements (e.g. As, Co, Cr) that are not precipitated by formation of sulphide minerals as well as phosphate.

Trace elements that form insoluble sulphide minerals at very low sulphide concentrations (≤ 10 nmol/l) will be removed from the water phase.

Following re-oxidation of the water phase the removal of Fe(II) from the water phase is strongly correlated with the removal of arsenic, cobalt, chromium and phosphate indicating a co-precipitation with, or an adsorption on freshly formed hydrous ferric oxides.

Acknowledgements. This work was supported by the Ministry of Research and Technology. We would like to thank Ms. V. Felsen for heavy metal determination

with AAS, Mr. M. Bormacher for nutrient analysis, Mr. H. Gramm for microelectrode measurements, Mr. K. Hennies for determining the "reactive" iron and AVS, and the crew of the research vessel MS Ludwig Prandtl for their assistance in sample collection.

References

Emerson S, Cranston RE, Liss PS (1979) Redox species in a reducing fjord: equilibrium and kinetic considerations. Deep-Sea Research 26A: 859–878

Heggie D, Lewis T (1984) Cobalt in porewaters of marine sediments. Nature 311: 453–455

Li Y-H, Gregory S (1974) Diffusion of ions in sea water and in deep-sea sediments. Geochim Cosmochim Acta 38: 703–714

Petersen W, Knauth H-D, Pepelnik R, Bendler I (1990) Vertical distribution of Chernobyl isotopes and their correlation with heavy metals and organic carbon in sediment cores of the Elbe estuary. The Science of the Total Environm 97/98: 531–547

Petersen W, Wallmann K, Schröer S, Schroeder F (1993) Studies on the adsorption of cadmium on hydrous iron(III) oxides in oxic sediments. Anal Chim Acta 273: 323–327

Revsbech NP, Joergensen BB, (1986) Microelectrodes: Their use in microbiological ecology. In: Marshall KC (ed) Advances in microbiol ecology, vol 9. Plenum, New York, pp 293–352

Schroeder F, Klages D, Blöcker G, Vajen-Finnern H Knauth H-D (1992) The application of a laboratory apparatus for the study of nutrient fluxes between sediment and water. Hydrobiologia 235/236: 54–552

Schroeder F, Klages D, Knauth H-D (1990) Bell-jar experiments and suspension measurements for the quantification of oxygen consumption and nitrogen conversion in sediments of the Elbe estuary. In: Michaelis W (ed) Coastal and Estuarine Studies 36, Conference Proceedings of "Estuarine Water Quality-Management", Springer, Berlin Heidelberg New York, pp 253–258

Schwedhelm E, Vollmer M, Kersten M (1988) Bestimmung von Konzentrationsgradienten gelöster Schwermetalle an der Sediment/Wasseroberfläche mit Hilfe der Dialysetechnik. Fresenius Z Anal Chem 332: 756–763

Sundby B, Anderson LG, Hall POJ, Iverfeldt A, van der Loeff MR, Westerlund SFG (1986) The effect of oxygen on release and uptake of cobalt, manganese, iron and phosphate at the sediment–water interface. Geochim Cosmochim Acta 46: 1281–1288

Wallmann K (1992) Die Löslichkeit und die Bindungsformen von Spurenmetallen in anaeroben Sedimenten. Vom Wasser 78: 1–20

Wallmann K, Hennies K, König I, Petersen W, Knauth H-D (1993) New procedure for determining reactive Fe(III) and Fe(II) minerals in sediments. Limnology Oceanogr. 38(8): 1803–1812

Westerlund SFG, Anderson LG, Hall POJ, Iverfeldt A, van der Loeff MR, Sundby B (1986) Benthic fluxes of cadmium, copper, nickel, zinc and lead in the coastal environment. Geochim Cosmochim Acta 46: 1289–1296

3 Interaction of Pesticides with River Sediments and Characterization of Organic Matter of the Sediments

G. Abbt-Braun and F.H. Frimmel

3.1 Introduction

Organic contaminants are widespread in the aquatic environment. As the solubility of hydrophobic organic chemicals is limited in water, their accumulation in the liquid phase of natural waters will be low compared to the particulate phase. The chemicals can be sorbed on the sediments as well as on the suspended matter (colloids). A short review about the concentration of typical organic pollutants in sediments is given by Förstner (1989). The mechanisms of the sorption are complex and not yet well understood. From the literature there is considerable data concerning the adsorption on soils (e.g. Calvet 1989 and Khan 1980). The sorption of the organics on the solids is dependent on the properties of the chemicals (e.g. functional groups, size and shape of the molecules) and of the adsorbents (e.g. size distribution, mineral composition, C_{org} content). Several attempts have been made to develop methods of describing the sorption behaviour. Karickhoff et al. (1979) suggested calculation of the distribution of organic chemicals between water and sediment. The partition coefficient (or distribution, sorption coefficient; K_D) is the ratio of the specific concentrations of the organic substances in the dissolved and adsorbed form.

The sorption behaviour of organophosphorus pesticides in natural sediments was studied by Fröbe et al. (1989). The sorption coefficient K_D was found to follow the same sequence as the lipophilicity of the compounds expressed as their n-octanol–water partition coefficient. However, it has been indicated that a simple partition model does not offer an adequate visualization of the sorption behaviour. The sorption of non-polar organic chemicals on aquifer material with low-carbon content was investigated by Stauffer et al. (1989). Neither a predictive

relationship between sorption coefficient and type of aquifer material, nor a correlation with any properties could be found.

Quality criteria for sediments are of fundamental need. Several approaches have been proposed (Chapman 1989; Förstner et al. 1987) and were reviewed by Adams and others (Adams et al. 1992; Burton and Scott 1992). The sediment–water equilibrium partitioning approach uses water quality criteria and a correction factor for the effect of organic carbon (Pavlou 1987; Shea 1988). The porewater concept uses the distribution equilibria of organic contaminants (non-polar or non-ionic ones) between sedimentary organic matter and interstitial water. The distribution coefficients (K_D) for the individual organic contaminants are used to describe the equilibrium concentration in the liquid (porewater) and in the solid phase (sediment). The contaminants are divided into several classes according to their specific properties. For the non-polar organic contaminants the n-octanol–water partition coefficient is related to the K_D value (Chiou et al. 1982; Karickhoff 1981). There is a great lack of data to describe the more polar contaminants.

The difficulties in the description arise also from the fact that pesticides form associates with dissolved organic carbon (Choudry 1980; Huber et al. 1992; Wershaw 1986). Therefore, influences on the adsorption, bioavailability and mobility compared to the free pesticides have to be expected. Lee and Farmer (1989) proposed an interaction of the bulk of organic matter with non-ionic pesticides. They found that the source of the dissolved organic carbon (DOC), the ionic strength and pH of the solution have a special influence on the interaction. A membrane model describing the sorptive interaction of hydrophobic pollutants with humic substances (HS) is given by Wershaw (1989). The binding mechanisms are found to be induced by enzymes (Bollag and Myers 1992).

In addition, the adsorption of the organic chemicals on the sediments is influenced by the DOC of the porewater. Capel and Eisenreich (1990) described the distribution of chlorinated hydrocarbons in sediment and porewater. They found that the affinity of the contaminants was different for several organic pools. It was relatively high for the bulk sediments but only about half the amount for the porewater compared with the solid phases (larvae or sediment).

Colloidal organic matter plays a specific role in binding organic compounds. It seems that the interaction of contaminants with the dispersed organic matter has similarities to the interaction with sediments and appears to be closely correlated to fundamental properties such as solubility and n-octanol–water partition coefficient (Sigleo and Means 1990). An equilibrium partitioning model which includes the organic load of the aqueous phase and the interstitial porewater is described by Hunchak-Kariouk and Suffet (1991).

Two general types of organic material were found to be present in the sediment samples (Evans et al. 1990): fragmentary plant material and aged condensed organic material coated on the sediment particles. It was obvious that the two types of organics showed different affinities for polycyclic aromatic hydrocarbons (PAH) due to an increasing aromaticity with increasing ageing of the organic ma-

terial. Humic substances (HS), the main fraction of this organic material, play an important part in the distribution of pollutants in the liquid and solid phase (Frimmel and Christman 1988).

Detailed investigations of the porewater concerning oxidation reactions were done by Orem and Gaudette (1984). As the anoxic material is oxidized, DOC was lost due to the coprecipitation of ferric oxihydrate. Structural changes in the dissolved organic matter due to the oxidation have also been indicated. Evidence is presented for the partitioning of hydrophobic chemicals in anoxic porewater, being significantly different to the partitioning measured in porewater saturated with oxygen (Hunchak-Kariouk and Suffet 1991).

The results from several approaches clearly show the need for further research. The objectives of this work were:

1. To give typical adsorption isotherms for the interaction of polar and non-polar pesticides with river sediments
2. To characterize the organic material in the porewater and the C_{org} sorbed on the sediments
3. To correlate the adsorption data with the C_{org} content and its character of the different sediment and soil materials

By this, a contribution to the understanding of the complex mixture of organic pollutants and sediments is made.

3.2 Experimental

Sediment-Material and Characterization. As adsorbents, sediments of the rivers Danube (DO1-S), Elbe (EL1-S and EL2-S), Main (MA1-S), Neckar (NE1-S), Rhine (RH3-S) and Weser (WE1-S) were used. In addition, sand of the aquifer from a bankfiltration site near the Rhine River at Düsseldorf (GR2-C-S), soil samples from a rendzina (E3) and sediment from a bog lake (HO6-S) were investigated. The mineral compositions of the sediment samples were determined by AAS and ICP / AES after pressure digestion with H_2SO_4 / HNO_3. The particle size distribution of the solid samples was estimated by wet sieving (> 6 μm) and by the method of Coulter (< 6 μm; Coulter 1953, 1956). The data characterizing the adsorbents are shown in Tables 3.1 and 3.2.

Pesticides. 2.4-Dichlorophenoxyacetic acid (2.4-D, Merck), atrazine (Riedel de Häen) were used without further purification. The same applies to metazachlor, metolachlor and simazin (all purchased by Labor Dr. Ehrenstorfer; see Table 3.3).

Batch Experiments. The adsorption experiments were performed as batch reactions under controlled temperature conditions (13°–15°C) in the dark. For the experiments 50- or 100-ml aqueous solution and 5 g or 10 g of the solid sample (dry

Table 3.1. Characteristic data for the sediments (C_{org} content was measured by elemental analysis)

Sample code	Origin	Fraction of the sample	Weight loss at 700 °C [%]	Weight loss at 550 °C [%]	C_{org} [%]	Particle size distribution
DO1-S	Sediment; Danube (damming, Ulm)	Total, moist	31.9	27.2	14.8	10% sand 82.5% silt 7.5% clay
E3	Soil; (Schloßpark, Karlsruhe)	< 0.6 mm, dry	16.0	16.0	8.0	–
EL1-S	Sediment; Elbe (Mühlenberger Loch, Hamburg)	Total, moist	5.2	5.0	2.5	–
EL2-S	Sediment, Elbe (Mühlenberger Loch, Hamburg)	Total, moist				
core:						
1– 2 cm			8.4	9.0	3.4	–
2– 3 cm			9.6	9.4	3.6	–
3– 4 cm			7.6	9.0	3.3	–
5– 6 cm			4.2	4.4	1.9	–
6– 7 cm			3.4	4.2	1.7	–
8– 9 cm			1.8	2.4	1.2	–
10–11 cm			3.0	4.2	1.7	–
11–13 cm			4.6	5.4	2.4	–
GR2-C-S	Sediment; Rhine (bank-filtration site, Düsseldorf)	< 0.6 mm, dry	0.2	0.2	0.1	–
HO6-S	Sediment; Hohlohsee (boglake, Schwarzwald)	Total, moist	85.0	–	–	–
MA1-S	Sediment; Main (lock, Besigheim)	Total, moist	–	15.2	4.9	22% sand 60% silt 18% clay
NE1-S	Sediment; Neckar (lock, Kleinkostheim)	Total, moist	–	15.6	5.7	13% sand 75% silt 12% clay
RH3-S	Sediment; Rhine (damming, Iffezheim)	Total, moist	18.0	10.8	4.6	8% sand 84% silt 8% clay
WE1-S	Sediment; Weser (harbour-basin, Bremen)	Total, moist	7.0	6.6	2.8	58% sand 40% silt 2% clay

Table 3.2. Metal composition of sediments after pressure digestion by H_2SO_4/HNO_3 (main components, calculated as oxides, % of dry weight)

Sediment	Fe_2O_3	MnO	Al_2O_3	CaO	MgO	Ash
EL2(1-2)-S	4.3	0.3	6.2	5.1	1.3	67
EL2(6-7)-S	2.1	0.1	2.9	3.7	0.8	81
MA1-S	4.1	0.1	8.9	4.0	1.4	58
NE1-S	4.3	0.1	9.1	9.0	2.6	49
RH3-S	4.4	–	9.1	11.1	3.0	49
WE1-S	3.9	0.2	4.9	1.6	1.0	76

weight, dr. wt.) were shaken for equilibration in a 100 or 250 ml glass-stoppered flask. The time of agitation was different for the rate experiments (1–192 h). The following matrices were applied: H_2O; 0.01 M $CaCl_2$; 0.34 M NaCl and a phosphate buffer (100 ml containing 61 ml 0.066 M KH_2PO_4; 39 ml 0.066 M $Na_2HPO_4 \cdot 2H_2O$). After the equilibration time, the supernatant was centrifuged for approximately 30 min (3600 RPM). The centrifuged solution was filtered through a membrane (cellulose nitrate with a pore size 0.45 μm for 2.4-D, and polycarbonate with a pore size 0.45 μm for atrazine, simazine, metolachlor and metazachlor). With the loaded solid phase the desorption experiments were done after adsorption of the chemicals, as mentioned above. The aqueous solution was changed for pesticide-free water, and the mixture was agitated. The dissolved pesticides were analyzed after different contact times of water and sediment.

Pesticide Analysis. The determination of the pesticides was done by HPLC (Table 3.3). As the initial concentrations were high enough, mostly no preconcentration steps were required. Wherever necessary, the pesticides were preconcentrated using C-18 cartridges.

Estimation of the C_{org} Content. The solid samples were dried at 105°C to constant weight. The residue after ignition and the loss of ignition were determined at 550°C according to DIN 38414 (DIN 1992) and at 700°C. The organic carbon content was calculated from the weight loss (x 0.5). In addition, elemental analysis was done according to Abbt-Braun et al. (1990). The inorganic carbon was volatilized by acidification, HCl (1:1).

Determination of C_{org}: Isolation of C_{org}. Soil and sediment samples were extracted with bidistilled water or aqueous pyrophosphate (0.1 M; pH = 7) for 48 h in the dark. The ratio of the solid sample (g) to the aqueous solution volume (ml) was different for the sediment samples – 1:1 (GR2-C-S), 1:7 (E3), 1:9 (WE1-S), 1:10 (EL1-S), 1:12 (RH3-S), 1:15 (MA1-S), 1:15 (NE1-S) and 1:80 (HO6-S). The extract was centrifuged before filtration (cellulose nitrate membrane 0.45 μm).

Table 3.3. Physical and chemical data of the pesticides and the experimental set-up for their analytical determination

Pesticide	2.4 - D	Atrazine	Metazachlor	Metolachlor	Simazine
Solubility in water [mg/l]	500 (25°C)	30 (22°C)	17 (20°C)	530 (20°C)	5 (20°C)
Analysis (HPLC)					
Column	RP-C-18	RP-C-18	RP-C-18	RP-C-18	RP-C-18
Injected solution	Aqueous	Aqueous	Aqueous	Aqueous	Aqueous
Eluent	MeOH/ AcOH/ H$_2$O (29 : 3 : 26)	H$_2$O/ Acetonitrile (gradient)	H$_2$O/ Acetonitrile (gradient)	H$_2$O/ Acetonitrile (gradient)	H$_2$O/ Acetonitrile (gradient)
Flow	1 ml/min	0.5 ml/min	0.5 ml/min	0.5 ml/min	0.5 ml/min
Detection (UV)	280 nm	241 / 230 / 254 nm	220 nm	220 nm	220 nm
Calibration	0.4 - 4.0 mg/l	0.025–5.0 mg/l	0.1–2.0 mg/l	0.1–2.0 mg/l	0.05–10 mg/l
Relative SD[a]	5%	10%	7%	10%	10%

[a] Relative standard deviation

From aqueous samples, the organic matter was isolated according to the XAD-2 method described elsewhere. (Abbt-Braun et al. 1991). The samples were kept in the protonated form as freeze-dried material.

Characterization Methods. Dissolved organic carbon (DOC) was determined after membrane filtration (cellulose nitrate 0.45 μm) using a UV-DOC analyzer (Dohrmann). C, H, N, O and S were determined in these freeze-dried samples (Abbt-Braun et al. 1990). The spectral absorbances at $\lambda = 254$ nm and $\lambda = 436$ nm were measured with a spectrophotometer (Lambda 5, Perkin Elmer) using 1-cm and 5-cm quartz cells and bidistilled water as a reference. IR-spectra were obtained using KBr-disks and a FT-IR-spectrometer (FTS-50, Digilab). The molecular weight distribution (analytical scale) was determined according to Fuchs (1985, 1986) using a TSK-HW-40(S) gel and DOC detection (DOC Analyzer, Gräntzel). In addition, HPLC experiments were run according to Gremm et al. (1991) using a C18-resin and on aqueous mobile phase with a flux of 0.3 ml/min (0.01% AcH (A); CH$_3$CN (B); gradient: 0 min 99% A; 2 min 70% A; 16 min 15% A; B = 100% A). In addition, a TSK-resin (G/2000 SWXL) was used as stationary phase and a phosphate buffer (2.5 mmol/l, pH = 7) as mobile phase with a flux of 1 ml/min. The proton capacity (H-Cap) was determined by titration of air free solution with an autotitrator (titrator DL 25, Mettler) under nitrogen atmosphere (Frimmel et al. 1985). The anionic particle charge (aPC) was measured by titration with 10^{-4} M poly DADMAC (diallyldimethyl-ammoniumchloride) and streaming current detection (PCD 02, Mütec) as described by Weis and Frimmel (1989).

3.3 Results and Discussion

3.3.1 Adsorption Behaviour

Sediment Characteristics. In order to get a broader view, different sediment samples were taken from several great rivers in Germany. In addition to the rivers, there was a variation in the detailed origin of the samples (harbour basin, damming, lock and bankfiltration site).The chemical and physical characteristics of the sediment material were clearly different concerning C_{org} content, particle size distribution and mineral composition. The C_{org} content (dr. wt.) of the river sediments varied between 2.8 (WE1-S) and 14.8% (DO1-S). The mean values were around 5% (MA1-S; NE1-S; RH3-S). The C_{org} content of the core (EL2-S) was at first decreasing with depth and was increasing at a 10-cm depth. The silt content was higher than 50% for all samples with the exception of the WE1-S sample, and the sand and clay content was quite different. The remaining ash content after wet decomposition was high (50–80% of dr. wt.). This was due to a high amount of silicates in the sediments. In case of MA1-S and NE1-S, quartz and feldspar were the main mineral components as shown by difractometric measurements.

Sediment Preparation for the Estimation of the Adsorption Behaviour of Organic Chemicals. In order to get the required validity and precision for the prediction of the sorption capacity of organic chemicals on sediments, some points concerning the pretreatment of the solid samples have to be considered. For soil samples, detailed instructions for the preparation are given by the OECD guideline (OECD 1981). A comparison of different pretreatment procedures was also described by Oepen et al. (1989). In handling the sediment samples, the possibility of alterations of the chemical forms of inorganic and organic components and of the biological potential of the samples have to be considered.

Standard analytical techniques for sampling, pretreatment and analytical requirements are less available for sediment studies than for water and soil studies. Some procedures were described by DIN (1993) and Kersten and Förstner (1987).

The chemical and physical characteristics of a sediment sample are changing during drying. Therefore, there is an influence on the sorption behaviour of organic chemicals as shown in Fig. 3.1.

The solid phase concentration of 2.4-D was higher for the sediments used in the original wet form as compared to the sediment sample dried at room temperature, rewatered and equilibrated by an aqueous solution. The adsorption ability of sediments for 2.4-D changed with storage time. Keeping the sediment samples at low temperature (refrigerated) for three months led to a higher adsorption of 2.4-D. The results were in the range of the results gained by the comparison of the dried and wet sediments.

To obtain meaningful results from laboratory experiments, it is recommended

Fig. 3.1. Comparison of the adsorption isotherms of 2.4-D on the Neckar River sediment (NE1-S wet and NE1-S dried and rewatered; sediment:solution = 1:10 / w:w)

to keep the sediment samples in the original aqueous matrix, and the experiments have to be done soon after sampling to minimize alteration of the material by chemical and biological processes.

Fig. 3.2. Adsorption kinetics of atrazine on the Main River (MA1-S) sediment (sediment:solution = 1:10 / w:w; c(0) = 0.5 mg/l)

Adsorption Rates. According to the OECD guideline for testing the adsorption and desorption of chemicals on soils, the time for equilibration was set up to 16 h. The adsorption rates were determined for the different types of sediments and organic chemicals. For atrazine and 2.4-D, initial concentrations of 0.5 mg/l and 4 mg/l, dissolved pesticides were used. The kinetics of the adsorption of atrazine and 2.4-D on sediment MA1-S are shown in Figs. 3.2 and 3.3.

The data show that the equilibrium is reached after 72 h for atrazine. There is no significant influence of the aqueous matrix. In case of 2.4-D, an equilibration time cannot be found in the different solutions. It seems that two different plateaus for equilibration are obtained. Only in the 0.34 *M* NaCl solution equilibration was reached after 24 h. The results can be explained with the fast biological degradation of 2.4-D. In case of the 0.34 *M* NaCl solution, the biological activity seems to be reduced to a far extent, as the cell membranes of the bacteria are influenced by the high salt concentration (osmosis). To investigate the possibility of avoiding biotic degradation of the added chemicals during sorption tests, several chemical methods of sediment sterilization were tested, as shown in Figs. 3.4 and 3.5.

The results for the system control experiment which had no sediment material, show clearly that there was no influence of the added toxic chemicals on the analytical determination and on the stability of 2.4-D (Fig. 3.4). The remaining liquid concentration in the samples not sterilized, was much lower after 192 h compared to 24 h. Biotic degradation can again explain the decrease of 2.4-D after 192 h. The adsorption rates of 2.4-D on sediments in the presence of NaN_3, formaldehyde and $HgCl_2$ showed the same tendency. The concentration of 2.4-D in the liquid phase was in the same range after 24 h and 192 h. This leads to the conclu-

Fig. 3.3. Adsorption kinetics of 2.4-D on the Main River (MA1-S) sediment (sediment:solution = 1:48 / w:w; c(0) = 4 mg/l)

Fig. 3.4. Influence of several sterilization methods on the adsorption of 2.4-D on the Neckar River (NE1-S) sediment (1% NaN₃; 10% formaldehyde; 500 mg/l HgCl₂; 0.4% peracetic acid; in 0.01 M CaCl₂ solution)

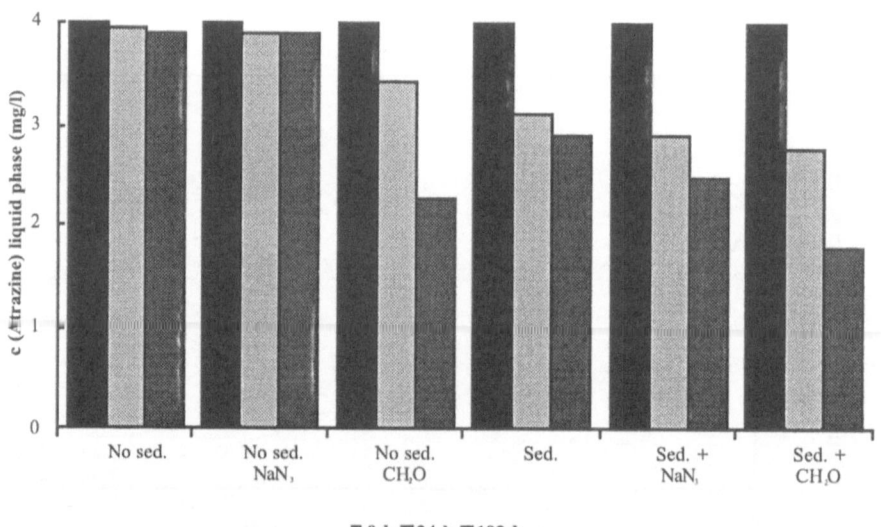

Fig. 3.5. Influence of several sterilization methods on the adsorption of atrazine on the Neckar River (NE1-S) sediment (1% NaN₃; 10% formaldehyde; in 0.01 M CaCl₂ solution)

sion that the sorption behaviour was not influenced by the toxic chemicals added for stabilization, and that the equilibration was reached after 24 h. The inhibition by peracetic acid turned out to be unsuccessful, as the remaining concentration of 2.4-D after 192 h in solution was much lower compared to the other chemicals. Therefore, it is useful in the case of 2.4-D, to take 24 h as equilibration time for the different kinds of aqueous solutions.

As the biotic degradation of atrazine is much slower, and as the equilibration is reached after 72 h, there is no need for sterilization in setting up adsorption isotherms. The addition of formaldehyde led anyway to errors in the analytical determination of atrazine (Fig. 3.5). It is also obvious that the sorption behaviour of the sediments is influenced by the toxic chemicals. The remaining liquid phase concentration of atrazine was much higher in the untreated samples than in the samples with NaN_3 or formaldehyde.

In contrast to the OECD guideline, the equilibration time has to be extended for sediment samples. The data show that equilibrium was reached for the sediment samples after 72 h for atrazine and after 24 h for 2.4-D. For atrazine and the soil sample, equilibration time was 96 h.

Adsorption Equilibria. As the chemical composition of the aqueous phase (river-water) is dependent on individual effluents in rivers or estuarine regions, the specific influence of the matrix of the solution on the adsorption/desorption behaviour of pesticides is of fundamental interest. Additional effects on the sorption capacity may also arise from the concentrations of different inorganic and organic constituents in the porewater.

To investigate the influence of the matrix of the solution, the adsorption of atrazine and 2.4-D on the sediments was studied using bidistilled water, 0.01 M $CaCl_2$ (according to the OECD guideline for testing of chemicals; OECD 1981), 0.34 M NaCl, a phosphate buffer (pH = 7) and the corresponding water of the river. Some results are shown in Table 3.4. In general, the adsorption of atrazine was higher than the one of 2.4-D. This might be due to the less polar character of atrazine.

Table 3.4. Influence of the matrix of the solution on the adsorption behaviour of 2.4-D (c(0) = 1 mg/ and atrazine (c(0)= 1 mg/l) on sediments and soil (soil:solution = 1:10; 1:5*; w:w)

System	Dissolved concentration after equilibration (% of the initial concentration)									
	2.4 - D					Atrazine				
	E3	MA1-S	NE1-S	RH3-S	WE1-S	E3	MA1-S	NE1-S	RH3-S	WE1-S EL1-
H_2O	74	80	90	93	98	38	74	78	76	80 55
0.01 M $CaCl_2$	40	80	80	72	91	41	69	62	70	79 67
0.34 M NaCl	52	90	90	-	-	31	73	76	78	80 79
Buffer (pH=7.0)	89*	80	90	-	-	-	81	67	-	77 70
River water	-	80	90	-	-	-	80	83	-	- -

The electrolyte in the liquid sample did not significantly influence the adsorption of atrazine. Compared to the soil samples, the sediments show lower adsorption capacities and were only little influenced by different solutions. It is obvious that the composition of the aqueous phase influenced the amount of 2.4-D adsorbed on the soil sample. This may be explained by the adsorption of cations on the solid phase which leads to a positive charge and as a consequence to a preferred adsorption of the 2.4-D anion. For sediment samples, the influence of the composition of the solution to the adsorption behaviour is not so obvious. The adsorption of the two pesticides on sediments was generally higher in the presence of 0.01 M $CaCl_2$ as compared to the other matrices. This might be due to ion pair formation in case of 2.4-D leading partly to uncharged molecules.

Desorption. In addition to the adsorption behaviour of chemicals on sediments, the knowledge of the desorption kinetics is a powerful tool for the evaluation of the quality of sediment samples. The investigations show that the desorption behaviour for two kinds of chemicals is different (Figs. 3.6 and 3.7).

The kinetics of the desorption of atrazine is faster than for 2.4-D. But the amount of the pesticides desorbed is different for several sediments. Within 1 h 60–70% of atrazine is desorbed from the sediments of NE1-S and MA1-S, whereas, only 30% is desorbed from RH3-S. Within 1 and 2 days the concentra-

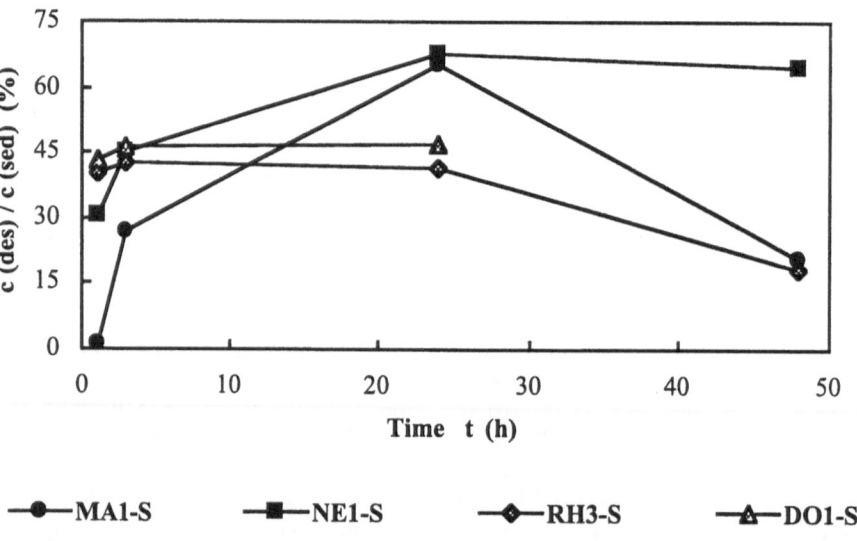

Fig. 3.6. Desorption of 2.4-D from sediments of the Main River (MA1-S), Neckar River (NE1-S), Rhine River (RH3-S) and the Danube River (DO1-S); sediment:solution=1:10/ w:w; 0.01 M $CaCl_2$; c(0) (initial concentration, solution before adsorption) = 16.0 mg/l; c(sed) = c(0) - c (liquid phase concentration after adsorption); c(des) = c (liquid phase concentration after desorption)

tion in the liquid phase is increasing slowly (10%). The desorption of 2.4-D is increasing within 1 day; 60–70% of 2.4-D is desorbed from sediments of NE1-S and MA1-S, but only 40% in case of RH3-S.

The different desorption behaviour of atrazine and 2.4-D is due to the different binding of the chemicals on the sediments as already shown by the adsorption equilibria. The ionic character of 2.4-D can lead to hydrogen bonding, whereas van der Waals-bonding will dominate the sorption of atrazine. The mechanical treatment of the sediments shows no influence on the desorption behaviour of the pesticides. By using ultrasonic vibration for 15 min the concentration of 2.4-D in the liquid phase is in the same range as obtained by shaking for 1 day (60–70% desorption). The low amount of the pesticides desorbed in case of RH3-S is striking, as the main physical and chemical character of all sediments is almost the same (mineral composition, see Table 3.2; C_{org} content, see Table 3.1). Differences are shown by the particle size distribution (clay: 8% (RH3-S); 18% (MA1-S); 12% (NE1-S)) and the chemical characteristic of the aqueous extracts of C_{org} (see Sect.3.3.2). As the adsorption capacity of the clay components is higher than for silt and sand particles the organic substances sorbed on the sediments and the DOC of the porewater must play a dominant part in the adsorption–desorption kinetic of the pesticides on sediments. As shown in the following (see Sect 3.3.2) the DOC of the porewater of RH3-S is much lower and the molecular size distribution is higher than for MA1-S and NE1-S. Therefore, high DOC values of the porewater can enhance the solubility of the pesticides.

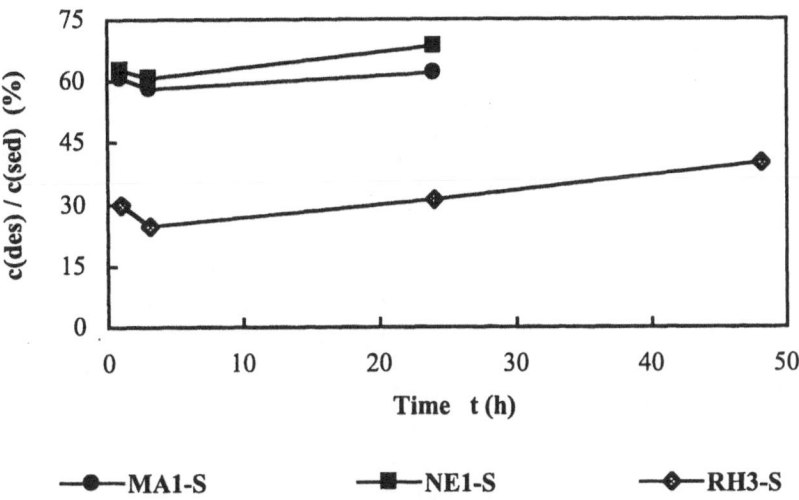

Fig. 3.7. Desorption of atrazine from sediments of the Main River(MA1-S), Neckar River (NE1-S) and Rhine River (RH3-S); sediment: solution = 1:10 / w:w; 0.01 M CaCl$_2$; c(0) (initial concentration, solution before adsorption) = 2.0 mg/l; c(sed) = c(0) - c (liquid phase concentration after adsorption); c(des) = c (liquid phase concentration after desorption)

Adsorption Isotherms. These were experimentally determined by variation of the concentration of the pesticides in a suspension containing a constant amount of adsorbents. Adsorption isotherms were also calculated using the linearized form of the Freundlich equation (Eq. 1):

$$\log (x/m) = \log K_F + (1/n)\log c \qquad (1)$$

where x/m is the amount q (μg/g; for $n = 1$) of pesticide adsorbed on the adsorbent, c (μg/ml; mg/l) is the concentration of pesticide in the supernatant solutions after equilibration, K_F and $(1/n)$ are the empirical Freundlich constants.

The best adsorption was found for the bog lake sediment. A relatively high adsorption was also found for the soil sample, whereas the river sediments showed fairly low adsorption capacities. The decreasing adsorption capacities can be related to the organic carbon content of the samples: bog lake> soil > river sediment. However, a correlation within the different samples of the river sediments is not obvious. For the sediments the influence of the matrix of the solution was not as high as for the soil sample in case of 2.4-D. The amount of the adsorption of atrazine was only little influenced by the different composition of the solution. For 2.4-D no measurable adsorption was found for the sediment of the bank-filtration site. It is known from literature that the carbon content of the adsorbents influences the adsorption behaviour. Low adsorption of 2.4-D was found by using

Fig. 3.8 a . Adsorption isotherms of 2.4-D (represented by *hatched areas*) on soil and sediments

Fig. 3.8 b. Adsorption isotherms of atrazine (represented by *hatched areas*) on soil and sediments

sediments of the bankfiltration site, which were preloaded with DOC from soil (E3) and extracted by 0.01 M CaCl$_2$. The solid phase concentration q was in the range of 0.5 to 2.0 μg/g. In case of atrazine measurable adsorption was found neither for the sediment of the bankfiltration site nor for the sediment preloaded with DOC.

Fig. 3.9. Adsorption isotherms of atrazine on sediments of the Neckar River (NE1-S) and the Rhine River (RH3-S)

To reach the real concentration range of atrazine in rivers, the adsorption iso-
therms were also set up in the low concentration range. As shown in Fig. 3.9 the
adsorption behaviour in the high and the low concentration ranges of the pesti-
cide is the same. Therefore, the experiments set up with the higher concentration
range do not have only model character as the adsorption behaviour is the same.

Fig. 3.10. Adsorption isotherms of metazachlor on sediments

The adsorption of metazachlor on DO1-S, EL2(5-6)-S and RH3-S is shown in
Fig. 3.10. In contrast to the investigations for 2.4-D and atrazine, the adsorption
experiments were only set up in some selected solutions (0.01 M CaCl$_2$, 0.34 M
NaCl and river-water). But as shown by the hatched areas the influence of the ma-
trix of the solution to the adsorption of metazachlor is much higher in case of
DO1-S and RH3-S as compared to 2.4-D and atrazine. The lower solubility of
metazachlor, as compared to 2.4-D and atrazine in aqueous systems and the spe-
cific functional groups of metazachlor which are influenced by the matrix of the
aqueous systems, may be an explanation.

The adsorption capacity for the three sediments is different (DO1-S > RH3-S >
EL2(5-6)-S). This goes along with the increasing amount of C$_{org}$ of the sediments
(C$_{org}$: 14.8% > 4.6% > 1.9%). Compared to the isotherms of atrazine and 2.4-D
the adsorption capacity of metazachlor is much higher on the sediments. This can
be related to the more hydrophobic character of metazachlor.

The highest adsorption capacities on sediments are found for metolachlor
(Fig. 3.11). The differences for the sediments DO1-S, EL2(8-9)-S and RH3-S are

due to the C_{org} content of the samples. The influence of the matrix of the solution is not as high as for metazachlor. The high adsorption capacity of metolachlor is striking, as the solubility is in the same range as 2.4-D. It is attractive to assume,

Fig. 3.11. Adsorption isotherms of metolachlor on sediments

that specific functional groups of metolachlor increase the sorption on sediments as compared to 2.4-D. The interaction of NH functions with heavy metals sorbed on the sediment may be an explanation (chelate complexation).

Adsorption isotherms were also determined for the different fractions (profile) of a sediment core (Figs. 3.12 a and 3.12 b).

The core was sampled at the Mühlenberger Loch in the River Elbe (EL2-S). The vertical length of the core was 14 cm. For the adsorption experiments, a 0.01 M CaCl$_2$ solution was used. It can be derived from Table 3.1 that the organic carbon content showed a minimum at the depth of 8–9 cm and is much greater at the surface and the bottom (1–2 cm: 3.4% C_{org}; 8–9 cm: 1.2% C_{org}; 10–11 cm: 1.7% C_{org}; 13–14 cm: 2.4% C_{org}). The adsorption decreased as the organic carbon content decreased. The same effects were found for both pesticides. The correlation of organic pollutants and the organic content of particle size fractions of sediments was shown by Evans et al. (1990). They found a positive linear relationship between PAH-concentration and organic matter content of sediments. In addition, they also showed that the organic matter content had a bimodal distribution through the sediment size fraction with maxima in the two largest and in the smallest fractions.

Fig. 3.12 a. Adsorption isotherms of 2.4-D on sediment of the Mühlenberger Loch (EL2-S, sediment core)

Fig. 3.12 b. Adsorption isotherms of atrazine on sediment of the Mühlenberger Loch (EL2-S, sediment core)

The results of the calculation of isotherms are summarized in Table 3.5. The K_F values of the sediments range up to 24. Again it is obvious that on average the adsorption of atrazine, metazachlor, metolachlor and simazine is higher than the one of 2.4-D. The K_F and $1/n$ values are not clearly correlated with the matrix of the solution. The K_F values of the soil samples are the highest values concerning the adsorption of the single chemicals (2.4-D and atrazine).

The dependence of the adsorption behaviour on the organic carbon content can be expressed as K_{OC} values according to Eq. 2.

$$K_{OC} = K_F \; / \; C_{org} \; * \; 100 \tag{2}$$

For the core sample (EL2-S) the highest K_{OC} values result for the fractions with the highest organic carbon content. A clear correlation within the different sediment samples was not obtained. The K_{OC} values of the soil samples were within the range of the samples EL1-S, NE1-S and WE1-S. As the adsorption isotherms of the five pesticides could not be prepared for the whole group of sediments and for all types of aqueous solutions, the following Figs. 3.13 a–c (see pages 71–72) can only show a trend in the different adsorption behaviour .

Table 3.5 a. Sorption coefficient K_F (ml/g) and $1/n$ values of the isotherms calculated with the Freundlich equation and the K_{OC} values

		Metazachlor			Metolachlor			Simazine		
		K_F	$1/n$	K_{OC}	K_F	$1/n$	K_{OC}	K_F	$1/n$	K_{OC}
Sediment (DO1-S)										
0.01 M CaCl$_2$		8.0	0.7	54	24.3	0.7	164	–	–	–
0.34 M NaCl		2.8	2.6	191	17.2	0.4	116	–	–	–
Danube - water		5.5	0.6	371	–	–	–	–	–	–
Sediment (EL2-S)										
0.01 M CaCl$_2$										
5–6	cm	1.8	0.4	95	–	–	–	–	–	–
8–9	cm	–	–	–	2.2	0.3	183	–	–	–
0.34 M NaCl										
5–6	cm	1.9	0.3	100	–	–	–	–	–	–
8–9	cm	–	–	–	3.5	0.5	292	–	–	–
Sediment (RH3-S)										
0.01 M CaCl$_2$		2.1	0.4	46	15.0	0.7	326	2.4	0.2	52
0.34 M NaCl		4.2	0.4	91	10.0	0.6	217	–	–	–
H$_2$O		–	–	–	–	–	–	–	–	–
Rhine - water		7.5	0.7	149	11.7	0.6	552	3.7	0.5	80

Table 3.5 b. Sorption coefficient K_F (ml/g) and 1/n values of the isotherms calculated with the Freundlich equation and the K_{OC} values

	2.4-D			Atrazine		
	K_F	1/n	K_{OC}	K_F	1/n	K_{OC}
Soil (E3)						
0.01 M CaCl$_2$	13.0	0.8	163	14.0	1.1	175
0.34 M NaCl	9.6	0.8	120	18.5	1.3	231
H$_2$O	3.4	0.8	43	15.5	1.0	194
Sediment (EL1-S)						
0.01 M CaCl$_2$	–	–	–	5.4	2.1	216
H$_2$O	–	–	–	4.9	1.0	196
Sediment (EL2-S)						
0.01 M CaCl$_2$						
1–2 cm	–	–	–	3.0	0.8	88
2–3 cm	3.0	0.9	105	–	–	–
3–4 cm	–	–	–	2.4	0.5	73
5–6 cm	1.6	0.8	84	–	–	–
6–7 cm	–	–	–	(<1)	a	(< 59)
8–9 cm	0.5	0.9	42	–	–	–
10–11 cm	–	–	–	0.7	1.0	41
13–14 cm	1.7	0.5	71	–	–	–
Sediment (MA1-S)						
0.01 M CaCl$_2$	0.9	1.1	18	0.9	1.2	19
0.34 M NaCl	0.6	1.1	12	0.1	1.5	2
H$_2$O	6.3	0.8	13	2.2	1.0	45
Buffer (pH = 7)	1.7	1.0	34	1.1	0.8	222
Main - water	2.9	0.9	59	0.9	1.1	18
Sediment (NE1-S)						
0.01 M CaCl$_2$	8.7	0.8	154	9.9	0.9	175
0.34 M NaCl	8.9	0.8	158	7.2	0.9	127
H$_2$O	2.7	0.9	48	0	2.3	0
Buffer (pH = 7)	1.9	1.0	34	0.4	1.4	6
Neckar - water	1.5	1.0	27	0.3	1.3	5
Sediment (RH3-S)						
0.01 M CaCl$_2$	3.0	0.7	64	3.5	0.8	76
H$_2$O	0.9	0.7	19	3.4	0.9	74
Sediment (WE1-S)						
0.01 M CaCl$_2$	0.9	0.5	321	3.7	1.2	132
0.34 M NaCl	–	–	–	1.5	0.5	54
H$_2$O	b	b	–	2.4	1.5	69
Buffer (pH = 7)	–	–	–	2.9	0.7	104

a no linear isotherm, not determinable
b no adsorption

Fig. 3.13 a. Correlation of the K_F values with the C_{org} content of the sediments

Fig. 3.13 b. Correlation of the K_{OC} values with the C_{org} content of the sediments

Fig. 3.13 c. Correlation of the K_{OC} values with the solubility of the pesticides

It was shown in the literature (see Sect 3.1) that the K_D values (see Eq. 3) are used to describe the equilibrium concentration in the liquid and in the solid phase.

$$K_D = (\mu g \; \text{pesticide/g sediment}) \; / \; (\mu g \; \text{pesticide/ml solution}) \qquad (3)$$

The values can only lead to a rough estimation in case of river sediments. The curves of the adsorption isotherms show no linear behaviour in all cases and therefore, the K_D values are different over the whole range of the liquid pesticide concentration. K_D values for the adsorption of 2.4-D and atrazine on MA1-S and NE1-S are shown in Table 3.6.

Table 3.6. K_D (mean values) of the adsorption of 2.4-D and atrazine on sediments from the Main (MA1-S) River and the Neckar (NE1-S) River. Values in parentheses r^2 give variance

Solution	2.4-D		Atrazine	
	MA1-S	NE1-S	MA1-S	NE1-S
0.01 M CaCl$_2$	2.4 (0.5)	1.9 (0.4)	3.4 (0.8)	4.0 (1.2)
0.34 M NaCl	1.5 (0.4)	1.4 (0.4)	3.1 (1.0)	3.7 (0.5)
H$_2$O	1.8 (0.4)	0.9 (0.2)	2.8 (0.4)	2.1 (1.4)
Buffer	1.9 (0.5)	1.5 (0.5)	3.3 (1.3)	4.0 (1.0)
River water	2.0 (0.4)	1.5 (0.3)	1.8 (0.5)	1.8 (0.5)

As already shown by the K_F, $1/n$ and K_{OC} values, the adsorption on sediments is higher for atrazine than for 2.4-D. The K_D values of atrazine are higher than the values of 2.4-D. Only in case of the river-water the values are in the same range. The individual values for the concentration range show a broad variability (for example: MA1-S, atrazine: 2.2–4.5 in 0.01 M CaCl$_2$; and 1.1–3.9 in 0.34 M NaCl). It has become obvious, that additional factors like particle size distribution, inorganic sediment composition and the character of the organic material also influence the adsorption of pesticides on sediments.

Parameters Influencing the Adsorption Behaviour.

DOC. Mechanisms which influence the interaction of organic chemicals and natural organic matter include ion exchange, hydrogen bonding, van der Waals forces and coordination through an attached metal ion (ligand exchange; Stevenson 1982). The mechanisms differ according to the type of pesticide. In addition to the interaction of the dissolved molecules with particulate organic matter, there is also an interaction of organic chemicals with the dissolved organic matter. It can be assumed from certain principles, that these interactions are the same.

As the sediment samples were used in the wet form, the DOC-concentration of the supernatant solution is increasing during the adsorption experiments. This is due to the distribution of the DOC of the porewater and to the mobilization of particulate organic matter from the sediments. The DOC-bleeding of the organic matter is different depending on the matrix of the solution. The concentration ranges from 5 to 15 mg/l DOC for the CaCl$_2$, H$_2$O or NaCl solution. In case of the pyrophosphate solution the DOC increases up to about 100 mg/l.

To investigate the influence of the DOC on the adsorption behaviour of the pesticides, the aqueous 0.01 M CaCl$_2$ solution was prepared from DOC-rich water (bog lake water). The adsorption of 2.4-D on sediment was higher in the presence of DOC-rich water (12.3 and 14.2 mg/l DOC from bog lake water and 6.3 mg/l DOC from the porewater) as compared to the aqueous solution (6.3 mg/l DOC from the porewater; see Fig. 3.14 a). Whereas, the adsorption of the more hydrophobic pesticide, atrazine, is not influenced (Fig. 3.14 b). This effect might be due to an interaction of atrazine and DOC. As there is no additional effect in DOC-rich water, it is obvious that there must be an interaction even at lower DOC-concentrations (DOC-bleeding of the porewater).

Fig. 3.14 a. Influence of DOC-rich water on the adsorption of 2.4-D on sediment from the Weser River, WE1-S (sediment:solution = 1:10 / w:w; 0.01 M CaCl$_2$, 12.3 and 14.2 mg/l DOC; N$_2$: anoxic conditions)

Fig. 3.14 b. Influence of DOC-rich water on the adsorption of atrazine on sediment from the Weser River, WE1-S (sediment:solution = 1:10 / w:w; 0.01 M CaCl$_2$, 12.9 and 14.1 mg/l DOC; N$_2$: anoxic conditions)

Sulphonic Acid. The relatively polar chemical 2.4-D was better adsorbed on the sediments in the presence of naphthaline-1.5-disulphonic acid (Fig. 3.15 a). The adsorption behaviour of atrazine was influenced in a complex form (Fig. 3.15 b).

Fig. 3.15 a. Influence of naphthaline-1.5-disulphonic acid (1 mg/l) on the adsorption of 2.4-D on sediment from the Weser River, WE1-S (sediment:solution = 1:10 / w:w; 0.01 M CaCl$_2$; N$_2$: anoxic conditions)

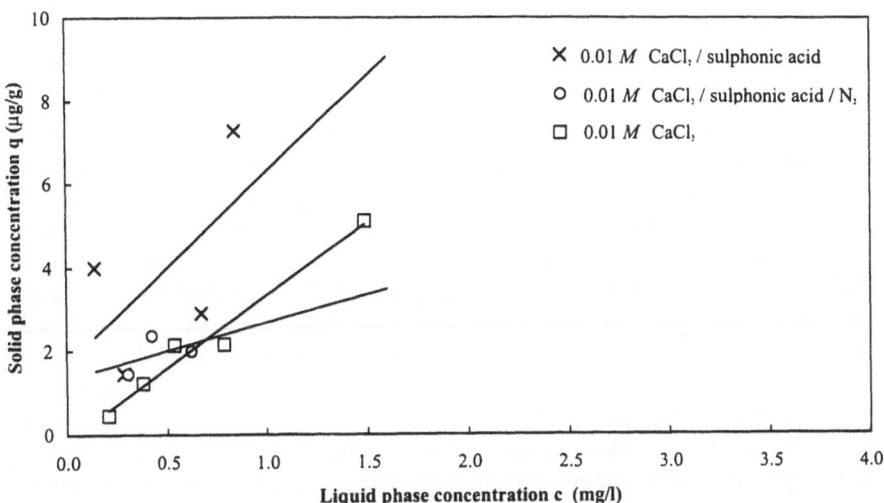

Fig. 3.15 b. Influence of naphthaline-1.5-disulphonic acid (1 mg/l) on the adsorption of atrazine on WE1-S sediment from the Weser River (sediment:solution = 1:10 / w:w; 0.01 M CaCl$_2$; N$_2$: anoxic conditions)

Redox Situation. The standard adsorption experiments of 2.4-D and atrazine were done under atmospheric conditions. The wet samples were originally obtained from anoxic sediments as shown by the low redox potential of the

porewater. During the experiments the reaction solution became more and more oxic, as shown by an increasing value of the redox potential. This had a strong influence on the adsorption–desorption behaviour of minerals and on the precipitation of heavy metals. As a consequence, the surface of the sediments may change. In Figs. 3.14 a, b and 3.15 a, b there was no influence concerning the adsorption behaviour of 2.4-D on sediments in nitrogen atmosphere. However, the adsorption behaviour of atrazine is slightly influenced, showing less adsorption in anoxic atmosphere.

3.3.2 Characterization of the Organic Matter of the Sediments and of the Porewater

It was shown that the organic components of the sediment have a great influence on the adsorption reactions. For an understanding of the sorption reactions detailed information of the character of the DOC of the porewater and of the particulate organic material is needed. Therefore, the sediments were extracted with bidistilled water or by a solution of pyrophosphate and the extracts were characterized by different spectroscopical and chromatographical methods. In addition, humic substances were isolated from the aqueous solution (0.1 M pyrophosphate) by the common XAD-procedure (Abbt-Braun et al. 1991). The data obtained from the aqueous and extracted solid organic material were compared.

DOC of the Porewater and of the Extracts. As the water content of the river sediments (MA1-S; NE1-S; RH3-S; WE1-S) was very high (50–70%) the water phase was separated by centrifuging the sediment material. The liquid phase is referred to as porewater in the following context. The DOC of the porewater ranged from 30 mg/l to 150 mg/l DOC (DOC in mg/l: MA1-S = 92.0; NE1-S = 57.6; RH3-S = 29.3; and WE1-S = 152.3). As known from literature (Piemontesi and Baccini 1986), the amount of the DOC of the porewater is dependant on the seasons. As in winter the biological activity is reduced, the DOC values are much lower. In summer and autumn the production of low molecular organic acids, alcohols and amines is high by fermentative processes leading to higher DOC values compared to the winter season. The sediment samples investigated were gained in summer and autumn. The varying DOC values reflect the different biological activity of the sampling sites. There is no correlation with the different C_{org} content of the sediments.

The specific spectral absorbances, UV (254 nm)/DOC, of the porewater from RH3-S and WE1-S were practically the same (0.6–0.7 l/mg·m). Samples MA1-S and NE1-S showed higher values (2.0–2.4 l/mg·m). From these results it can be concluded that the character of the organic substances is similar within the two "couples" Rhine/Weser and Main/Neckar. High values of UV-absorbance go along with molecules containing conjugated ketone structures and aromatic or unsaturated systems.

After centrifuging, the sediments were extracted with water. The yield of DOC was in the same range for the sediments of the rivers Rhine and Weser (DOC in mg/l: RH3-S = 46.3; WE1-S = 47.5; extract of 50 ml). The DOC-concentration of the aqueous extracts of MA1-S and NE1-S was much higher (DOC in mg/l: MA1-S = 156 mg/l; NE1-S = 153 mg/l; extract of 50 ml). The yield of the extractions was less than 1% concerning the C_{org} content of the samples. A correlation with the DOC of the porewater and the C_{org} content could not be established.

The specific spectral absorbance is different in the water extracts. Compared to the porewater, there is a decrease for the samples MA1-S and NE1-S, UV (254 nm)/DOC; l/mg·m; 2.1 and 1.3), the values are higher for RH3-S and WE1-S (UV (254 nm)/DOC; l/mg·m; 1.8 and 3.6) are higher. It seems that the organic substances of the aqueous extracts show approximately the same functional groups in case of MA1-S and NE1-S. However, organic material extracted from RH3-S and WE1-S has higher amounts of substances showing ketone groups and double bonds than the molecules in the porewater.

The samples were also extracted using a solution of pyrophosphate. The yield of the DOC which was extracted is about 10–15% of the C_{org} content.

HS were isolated of the pyrophosphate solution with the XAD-2 method. The percentages of the DOC-fractions of the extract are given in Table 3.7. The different treatment steps of the XAD-procedure lead to the different fractions: HS (humic substances), FA (fulvic acids), NHS (non-humic substances, not adsorbable on the XAD-2 resin at pH = 2). The Δ values give the amount of DOC in the original extract which was neither recovered as HS nor as NHS. This part was irreversibly adsorbed on the XAD-resin. As the extraction procedure is done at pH = 7, the HA-fraction cannot be isolated totally. The insoluble fraction humin of sedimentary humic substances was not extracted.

Relatively high amounts of HS (calculated by DOC balancing; >60% of the DOC of the pyrophosphate solution) are typical for the soil sample. The amount of HS of the sediments gained from rivers were in the same range (MA1-S, NE1-S)

Table 3.7. DOC balance of the XAD-isolation from the extracts of sediment samples (% DOC of the 0.1 M pyrophosphate solution)

Sample	HS % DOC	FA % DOC	NHS % DOC	Δ (100%-HS-NHS)
E3	64	37	23	13
GR2-C-S	37	35	50	13
HO6-S	39	32	26	35
MA1-S	58	29	37	5
NE1-S	72	30	31	–
RH3-S	48	22	22	30
WE1-S	50	15	20	30

or lower (~50% for RH3-S and WE1-S). The yield for the samples of the bog lake and of the bankfiltration site was around 40%.

The amount of the FA-fractions was different. Only for the soil and sediment samples it was in the same range (50% of HS; DOC balancing). It is striking, that the FA-fraction of WE1-S is much lower (15%). On the contrary, HS are nearly equally proportional to the FA-fraction in case of the river sediment with the low organic carbon content and the bog lake sample. This shows that no HA-fractions can be isolated at pH = 7. Whereas, in case of MA1-S, NE1-S and RH3-S around 50% and for the WE1-S sample ca. 70% of the HS are attributed to the HA-fraction. As known from literature, HA have a higher amount of aromaticity and less oxygen-containing functional groups than FA (Schnitzer and Khan 1972). As a consequence of the higher hydrophobic character of HA, the yield of aqueous extraction at pH = 7 was much less. The yield of HA can also reflect the stages of ageing of the organic material coated on the sediment (Frimmel and Weis 1991; Weis and Frimmel 1991). In the case of the bog lake sample and the bankfiltration sediment, it is attractive to assume that there were two different types of particulate organic material. One fraction could be extracted fairly easily and the other one (HA) seemed to be more hydrophobic. Here the long sojourn time of organic material in the underground and the high adsorption–desorption capacities of the system give reasonable explanations. In contrast, the river sediments showed different fractions of particulate organic matter. This is accompanied by a higher biological activity in these locations leading to a broad spectrum of organic material.

The NHS-fraction had in general a very hydrophilic character. The amount was ca. 20–30% of DOC for all samples, with the exception of GR2-C-S, where the NHS were about 50% of DOC.

It is obvious that there is a missing amount of DOC (Δ) in the mass balance. This DOC-fraction was adsorbed on the XAD-2 resin at pH = 2 and could not be desorbed at high pH values. Therefore, the character of the fraction must have been very hydrophobic. The amount of this type of DOC was relatively high for the extracts of the sediments of the rivers Rhine and Weser and of the bog lake.

The data of the DOC balance show that each sediment sample had its own individual character reflecting the specific situation and location of the system. No prediction of the amount of different organic fractions sorbed on the material could be made.

Chromatographic Behaviour. The substances were separated chromatographically according to their molecular size: the higher the retention time, the lower the molecular size. Even though there is a fair amount of randomness by assigning molecular weights to the fractions at different retention times, retention times are suited for a finger-print characterization of the fractionated matter on the basis of relative comparison.

The samples were also investigated using reversed phase material (C18-resins). In reversed phase chromatography hydrophilic substances are eluted first, where-

as, hydrophobic substances show higher retention times. The chromatograms were divided into two sections: a hydrophilic fraction ($t < 3$ min) and a hydrophobic fraction ($t > 3$ min).

The nominal molecular weight distribution, as analyzed by HPLC and UV detection ($\lambda = 254$ nm), is shown in Fig. 3.16 for the porewater of RH3-S. More than 50% of the peak area is attributed to the high molecular weight fraction ($t < 6$ min). This sample showed a higher amount of larger molecules than MA1-S and WE1-S (90% of the peak area at $t > 6$ min). The substances of these porewaters had a lower nominal molecular weight and the distribution of the molecules was more inhomogenious. It is obvious that the DOC of the porewater of RH3-S had the lowest molecular size. The amount of the hydrophilic fraction in the porewater was high (40–60% of the total area). The fractions obtained at higher retention times showed lower relative abundances and a broad distribution (Fig. 3.16 b).

Comparing the results, it can be concluded that the organic substances in the porewater were fairly hydrophilic and high in molecular weight. The lower the molecular weight, the more hydrophobic the character of the substances becomes.

Fig. 3.16 a. Gelchromatograms of porewater samples of the rivers Main (MA1-S), Rhine (RH3-S) and Weser (WE1-S)

Fig. 3.16 b. RP-C18 chromatograms of porewater samples of the rivers Main (MA1-S), Rhine (RH3-S) and Weser (WE1-S)

The chromatograms gained by gelchromatography are different for the pore-water samples and the substances extracted with water (Fig. 3.17 a). The chromatograms of the porewater and of the aqueous extract of Rhine sediment show the same shape but the concentration of the substances in the lower mass range was obviously higher as the amount of the high molecular fraction (Fig. 3.16). In the case of the Weser sediment, the distribution of the extracted material is broad but the distribution in the high and low mass range is the same. The same trend is obvious for MA1-S (increase in the high molecular mass range, 20% of the peak area at $t > 6$ min)

In general the porewaters are less structured than the water extracts in the case of the gelchromatograms. Also the RP-chromatograms show significant differences for the different samples from the same origin (Fig. 3.17 b). Only the samples from the RH3-sediment show practically identical RP-chromatograms.

Fig. 3.17 a. Gelchromatograms of aqueous extracts of sediments of the rivers Main (MA1-S), Rhine (RH3-S) and Weser (WE1-S)

Fig. 3.17 b. RP-C18 chromatograms of aqueous extracts of sediments of the rivers Main (MA1-S), Rhine (RH3-S) and Weser (WE1-S)

Fig. 3.18 a. Gelchromatograms of DOC-fractions (pyrophosphate extracts, NHS, FA) of the Rhine River RH3-sediment

The gelchromatographic behaviour of the DOC-fractions gained by the XAD-procedure is shown in Figs. 3.18 a and 3.18 b.

In contrast to Figs. 3.16 and 3.17, the detection was done by DOC. The chromatograms gained by UV detection show the same behaviour. The nominal molecular weight distribution was different for the DOC-fractions as shown for EL1-S and RH3-S. The chromatograms of RH3-S are typical for sediment FA. The same shape was found for GR2-C-S, HO6-S, MA1-S, NE1-S and WE1-S and for the soil sample E3. The NHS-fraction shows substances in the high and low molecular weight range, whereas the chromatogram of the FA-fraction shows a relatively broad and unstructured distribution of material in between. The shape of the chromatograms for EL1-S is entirely different. The NHS-fraction has a high amount of low molecular weight substances. The chromatogram of the FA presents a broad distribution which appears at higher retention times compared to RH3-S FA.

Fig. 3.18 b. Gelchromatograms of DOC-fractions (pyrophosphate extract, NHS, FA) of the Elbe River EL1-sediment

The chromatograms obtained by C18-reversed phase resins are totally different for the FA- and the NHS-fractions. In general all samples showed the same behaviour. The hydrophilic fraction of the NHS is high (80–90% of the total area), whereas the hydrophobic fraction of the FA is around 60–70% of the total area.

The organic substances of the porewater seem to be different. Therefore, it is possible that they reflect the more actual biological and chemical status of the liquid phase of the sediment-only system. As the organic material already incorporated into the solid phase of the sediment (C_{org}) can be aged, the extracts of different sediments show more similar chromatographic behaviour.

Fig. 3.19 a. RP-C18 chromatograms of DOC-fractions (pyrophosphate extract, NHS, FA) of the Rhine River RH3-sediment

Fig. 3.19 b. RP-C18 chromatograms of DOC-fractions (NHS, FA) of the Elbe River EL1-sediment

Spectroscopical and Chemical Parameters. The FA-fractions were character-
ized by several spectroscopical and chemical methods (Table 3.8).

The specific spectral absorption coefficients at $\lambda = 254$ nm of the sediment FAs
are significantly different. Higher values are found for the soil, bog lake and
bankfiltration site samples. The sediment samples show lower values, whereas
RH3-S and EL1-S are between the samples of soil and bankfiltration site. This
might be due to the different origin of the organic materials. The starting material
of organic sediment material is different, as compared to soil material. It is
derived from algae and larvae for sediment samples leading to a higher amount of
aliphatic chains, whereas lignins lead to aromatic structures in case of soil and bog
lake samples. However, higher values indicate also a higher status of
condensation and ageing of the organic material (EL1-S and RH3-S). The data at
$\lambda = 436$ nm show that the sediment samples have a higher amount of functional
groups responsible for the yellow brown colour. EL1-S FA indicates the highest
value. This agrees with the results obtained for the UV-range.

The elemental composition varies for the different FA samples. The higher the
aromatic character is, the lower the H/C values are. This parameter can, therefore,
stand for the different stages of ageing. The data agree with the specific spectral
absorbance. It seems that the soil sample (E3) and the river sample (RH3-S FA)
are more aromatic than the other samples. The O/C values are in the same range.
It is obvious that the river sediment samples have higher amounts of nitrogen
when compared to the bog lake and soil sample. This indicates differences in the

Table 3.8. Physical, chemical and spectroscopical data of isolated FA from sediment

| | Sample | | | | | | | |
	E3-FA	EL1-S-FA	GR2-C-S-FA	HO6-S-FA	MA1-S-FA	NE1-S-FA	RH3-S-FA	WE1-S-FA
A(254)/DOC; pH=7 l/(mg*m)	4.40	5.34	5.42	5.60	3.56	4.29	4.96	3.63
A(436)/DOC; pH=7 l/(mg*m)	0.33	0.71	0.51	0.47	0.29	0.43	0.40	0.39
A(254)/A(436)	13.50	7.52	10.71	11.97	12.39	9.99	12.54	9.23
H/C	0.827	–	0.972	0.928	0.980	0.882	0.777	0.980
O/C	0.573	–	0.520	0.534	0.582	0.509	0.531	0.671
N/C	0.027	–	0.043	0.021	0.097	0.052	0.041	0.052
S/C	0.009	–	–	–	0.028	0.020	0.040	0.021
H-Cap. μmol/mg DOC								
Σ	13.0	–	17.5	16.7	–	14.2	–	–
< pH 7	11.0	–	10.9	9.5	–	12.1	11.1	9.6
> pH 7	2.0	–	6.7	7.2	–	2.1	a	a
aPC μmol/mg DOC	4.37	6.61	6.74	5.88	3.94	6.89	7.31	5.36

a not determined

biochemical reactions in the sediment systems depending on the sources of organic material. As shown in the literature, the elemental composition of humic substances in lake sediments, as shown by the H/C, O/C and N/C show the same range as soil humic substances (Ishiwatari 1985; Steelink 1985). For all sediment samples a trend with depth was shown: an increase in carbon content, in H/C-ratio, in C/N-ratio and in colour intensity; and a decrease in proteinaceous material and in carbohydrate-like material. Such changes are due to different degradation reactions in an anaerobic environment and the diagenetic changes with burial of the sediment. The high contents of sulphur are also typical for the sediment FAs.

The IR spectra are shown in Fig. 3.20. A typical IR spectrum of FA is represented by E3 FA. The shape of the FA spectra from the bog lake sediment and the bankfiltration site is similar. The distinct band in the range between 2990 and 3000 cm^{-1} is striking. These wavenumbers are caused by aliphatic groups. The spectra of the river sediments FA (MA1-S, NE1-S, RH3-S and WE1-S) are similar to the spectrum of GR2-C-S FA. The IR-spectrum of EL1-S FA is different. The relative abundances of the bands are reversed. This shape is often shown by HA samples (Abbt-Braun and Frimmel 1991). The range between 1100 up to 1300 cm^{-1} can be assigned to C-O vibrations of polysaccharide structures.

Acid functional groups can be determined by titration. The proton capacities

Fig. 3.20. FT-IR-spectra of fulvic acids (KBr-pellets, 256 scans, resolution 2 cm^{-1}; *1* E3 FA; *2* GR2-C-S FA; *3* HO6-S FA; *4* EL1-S FA)

(H-cap.) are higher for the soil sample compared to the FA of the sediments (Table 3.8). The pH-range can be divided into a range below (carboxylic groups) and above (phenolic groups) pH = 7. It is obvious that some FAs isolated from sediments have higher amounts of phenolic groups, whereas the carboxylic groups show the same concentration.

In addition, the anionic particle charge (aPC) is higher for FA isolated from sediments than for FA isolated from soil (see Table 3.8).

3.4 Conclusion

Sediments are highly complex systems. Therefore, it is difficult to obtain meaningful data with high reproducability. Working under defined conditions and observing the interaction with clearly identifiable substances is promising.

The preparation of the sediments and the experimental procedures of the adsorption investigations have great influence on the study results of the sorption capacities. There was little influence of the matrix of the solution on the adsorption equilibria in the case of river sediments and 2.4-D, atrazine and metolachlor. The adsorption increased as follows: 2.4-D < atrazine < metazachlor < metolachlor. No clear correlation of the adsorption capacity with the C_{org} content of the river sediments was obtained. However, the detailed investigation of a core sample showed increasing adsorption with increasing C_{org} content. The data show that the adsorption behaviour of atrazine and 2.4-D on river sediment is different to the adsorption on soil and bog lake sediments. It seems that a generalization and prediction of the equilibration time for the adsorption behaviour of pesticides is not possible. It has to be experimentally determined for the individual chemicals and sediments. The K_F and 1/n values of the Freundlich equation gave no clear coherence with the C_{org} content and the K_{OC} values in case of river sediments. However, the results help to understand details of adsorption and the reactions in heterogeneous aquatic systems. From the results, it is obvious, that the adsorption of organic micro-pollutants on the different solid phases is the result of more complex interactions than has been assumed. Therefore, a detailed characterization of the particulate organic material and of the inorganic material of the river sediments has to be achieved with respect to the partitioning of the different types of xenobiotics. The knowledge of these data is of fundamental importance for the estimation of distribution, e. g. immobilization and bioavailability of chemicals.

Acknowledgement. The work was financially supported by the Bundesministerium für Forschung und Technologie (BMFT grant WT 90169). For the fine technical assistance, we thank S. Klose, J. Gandhi, D. Bandara, C. Eisenhauer, E. Miller, H. Schütz and J. Taherivand.

References

Adams WJ, Kimerle RA, Barnett JW Jr (1992) Sediment quality and aquatic life assessment. Environ Sci Technol 26:1864–1875

Abbt-Braun G, Schmiedel U, Frimmel FH (1990) Elementaranalytische Untersuchungen isolierter Fulvinsäuren unterschiedlichen Ursprungs. Vom Wasser 75:59–73

Abbt-Braun G, Frimmel FH, Lipp P (1991) Isolation of organic substances from aquatic and terrestrial systems – comparison of some methods. Z Wasser-Abwasser-Forsch 24:285–292

Bollag JM, Myers C (1992) Detoxification of aquatic and terrestrial sites through binding of pollutants to humic substances. Sci Total Environ 117/118:357–366

Burton GA Jr, Scott KJ (1992) Sediment toxicity evaluations – their niche in ecological assessments. Environ Sci Technol 26:2068–2075

Calvet R (1989) Adsorption of organic chemicals in soils. Environ Health Perspectives 83:145–177

Capel PD, Eisenreich SJ (1990) Relationship betweeen chlorinated hydrocarbons and organic carbon in sediment and porewater. J Great Lakes Res 16 H:245–257

Chapman PM (1989) Current approaches to developing sediment quality criteria. Environ Toxicol Chem 8:589–599

Chiou CT, Schmedding DV, Manes M (1982) Partitioning of organic compounds in octanol–water systems. Environ Sci Technol 16:4–10

Choudry GG (1980) Interactions of humic substances with environmental chemicals. In: Hutzinger O (ed) (1980) The handbook of environmental chemistry, vol. 2 part B. Springer, Berlin Heidelberg New York, pp 103–128

Coulter WH (1953) US-Patent 2.656.508, issued Oct. 20. 1953, applied for Aug. 27. 1949

Coulter WH (1956) Proc Natl Electr Conf 12:1034

DIN (1993) Deutsche Einheitsverfahren zur Wasser-, Abwasser- und Schlammuntersuchung. Verlag Chemie,Weinheim

Evans KM, Gill RA, Robotham PWJ (1990) The PAH and organic content of sediment particle size fractions. Water Air Soil Pollut 51:13–31

Förstner U (1989) Contaminated soils.In: Lecture notes in earth science, vol. 21. Springer, Berlin Heidelberg New York

Förstner U, Ackermann F, Alberti J, Calmano W, Frimmel FH, Kornatzki KH, Leschber R, Roßknecht H, Schleichert U, Tent L (1987) Qualitätskriterien für Gewässersedimente – Allgemeine Problematik und internationaler Stand der Diskussion. Z Wasser-Abwasser-Forsch 20:54–59

Frimmel FH, Christman RF (eds) (1988) Humic substances and their role in the environment. Dahlem workshop reports. John Wiley & Sons, Chichester

Frimmel FH, Hopp W, Quentin KE (1985) Titration isolierter aquatischer Huminstoffe und ihrer Calcium-Komplexe mit starken Basen und Säuren. Z Wasser-Abwasser-Forsch 18:259–262

Frimmel FH, Weis M (1991) Ageing effects of highmolecular weight organic acids which can be isolated from landfill leachates. Wat Sci Tech 23:419–426

Fröbe Z, Drevenkar V, Stengl B, Juracic M (1989) Sorption behaviour of some organophosphorus pesticides in natural sediments. Toxicol Environ Chem Rev 19:69–82

Fuchs F (1985, 1986) Gelchromatographische Trennung von organischen Wasserinhaltsstoffen. Vom Wasser 64:129–144; 65:93–105; 66:127–136

Gremm T, Abbt-Braun G, Frimmel FH (1991) Einfluß verschiedener Parameter auf die HPLC-Charakterisierung organischer Säuren. Vom Wasser 77:231–241

Huber SA, Scheunert J, Dörfler U, Frimmel FH (1992) Zum Einfluß des gelösten organischen Kohlenstoffs (DOC) auf das Mobilisierungsverhalten einiger Pestizide. Acta Hydrochim Hydrobiol 20:74–81

Hunchak-Kariouk K, Suffet JH (1991) Binding of hydrophobic organic pollutants to humic substances. Humus-uutiset, Finnish Humus News 3(3) – The Third International Nordic Symposium on Humic Substances:157–170

Ishiwatari R (1985) Geochemistry of humic substances in lake sediments. In: Aiken GR, McKnight DM, Wershaw RL (eds) (1985) Humic substances in soil, sediment, and water. John Wiley & sons, New York, pp 147–180

Karickhoff SW, Brown DS, Scott TA (1979) Sorption of hydrophobic pollutants on natural sediments. Water Res 13:241–248

Karickhoff SW (1981) Semi-empirical estimation of sorption of hydrophobic pollutants on natural sediments and soils. Chemosphere 10:833–846

Kersten M, Förstner U (1987) Effect of sample pretreatment on the reliability of solid speciation data of heavy metals-implication for the study of diagenetic processes. Mar Chem 22:299–312

Khan SU (1980) Physicochemical processes affecting pesticides in soil. In: Khan SU (ed) (1980) Pesticides in the soil environment. Elsevier Sci Publ Co, Amsterdam, pp 29–118

Lee D-Y, Farmer WJ (1989) Dissolved organic matter interaction with napropamide and four other nonionic pesticides. J Environ Qual 18:468–474

OECD(1981)Guideline for testing of chemicals (106) adsorption/desorption. May 12, 1981

Oepen B van, Kördel W, Klein W (1989) Soil preparation for the estimation of adsorption coefficients (koc) of organic chemicals. Chemosphere 18:1495–1511

Orem WH, Gaudette HE (1984) Organic matter in anoxic marine pore water: oxidation effects. Org Geochem 5:175–181

Pavlou SP (1987) The use of the equilibrium partitioning approach in determining safe levels of contaminants in marine sediments. In: Dickson KL, Maki AW, Brungs WA (eds) Fate and effects of sediment-bound chemicals in aquatic systems. Pergamon Press, Toronto, pp 388–412

Piemontesi D, Baccini P (1986) Geochemical characteristics if dissolved organic matter in interstitial waters of lacustrive sediments and its influence on copper and zinc transport. Environ Techn Let 7:577–592

Schnitzer FJ, Khan SU(1972)Humic substances in the environment.Marcel Dekker,New York

Shea D (1988) Developing national sediment quality criteria. Environ Sci Technol 22:1256–1261

Sigleo AC, Means JC (1990) Organic and inorganic components in estuarine colloids: implications for sorption and transport of pollutants. Reviews of Environmental Contamination and Toxicology 112:123–147

Stauffer TB, MacIntyre WG, Wickman DC (1989) Sorption of nonpolar organic chemicals on low-carbon-content aquifer materials. Environmental Toxicology and Chemistry 8:845–852

Steelink C (1985) Implications of elemental characteristics of humic substances. In: Aiken GR, McKnight DM, Wershaw RL (eds) (1985) Humic substances in soil, sediment, and water. John Wiley & sons, New York, pp 457–476

Stevenson FJ (1982) Humus chemistry; genesis, composition, reactions. Wiley-Interscience, New York

Weis M, Frimmel FH (1989) Electrochemical determination of the anionic particle charge of aquatic fulvic acids. Fresenius Z Anal Chem 335:927–930

Weis M, Frimmel FH (1991) Distribution of highmolecular weight organic acids from landfill leachate in the liquid / solid phases of an anaerobic model aquifer. Z Wasser-Abwasser-Forsch 24:113–118

Wershaw RL (1986) A new model for humic materials and their interactions with hydrophobic organic chemicals in soil–water or sediment–water systems. J Contaminant Hydrol 1:29–45

Wershaw RL (1989) Application of a membrane model to the sorptive interactions of humic substances. Environ Health Perspectives 83:191–203

4 Partitioning Behaviour of Phenols Between Water and Hydrophilic/Hydrophobic Interfaces

P. Belouschek, B. Gorny, K. Hof, S. Nowicki, U. Skiba ,V. Neitzel, and Z. Adamczyk

4.1 Introduction

In the actual discussion of sediment quality criteria the approaches of sediment–water equilibria being more oriented to basic science are of particular interest (Förstner 1989). Since sediments are increasingly recognized as both a carrier and a possible source of contaminants in aquatic systems, the partitioning mechanisms of chemicals between water and sediment play a major role for understanding the pollution of sediments (Förstner 1989; Förstner et al. 1987).

In sediments the partitioning of solute and solid–water constituents between the aqueous phase and the inorganic and organic particles may be described by adsorption processes and the Nernst partition coefficient, respectively. Although a lot of adsorption experiments (Karickhoff 1979, 1984; Huang 1980) as well as investigations of partition mechanisms, especially of the n-octanol/water partition coefficient (Chiou and Schmedding 1982; Chiou et al. 1977; Mackay 1982) of different organic compounds are carried out, there exists no systematic views demonstrating the interdependence between the properties of the solute molecules and of the solid surfaces. But the understanding of these differentiated processes is the preposition for interpreting quantitatively the contamination of aquatic sediments and also for developing approaches of sediment–water equilibria.

Thus, it is the essential aim of this work to present the background of partition mechanisms of a typical organic molecule in aquatic systems on the basis of the hydrophilic–hydrophobic properties of the transferring organic molecule as well as of the accumulating solid or liquid phase. Phenol and its derivates are chosen as typical environmental molecules, because its hydrophilic–hydrophobic properties may be changed by distinct substitutes and their position in the molecule, respectively. The typical minerals of aquatic systems, like clay minerals or metal oxides,

are characterized by hydrophilic surface properties. The hydrophilic–hydrophobic balance of surface properties may be influenced especially by organic coating materials. A special case is the adsorption (deposition) of colloid-sized minerals as well as coated colloidal particles or organic macromolecules on surfaces (collectors). The influence of the hydrophilic–hydrophobic properties of the paraffinic phase on the partition coefficients of phenols is shown by using n-decane as well as n-decanol and mixtures of it.

4.2 Fundamentals

The partitioning effects of solutes between the liquid and solid (adsorbent) phase may be described theoretically by adsorption isotherms which can be classified into different major classes (Huang 1980; Giles et al. 1960; see Fig. 4.1).

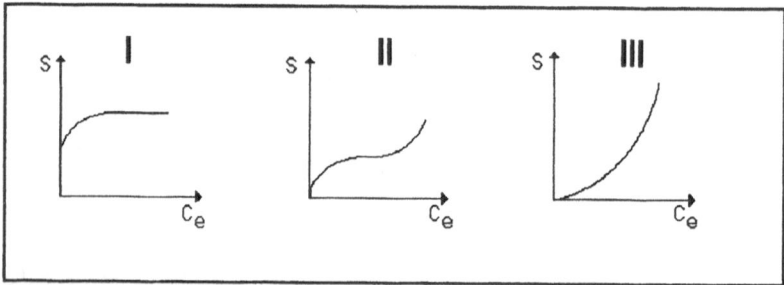

Fig. 4.1. Classification of adsorption isotherms; where **I** is monomolecular adsorption (Langmuir-type); **II** is multimolecular adsorption, typical for strong interaction between adsorbate and adsorbent; and **III** is multimolecular adsorption, typical for weak interaction between adsorbate and adsorbent

The Langmuir adsorption is described by

$$\frac{C}{x/m} = \frac{1}{kb} + \frac{C}{b} \tag{1}$$

with

C	=	The equilibrium concentration of adsorbate
x/m	=	The amount of adsorbate adsorbed per unit mass of adsorbent
b	=	The adsorption maxiumum
k	=	A term related to the bonding energy

To describe empirically adsorption isotherms the Freundlich-type is generally used:

$$\frac{x}{m} = kC^n \tag{2}$$

with C = The equilibrium concentration of adsorbate
 x/m = The amount of adsorbate adsorbed per unit mass of
 adsorbent
 k and n = Terms which may be related to the bonding energy

The distribution behaviour of organic molecules between an aqueous and a paraffinic phase (Sherwood et al. 1975) is generally given by the partition coefficient being defined as

$$K_D = \frac{C_p}{C_w} \tag{3}$$

with c_p, the concentration of the organic compound in the paraffinic phase and c_w, the corresponding in the aqueous phase.

In general, the concentrations have the dimension mg/l, but for determining the thermodynamic parameters of transfer, it is more convenient to express the concentrations as mole fractions x in moles/l.

The thermodynamic parameters of transfer are given as

$$\mu_w = \mu_p = {}^0\mu_w + RT\ln x_w + {}^0\mu_p + RT\ln x_p \tag{4}$$

and

$$^0\mu_w = {}^0\mu_p = RT\ln \frac{x_p}{x_w} = RT\ln {}^mk_w = \Delta G_{tr} \tag{5}$$

It is μ the chemical potential, ΔG_{tr} the free enthalpy of transfer, mk_w the partition coefficient determined by using mole fractions. The enthalpy ΔH_{tr} and the entropy ΔS_{tr} may be obtained according to

$$\Delta G_{tr} = \Delta H_{tr} - T\Delta S_{tr} \tag{6}$$

Furthermore, it is

$$\Delta G_{tr} = \Delta G_{s,w} - \Delta G_{s,p} \tag{7.1}$$

$$\Delta H_{tr} = \Delta H_{s,w} - \Delta H_{s,p} \tag{7.2}$$

$$\Delta S_{tr} = \Delta S_{s,w} - \Delta S_{s,p} \tag{7.3}$$

The index 's' correlates to the corresponding thermodynamic parameter of solution in water (w) or in the paraffinic phase (p).

The amount of partitioning effects is governed by the intrinsic interdependence of adsorbates and adsorbents as well as the transferring molecule and the accumulating lipophilic phase being characterized by the properties of the transferring chemicals and by the surface properties of the solid phases, respectively.

Typical minerals interacting with dissolved constituents in natural waters are clay minerals, carbonates, quartz and feldspar (Stumm and Morgan 1981; Mortland 1970). These solid materials have an amphoteric character and vary greatly in the nature and magnitude of the negative and positive charge. At usual pH-values of natural water, most of the mineral surfaces are negatively charged, except those of iron oxides, aluminium oxides and manganese oxides.

Typical chemicals for determining partition coefficients are liquid alcanes and alcanols.

Depending upon the kind of substituents and their position, the phenol molecules can be set up in the following categories:

1. Weak hydrophilic substance (phenol)
2. Non-polar (pentachlorophenol)
3. Anionic compound (phenolate at high pH-values)

Based on these properties the phenolic compounds may interact with the surface of the adsorbent as well as the alcanes or alcanols of the lipophilic phase by the following bonding mechanisms:

1. Van der Waals forces
2. Hydrogen bonds (hydrophobic interaction)
3. Ion-molecule and ion-dipole interactions
4. Coulomb forces.

4.3 Adsorption of Phenols on Hydrophilic Mineral Surfaces

The sorption behaviour of organic molecules on mineral surfaces from aqueous solutions is determined by the chemical and physical properties of the solutes, as well as by the ionic and hydrolysis of the inorganic solid surfaces. The consequences of these interdependences will be shown by the adsorption of substituted phenols on different clay minerals and colloidal silica particles.

Fig. 4.2 represents the adsorption isotherms of phenol, 4-chlorophenols and 2.6-dichlorophenols on kaolinite. In all cases the adsorption isotherms correspond to the III-type (Giles et al. 1960). Similar results are obtained for the adsorption of substituted phenols on other clay minerals as well as on colloidal oxidic particles. It is interesting to note that the adsorption isotherms on kaolinite and silica are al-

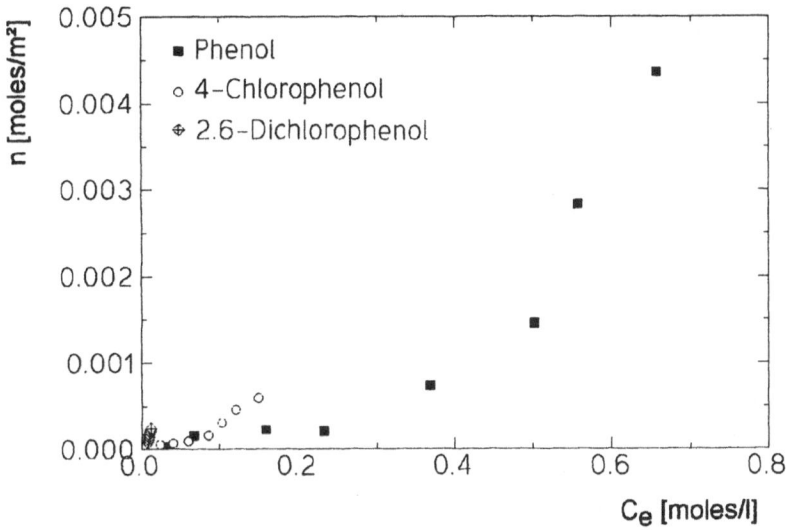

Fig. 4.2. Adsorptions isotherms of subtstituted phenols on kaolinite at pH=7 and T=293 K

most identical, a fact which indicates the non-specific adsorption behaviour of these charged, but strong hydrated surfaces.

When evaluating the adsorption isotherms according the Langmuir and Freundlich theory, two straight lines of different slopes are obtained demonstrating two different sorption mechanisms. At low phenol concentrations a step by step adsorption up to monomolecular coverage is given followed by multi-layer adsorption. The transition is at equilibrium concentration of about 0.15 moles/l.

This co-operative change of the surface properties from hydrophile to hydrophobe is clearly reflected in the amount of the adsorption capacities, e.g. 0.0028 g/m^2 and 0.18 g/m^2.

The adsorption of phenolic compounds on clay minerals and oxidic particles is markedly increased from aqueous electrolyte solutions at ionic concentrations of about 10^{-2} to 10^{-1} moles/l. It should be mentioned that similar results are obtained for the adsorption of phenols from aqueous electrolyte solutions on hydrophobic surfaces, e.g. activated carbon. This enhanced adsorption may be explained by so-called salting out effects (Peschel et al.1978).

The adsorption measurements of substituted phenols on the negatively charged clay minerals or silica particles in dependence of pH clearly demonstrate the influence of the acid–base character of the phenol compounds. Whereas the adsorption behaviour of e.g. 2.6-dichlorophenol is in dependance with pH, the adsorption of the phenol molecules is markedly reduced at pH \geq 7 due to the shift from phenol to phenolate ion.

Even these results support the view that in aqueous systems the adsorptions of phenols is non-specific.

Fig. 4.3. Adsorption of phenol on kaolinite, montmorillonite and Ca-bentonite

As may be drawn from Fig. 4.3, the adsorption capacity of kaolinite is more pronounced than that of the multi-layer swelling clay mineral montmorillonite which has about the tenfold total surface. Such results were be found for phenolic compounds as well as for volatile and non-volatile organic molecules. It follows from these results that weak hydrophilic organic molecules and apolar compounds are not (ad)sorbed into the inner surface of swelling clay minerals. Similar results are obtained by the adsorption of benzene, toluene, kaolinite and montmorillonite from water (Ziechmann and Müller-Wegener 1990). This view is supported by the fact that kaolinite and silica particles show similar adsorption capacities indicating also the non-specific properties of hydrophilic inorganic surfaces for adsorbing of weak hydrophilic and apolar organic molecules.

4.4 Sorption of Phenols on Coated Surfaces

The coating of mineral particles by organic material plays an important role in aquatic systems, especially in water sediments (Förstner 1989; Ziechmann and Müller-Wegener 1990). Typical coatings are humic substances. The significance of organic coatings for aquatic systems is based on their ability to increase the

adsorption capacities, as well as on their properties to influence the aggregation (flocculation) and disaggregation (steric stabilization) processes.

In order to get a deeper understanding for the consequences of coatings it seems to be useful to carry out model experiments.

Synthetic polymer molecules like polyacrylamide resemble humic substances (Dean 1969). The significance of organic coatings on adsorption and aggregation phenomena may be studied for the example of the adsorption of substituted phenols on well-defined silica particles coated by polyacryamid molecules having non-ionic as well as cationic or anionic properties. When using the method of dynamic light scattering for investigating the hydrodynamic particle diameter, it is not only possible to detect the radius of aggregated particles, but also the layer thickness of the adsorbed polyacryamide (Decora 1985). Thus, changes during the adsorption of phenol on the coated mineral particles of colloidal size can be directly determined.

But before presenting such results, it is necessary to point out the influence of adsorbed polymers on the aggregation and disaggregation behaviour of colloidal particles. This will be shown on the basis of the adsorption of polyacrylamides having non-ionic and ionic properties on colloidal silica particles in dependence of pH-value (Belouschek et al. in press; Belouschek et al. 1992). In Figs. 4.4 and 4.5 the diameters of single coated particles as well as of particle aggregates are presented as a function of the amount of adsorbed cationic polyacrylamides per

Fig. 4.4. Particle diameter as well as aggregation diameters (d represents particle layer thickness) as a function of the amount of adsorbed 45% cationic polyacrylamide (C_p) at ■ pH = 7 before 18 nm; and at ❑ pH = 5 after 25 nm

Fig. 4.5. Particle diameter as well as aggregation diameters (*d* represents particle layer thickness) as a function of the amount of adsorbed 25% cationic polyacrylamide (C_p) at ■ pH = 7 before 37 nm and after 55 nm; and at ◻ pH = 5 before 25 nm and after 70 nm

m^2. At a small concentration interval a flocculation maximum occurs followed by steric stabilization. There exists a direct correlation between the surface charge and the amount of cationic endowment of the polyacryamide. Interaction forces – distance measurements between macroscopic silica surfaces coated by poly-acryamides of high mole mass – clearly indicate the bridging mechanism as the rule of the flocculation process (Belouschek et al. in press).

From the described experiments the layer thickness of the adsorbed polyacryl-amides are determined. It is interesting to note that the half-layer thickness is found before the flocculation maximum in comparison to the steric stabilization after the flocculation. This result is supported by the theory of LA MER (LaMer and Healy 1963). Similar layer thicknesses (see Figs. 4.4 and 4.5) are also de-tected by other investigations (Lee and Somasundaran 1989; Kitchener 1972). When adsorbing non-ionic or anionic endowed polyacryamides on negatively charged silica particles, flocculation maxima can not be found. The determined layer thicknesses are also markedly smaller with d = 5–15 nm than the corresponding of cationic polymers.

In Fig. 4.6 the (ad)sorption isotherms of phenol on silica particles coated by different polyacryamides are presented.

It is clearly evident that the adsorption isotherms of phenol on uncoated as well as non-ionic and anionic polyacrylamides are congruent within the experimental error. The (ad)sorption capacities of the cationic polyacrylamides are markedly

Fig. 4.6. Adsorption isotherms of phenol on colloidal silica particles coated by polyacrylamides at pH = 7 and T = 298 K

more pronounced, whereby the 25% cationic polyacrylamide has the highest capacity. It is interesting to note that the layer thicknesses of the adsorbed polyacrylamides increase with growing equilibrium concentration of phenol in differentiated manner (see Fig. 4.7).

Fig. 4.7. Changes of layer thicknesses, d, of adsorbed 45% cationic polyacrylamides, $C_p = 10^{-5}$ g/m^2, as a function of phenol equilibrium concentration

This effect is especially marked in the case of cationic polyacrylamides. As may be drawn from Fig. 4.4, the enlargement of the layer thicknesses starts after the monomolecular coverages of phenols are reached (see also Chap. 2).

It is clearly evident from these results that the (ad)sorption of phenol on coated particles is characterized by synergistic effects of adsorption amounts and conformational changes of the organic polymer chain segments. These synergistic effects cause significant enlargement of the layer thickness which means that the number of adsorption sites is increased due to the conformational changes.

These results indicate the interrelation between the (ad)sorption amount of organic molecules and the content of total organic carbon of the considered adsorbent.

4.5 Adsorption of Colloidal Particles on Surfaces

The natural colloids may be of inorganic as well as of organic origin. These colloids being mostly negatively charged at natural pH-value interact with metals to form metal-organic complexes by ion exchange and chelation processes, and interact with organic chemicals by protonation, Van der Waals forces, coordination complexation, and hydrophobic bonding. Thus, colloidal particles are typical carriers of inorganic and organic pollutents in freshwater (Carter and Suffet 1982), estuarine (Poirrier et al. 1972), and seawater (Pierce et al. 1974). In this connection the interactions of colloid-sized particles with larger ones or with boundary surfaces (collectors) leading to adsorption and deposition is of particular interest in the contamination of aquatic systems.

The significance of particle deposition has been recognized long ago and many works have appeared concerning mostly kinetic aspects of this process (Gimbel 1984). On the other hand, important non-linear effects appearing at large surface concentrations of adsorbed particles are little studied either experimentally or theoretically mostly due to difficulties in considering properly the many-bodied problem arising. Based on Monte-Carlo simulations the localized, sequential adsorption of colloided particles interacting via the screened coulomb potential may be analyzed by taking into account the lateral interactions (Feder and Glaever 1989), especially the lateral double-layer interactions (Adamzcyk and Belouschek 1991).

By the latter theoretical method, the deposition of colloids onto plates, spheres and cylinders may be calculated in dependence of the ratio (A) of the radius of the colloidparticle (a) and of the collector (R). Furthermore, the parameter κa (k: Debye-Hückel parameter) is the master factor determining the lateral interactions and herewith the surface coverage. It results from the Monte-Carlo simulations that for colloid-sized particles smaller than 1 μm the lateral double-layer interac-

tions play an important role, whereas for diameters larger than $1\,\mu$m they may be neglected. In the latter case, the expansion of the electrical double-layer is extremely small in comparison to the radius of the particle ($\kappa a = \infty$). In this case the maximum surface coverage (Θ) may be described by the relation

$$\Theta = \eta a^2 Np/S^* \tag{8}$$

It follows from the Monte-Carlo simulations: a maximum value of $\Theta = 54{,}7\%$ related to the geometrical collector surface.

In Eq. 8 Np is the average number of the adsorbed particles. The effective collector surface (S^*) can be determined by the expression

$$S^* = S_c(1+A)^m \tag{9}$$

with S_c, the general collector surface, and $m = 0, 1, 2$ for plate, sphere and cylinder.

For colloid-sized particles having a diameter smaller than $1\ \mu$m, the really maximum surface coverage (Θ) corresponds to the following relation

$$\Theta_{mx} = \Theta(1+A)^m/(1+H^*) \tag{10}$$

with H^* the real expansion of the lateral double-layer interactions.

It may be derived from the Monte-Carlo simulations that Θ_{mx} is always smaller than Θ. For high electrolyte concentrations of the aqueous phase and particle diameters larger than $1\ \mu$m, the lateral double-layer effects may be neglected. These results are demonstrated in Fig. 4.8 showing exemplarily the dependence of the surface coverage from the parameters κa as a function of $R = a$, where R is a cylindrical collector.

It is clearly evident from Fig. 4.8 that the amount of adsorbed colloidal particles on surfaces is significantly determined by the electrolyte concentration of the aqueous phase. Thus, the amount of deposited particles on surfaces is much higher in estuary and sea than in lakes or rivers.

As may be drawn from Fig. 4.9, the kinetic of particle adsorption is also determined by the electrolyte concentration.

It is clearly evident that in the estuary and sea having high electrolyte concentrations, the adsorption ensures much quicker than in waters of low ionic strengths.

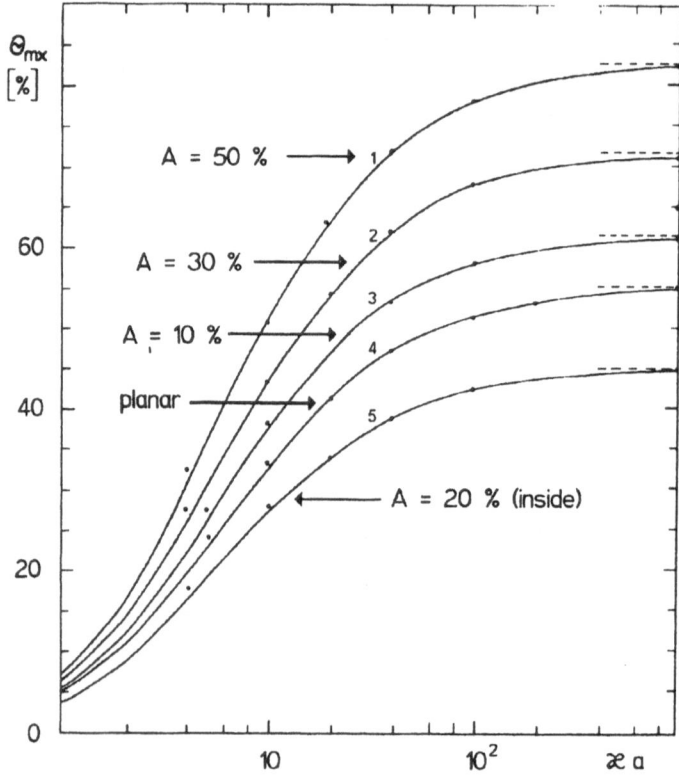

Fig. 4.8. Maximum surface coverage of a cylindrical particle as a function κa

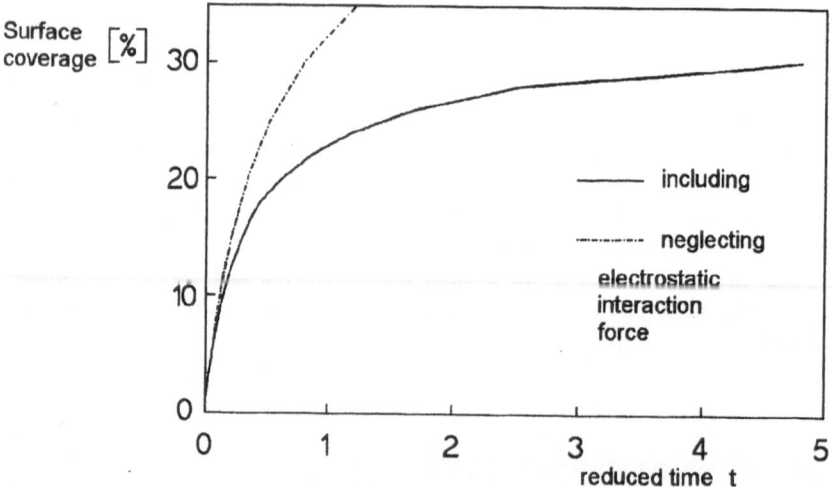

Fig. 4.9. The kinetic of particle adsorption as a function of low and high electrolyte concentrations

The significance of these results for aquatic systems lies in the accumulation of pollutants, e.g. heavy metals on the colloids and therefore on the collectors. Tables 4.1 and 4.2 show the calculated results of spherical and cylindrical collectors for two radii each. The heavy metal concentration and adsorption density (from adsorption experiments) is considered constant in all cases; the particle radius a, the collector radius R and the ionic strength c are variable. For the first example (small collector and particle radius) in Tables 4.1 and 4.2 it is obvious that the accumulation n increases with the ionic strength and the particle radius, whereas the second example exhibits for both collectors no dependence upon the ionic strength even by low aspect ratios. This indicates for small radii (particle and collector) a dependence on the ionic strength, so that in effect the calculation of the surface concentration must take the lateral interaction forces between the particles into account (soft sphere calculation). At a size greater 1 μm, the increasing ionic strength and a further particle enlargement show no significant influence on the accumulation of pollutants at not only spherical but also cylindrical collectors.

Table 4.1a. Calculated pollutant accumulation on a spherical collector. Example 1: spherical collector: $R = 1$ μm, $O = 1.257 \cdot 10^{-11}/m^2$; pollutant concentration: $5 \cdot 10^{-5}$ moles/l; pollutant adsorption: $6.28 \cdot 10^{-7}$ moles/m^2

a [μm]	c [moles/l]	κa	Θ_{mx}	$n \cdot 10^{17}$ [a]
0.1	10^{-4}	~5	~22%	0.69
A=10%	10^{-3}	~10	~40%	1.27
	10^{-1}	~10^2	~65%	2.06
0.2	10^{-4}	~10	~47%	1.48
A=20%	10^{-3}	~20	~60%	1.89
	10^{-1}	~$2 \cdot 10^2$	~88%	2.77

[a] [moles/collector]

Table 4.1b. Calculated pollutant accumulation on a spherical collector. Example 2: spherical collector: $R = 100$ μm, $O = 1.257 \cdot 10^{-7}/m^2$; pollutant concentration: $5 \cdot 10^{-5}$ molesmol/m^2; pollutant adsorption: $6.28 \cdot 10^{-7}$ moles/m^2

a [μm]	c [moles/l]	κa	Θ_{mx}	$n \cdot 10^{13}$ [a]
10	10^{-4}	~$5 \cdot 10^2$	~65%	2.04
A=10%	10^{-3}	~10^3	~65%	2.04
	10^{-1}	~10^4	~65%	2.04
20	10^{-4}	~10^3	~78%	2.46
A=20%	10^{-3}	~$2 \cdot 10^3$	~78%	2.46
	10^{-1}	~$2 \cdot 10^4$	~78%	2.46

[a] [moles/collector]

For the comparison of sphere and cylinder collector we assumed the same outer surface area, whereby the cylinder possesses in addition an internal interface where adsorption could take place. If one compares only the deposition of colloids on the external interfaces, the pollutant concentration on the sphere is always higher, whereas by possible internal deposition of colloids the accumulated heavy metal concentration is higher on the cylindrical collector.

Table 4.2 a. Calculated pollutant accumulation on a cylindrical collector. Example 1: cylindrical collector $R_a = 1$ μm, $O = 1.257 \cdot 10^{-11}/m^2$; $R_i = 0.8$ μm, $L = 2$ μm, $d = 0.2$ μm; pollutant concentration $5 \cdot 10^{-5}$ moles/l; pollutant adsorption $6.28 \cdot 10^{-7}$ moles/m^2

a [μm]	c [moles/l]	κa	Θ_{mxa}	Θ_{mxi}	$n \cdot 10^{17}$ [a]		
					n_a	n_i	n_{ges}
0.1	10^{-4}	~5	~26%	~16%	0.822	0.404	1.22
A=10%	10^{-3}	~10	~37%	~25%	1.17	0.632	1.80
	10^{-1}	~10^2	~57%	~37%	1.80	0.935	2.74
0.2	10^{-4}	~10	~40%	~27%	1.26	0.682	1.94
A=20%	10^{-3}	~20	~50%	~33%	1.58	0.834	2.41
	10^{-1}	~$2 \cdot 10^2$	~56%	~42%	1.77	1.06	2.83

[a] [moles/collector]

Table 4.2 b. Calculated pollutant accumulation on a cylindrical collector. Example 2: cylindrical collector $R_a = 1$ μm, $O = 1.257 \cdot 10^{-11}/m^2$; $R_i = 80$ μm, $L = 200$ μm, $d = 20$ μm; pollutant concentration $5 \cdot 1$ 0^{-5} moles/l; pollutant adsorption $6.28 \cdot 10^{-7}$ moles/m^2

a [μm]	c [moles/l]	κa	Θ_{mxa}	Θ_{mxi}	$n \cdot 10^{17}$ [a]		
					n_a	n_i	n_{ges}
10	10^{-4}	~$5 \cdot 10^2$	~62%	~40%	1.96	1.01	2.97
A=10%	10^{-3}	~10^3	~63%	~40%	1.99	1.01	3.00
	10^{-1}	~10^4	~63%	~40%	1.99	1.01	3.00
20	10^{-4}	~10^3	~66%	~45%	2.08	1.14	3.22
A=20%	10^{-3}	~$2 \cdot 10^3$	~66%	~45%	2.08	1.14	3.22
	10^{-1}	~$2 \cdot 10^4$	~66%	~45%	2.08	1.14	3.22

[a] [moles/collector]

These results demonstrate the importance of detailed consideration for the understanding and prediction of accumulation and deposition processes in natural aquatic systems. It is not sufficient to take account of the predominate ionic strength since the shape and above all of the particle and collector radius should also be considered.

4.6 The Partition Coefficient: k_D

Partition coefficients are usually determined between water and n-octanol (Chiou and Schmedding 1982; Chiou et al. 1977; Mackay 1982). In order to study the influence of the hydrophilic–hydrophobic properties of the paraffinic phase on the amount of the partition coefficient, we investigated the partition behaviour of phenol and substituted phenol compounds between water and n-decane/n-decanol and mixtures of it (Belouschek and Neitzel 1985; Belouschek et al. 1986). It should be mentioned that similar partition coefficients are obtained for n-octanol and n-decanol.

As may be drawn from Table 4.3, which enables a comparison of the partition coefficients of the phenols and some selected halobenzenes between water and n-decane/n-decanol, there exists a significant influence of the hydrophobic–hydrophilic balance of the organic chemicals and the paraffinic phase on the partition behaviour.

It is evident that the partition coefficients are markedly increased for polar organic chemicals, e.g. substituted phenols, between water and n-decanol in comparison to the paraffinic phase n-decane.

The differences are reduced for phenols being substituted in ortho-position. Similar partition coefficients are found for typical hydrophobic substances, e.g. chlorobenzene, when using n-decane or n-decanol as paraffinic phase.

The interdependence of hydrophilic and hydrophobic properties between organic compounds and the paraffinic phase is also clearly demonstrated by the temperature dependence of partition coefficients.

Table 4.3. Partition coefficients of polar organic compounds between water and n-decane/ n-decanol at 293 K

	$^m k_D$ Water / n-decanol	$^m k_D$ Water / n-decane
phenol	278	0.909
4-methylphenol	757	2.55
2-methylphenol	845	8.01
3-methylphenol	862	3.38
2-chlorophenol	1408	54.3
4-chlorophenol	2490	4.4
2-chlorophenol	2960	4.9
2,6-dimethylphenol	1802	101.2
3,5-dimethylphenol	2204	14.62
chlorobenzene	6700	7086

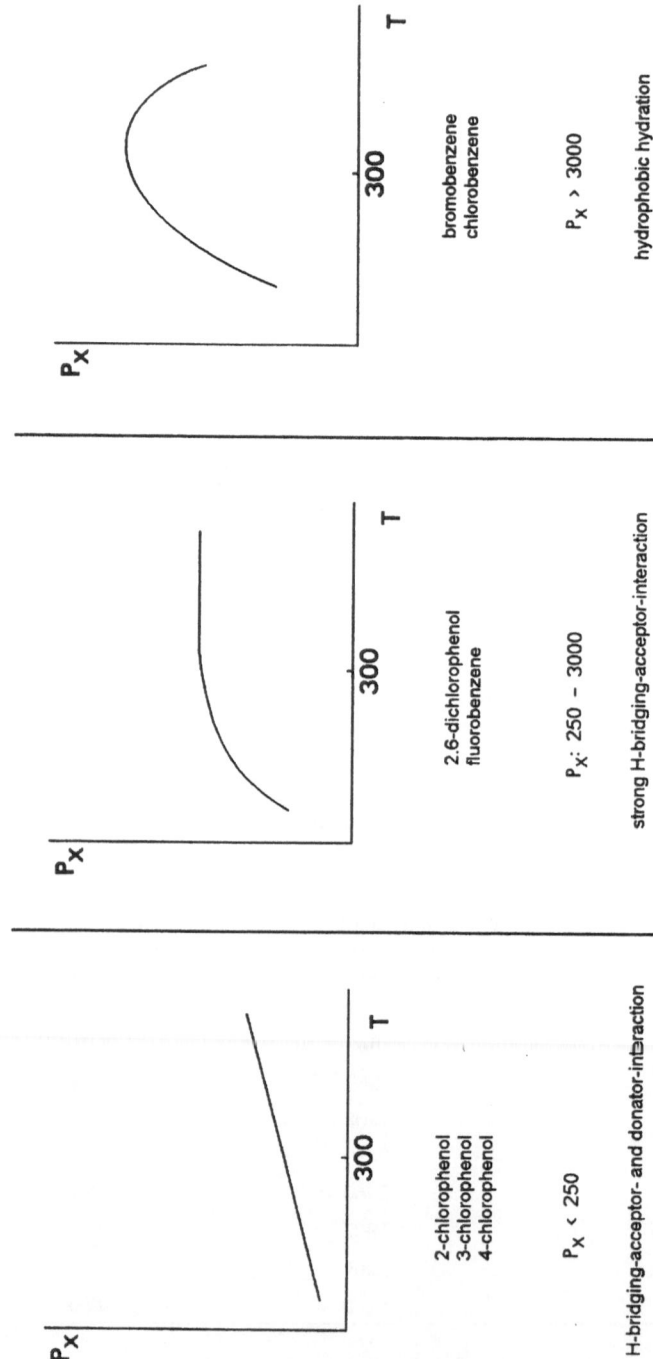

Fig. 4.10. Temperature dependence of partition coefficients of different organic compounds between water and n-decane; where P_X: partition coefficients and T (temperature): K

It is clearly evident from Fig. 4.10 that the temperature dependence of the partition behaviour between water and n-decane, an ideally hydrophobic paraffinic phase, may be divided into three groups:

1. Organic chemicals carrying out H-bridging- and donator-interactions show a continuous increase of the partition coefficient with temperature.
2. Organic compounds of H-bridging interactions, e.g. 2.6-dichlorophenol or fluorobenzene, have partition coefficients in dependence of temperature being characterized by an increase until about 300 K in order to remain constant thereafter.
3. The partition behaviour of hydrophobic organic chemicals is characterized by a maximum at about 300 K.

The percentage increase of the partition coefficients with temperature is presented for some typical organohalogenated compounds in Table 4.4. It is evident that a rise of about 15% of the partition coefficients is found for temperatures of about 303–313 K if the partition coefficients valid for 283 K are related to 100%. The growth of the partition coefficients with temperature is markedly diminished for the halobenzenes in comparison to the chlorophenols.

Table 4.4. Percentage increase of the partition coefficients with temperature for some typical organohalogenated compounds (283 K \cong 100%)

	Temperature (K)	Percentage of increase (%)
Fluorobenzene	313	25
Chlorobenzene	298–303	15
Bromobenzene	298–303	15
2-chlorophenol	303	35
	313	50
3-chlorophenol	303	50
	313	90
	303	60
	313	85
2,6-dimethylphenol	313	32

It is now interesting that the partition coefficient of hydrophobic organics show also a maximum at about 300 K when the paraffinic phase is n-decanol or mixtures of n-decane /n-decanol (see Figs. 4.11 and 4.12).

As may be drawn from Figs. 4.11 and 4.12 the amount of the partition coefficients of polar organic molecules with hydrophilic groups varies significantly with the composition of lipophilic phase. The partition coefficients diminish with temperature for n-decane as lipophilic phase, whereas an increase is found using

n-decanol. It is interesting that for the mixture: 99.5 vol% n-decane/0.5 vol% n-decanol the partition coefficients remain constant in the temperature interval investigated.

The reasons of this divergent behaviour of partitioning effects lie in the different solution behaviour of the transferred apolar or polar as well as hydrophobic or hydrophilic molecules in the aqueous or non-aqueous phase.

In the case of hydrophobic apolar molecules the partition effect is determined only by the solution behaviour in the aqueous phase. The solubility of hydrophobic, apolar molecules like benzene or chlorobenzene in water is charac-

Fig. 4.11. Temperature dependence of partition coefficients of hydrophobic molecules for n-decanol as well as mixtures of n-decane/n-decanol. The *closed squares* indicate: chlorobenzene between water and decane; *open circles*: 25 vol% decanol + 75 vol% decane; *closed circles*: 50 vol% decanol + 50 vol% decane; *open triangles*: 75 vol% decanol + 25 vol% decane; and *closed triangles*: decanol

Fig. 4.12. Temperature dependence of partition coefficients of hydrophobic molecules for n-decanol as well as mixtures of n-decane/n-decanol. The *x-symbols* indicate: 4-chlorophenol between water and decane; *open circles*: 0.5 vol% decanol + 99.5 vol% decane; *closed circles*: 2 vol% decanol + 98 vol% decane; *open triangles*: 5 vol% decanol + 95 vol% decane; *closed triangles*: 10 vol% decanol + 90 vol% decane; *open squares*: 50 vol% decanol + 50 vol% decane; and *closed squares* : decanol

Table 4.5. Thermodynamic parameters of solution in water (W) and n-decane (O) and of transfer: ΔG : free enthalpy, ΔH: enthalpy, and ΔS: entropy

System		ΔG [kJ/mol]	ΔH [kJ/mol]	ΔS [J/mol K]
4-chlorophenol	water	-13.8	- 4.0	33
	n-decane	- 9.6	-17.5	-27
Transfer W \rightarrow O		- 3.9	14.3	61
2,6-dichlorophenol	water	-20.8	-16.3	15
	n-decane	- 6.9	-22.4	-53
Transfer W \rightarrow O		-13.9	5.8	71

terized by a significant minimum at about 298 K (Belouschek and Neitzel 1985). Since hydrophobic apolar molecules may only interact by the weak dispersion or Keesom- and Debye-energies, respectively, the influence of the lipophilic phase is not so dominant for the partitioning effects. Thus, the transfer of apolar hydrophobic molecules between the aqueous and the paraffinic phase is mainly determined by the hydrophobic hydration (Tanford 1980).

Hydrophobic molecules are only weakly soluble in water and possess therefore, a positive enthalpy of solution. This means the solution of hydrophobic molecules in water is only due to entropic effects. It is interesting to note that the thermody-namic parameters of aqueous solution valid for chlorobenzene are almost identical with that of the transfer from water to n-decane. Thus, the temperature dependence of the partition coefficient may only be explained by the solution behaviour in water. There exists a minimum of solubility of chlorobenzene in water at about 300 K corresponding with a maximum of the partition coefficient. By increasing temperature the hydrophilic hydration is destroyed causing to an enhancement of interactions between the hydrophobic molecules.

In Table 4.5 the thermodynamic parameters of solutions consist of the transfer for 4-chlorophenol and 2.6-dichlorophenol. It is clearly evident that the differences between the thermodynamic parameters of water and n-decane coin-cide very well to that of the transfer from water to n-decane.

The solubility of 4-chlorophenol in water increases slowly with temperature in comparison to that of n-decane. Thus, the partition equilibrium is shifted to the lipophilic phase by increasing temperature. In the wise of 2.6-dichlorophenol the solubility in water is markedly increased above 300 K due to hydrophobic interac-tions, so that the solubility in the n-decane phase is compensated. This means the partition coefficient of 2.6-dichlorophenol between water and n-decane is almost unchanged above 300 K.

The partition behaviour of a chemical compound between two phases is deter-mined by energetic processes, e.g. changes of intermolecular interactions and of solubility behaviour, being characterized quantitatively by the thermodynamic transfer parameters shown in the Equations 7.1–7.3. The class and position of

Fig. 4.13. Correlation of the free enthalpy of transfer of selected organic chemicals from water to n-decane with the free enthalpy of solution in water at 293 K

functional groups in an organic molecule as well as the physical-chemical properties of solvents which influence the partition mechanisms in characteristic kind, changes the thermodynamic parameters of transfer and determine the amounts of ΔG_{tr}, ΔH_{tr} and ΔS_{tr}. For understanding the different temperature dependences of the partition coefficients of the investigated phenols and halobenzenes and the linear relationship of the free enthalpy of partition and of solution, it is necessary to compare the thermodynamic parameters of transfer with those of the aqueous and of the paraffinic solution (see Eqs. 7.1–7.3).

In the Figs. 4. 13 and 4.14 the free enthalpies of transfer valid for the system water/n-decane as well as water/n-decanol are plotted against the free energies of solubility in water for phenol, substituded methyl- and chlorophenols as well asbenzene and halobenzenes. Global correlation isn't obtained for n-decane and n-decanol, as may be drawn from Figs. 4.13 and 4.14 respectively. When using a global correlation, it results generally that in water, insoluble organic compounds are enriched in a lipophilic phase. Since the partition coefficients of the considered organic molecules are more different in the wise of n-decane than that of n-decanol, a global correlation seems to be more significant for n-decanol than for n-decane. This view is supported by the fact that the differences of the free en-thalpies of transfer are reduced to about 5 kJ/mol for n-decanol as lipophilic phase, in comparison to about 25 kJ/mol for n-decane.It is clearly evident from Fig. 4.13 that linear correlation exists for methylphenols and the para- and meta-chlorophenols when starting from the phenol molecule. On the other hand, benzene may directly be correlated with chloro- and bromobenzene, but not with

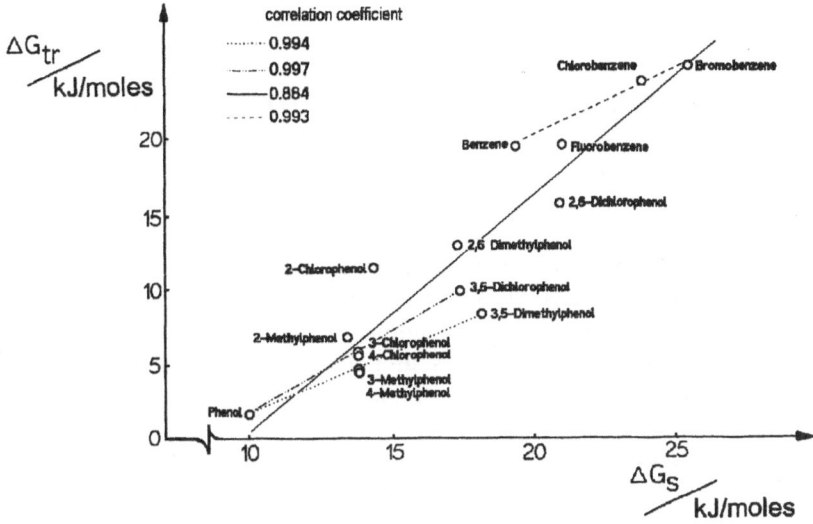

Fig. 4.14. Correlation of the free enthalpy of selected organic chemicals from water to n-decanol with the free enthalpy of solution in water at 293 K

fluorobenzene.

Regarding the chlorophenols, it may be seen that only the phenols having chlorine atoms in meta- and para-position present a satisfactory correlation. The ortho-substituted chlorophenols cannot fit into any straightline correlation.

In the case of n-decanol as lipophilic phase, three linear straightlines are found for the 3-chloro- and 4-chlorophenol: the 2-chloro- and 3,5-dichlorophenol and the methyl-, 2,6-dichlorophenol and the chlorobenzene.

The results clearly indicate that the contributions of the solution behaviour of the organic molecules in the aqueous phase to partition coefficients is reduced by a lipophilic phase like n-decanol being also able to form H-bridging bonds. This superposition of contributions of intermolecular force from the aqueous and the n-decanol phase reduces the possibility to interpret the linear relationships.

These results clearly indicate that a linear relation of the free enthalpy of transfer and of solution is only given for the case of additivity of increments (Leo et al. 1976). It should be considered that a correlation of the partition coefficient with the solubility, e.g. in water, is only given by a similar solution behaviour of the organic molecules in the paraffinic phase (Leo et al. 1976, 1980; Fujita et al. 1964).

The sensitivity of these boundary conditions on the correlation factors obtained by the linear relation of the free enthalpies of transfer and solution is intrinsically demonstrated by the results shown in Figs. 4.13 and 4.14.

4.7 Concluding Remarks

The differentiated knowledge of the partitioning behaviour is the essential preposition for developing sediment–water equilibrium criteria. The results presented clearly indicate that only for apolar organic compounds a simple correlation of the partition coefficient and the distribution behaviour between the aqueous phase and hydrophilic–hydrophobic phase can be derived. In the latter case, the amount of the partitioning coefficients is chiefly determined by the contribution of Van-der-Waals forces. The partition behaviour of polar molecules is ascertained by well-balanced contributions of intermolecular forces being influenced not only by the properties of polar molecules but also by the significance of the hydrophile and hydrophobe adsorbent or sorbent phase, respectively.

Thus, a simple procedure to develop a practicable sediment–water equilibria assessment which includes all possible contributions of intermolecular forces to the partition coefficients does not exist.

References

Adamzcyk Z, Belouschek P (1991) Localized adsorption of particles on spherical and cylindrical interfaces. J Colloidal and Interface Sci 146:123–136

Belouschek P, Gorny B, Lorenz D () Zur Aggregation von Silika-Teilchen in Gegenwart von Polyacrylamiden. being published in Vom Wasser

Belouschek P, Lorenz D, Adamzcyk Z (1992) Kriterien der Kolloidstabilität in Aquatischen Systemen. being published in Wasserkalender 93. E Schmid Verlag, Berlin

Belouschek P, Neitzel V (1985) Die Bedeutung der Solvatation für die Verteilung von halogenorganischenVerbindungen zwischen Wasser und einer Paraffinphase. Z Wasser-Abwasser-Forsch 18:51–55

Belouschek P, Neitzel V, Belouschek I (1986) Entfernung niedermolekularer halogenorganischer Verbindungen aus Abwassern mittels Sorption an temporärstabilen Paraffindispersionen – Modelluntersuchungen. Wasser - Abwasser - gwf 127:25–31

Belouschek P, Nowicki S, Skiba U () Adsorption of phenols on hydrophilic mineral surfaces from aqueous solutions being prepared.

Carter CW, Suffet IH (1982) Binding of DDT to dissolved humic materials. Env Sci Technol 16:735-740

Chiou CT, Freed VH, Schmedding DW, Kohnert R (1977) Partition coefficient and bioaccumulation of selected organic chemicals. Env Sci Technol 11:475–478

Chiou CT, Schmedding DW (1982) Partitioning of organic compounds in octanol–water systems. Environ Sci Technol 16:502–508

Dean RB (1969) Colloids complicate treatment processes. Env Sci Technol 3:820

Decora R (1985) Dynamic light scattering. Verlag, New York

Feder J, Giaever I (1989) J Colloid Interface Sci 78:144

Förstner M (1989) Contaminated sediments. In: Lecture Notes in Earth Sciences. Springer, Berlin Heidelberg new York

Förstner U et al. (1987) Qualitätskriterien für Gewässersedimente – Allgemeine Problematik und internationaler Stand der Diskussion Z Wasser-Abwasser-Forsch 20:54–59

Fujita T, Wasa J, Hansch C (1964) A new substituent constant, P, derived from partition coefficient. J Amer Chem Soc 86:5175

Giles CH, MacEwan TH, Nakhwa SN, Smith DJ (1960) Chem Soc 786:3973

Gimbel R (1984) Abscheidung von Trübstoffen aus Flüssigkeiten in Tiefenfiltern, Heft 25 der Veröffentlichung Wasserchemie der Universität Karlsruhe, Karlsruhe

Huang PM (1980) Adsorption processes in soil. In: The Handbook of Environmental Chemistry – Reactions and Processes, Vol 2, Part A. pp 48–59

Karickhof SW (1979) Sorption of hydrophobic pollutants on sediments. Water Research 13:241–248

Karickhoff SW (1984) Organic pollutant sorption in aquatic systems. J Hydraul Div Am Soc Civ Eng 110:707–735

Kitchener JA (1972) Principles of action of polymeric flocculants. Brit Polym J 4(3):217–229

LaMer VK, Healy TW (1963) Adsorption-flocculation reactions of macromolecules at the solid-liquid interface. Rev Pure Appl Chem 13:112–133

Lee LT, Somasundaran Pl (1989) Adsorption of polyacrylamide on oxide minerals. Langmuir 5:854–860

Leo A, Hansch C, Elkins D (1971) Partition coefficients and their uses. Chemical Reviews 71:525–616

Leo A, Hansch C, Jow PYC (1976) Dependence of hydrophobicity of apolar molecules on their molecular volume. J Medicinal Chem 19:611–615

Mackay D (1982) The correlation of bioconcentration factors. Env Sci Technol 16:274–277

Mortland MM (1970) Adv Agron 22

Peschel G, Belouschek P, Kress B, Reinhard R (1978) Zur Adsorption von Phenolen aus wäßriger Elektrolytlösung an Aktivkohle. Progr Colloid & Polyme Sci 65:83

Pierce RH, Olney CE, Felbeck GT (1974) pp-DDt adsorption to suspended particulate matter in sea water. Geochim Cosmochim Acta 38:1061–1073

Poirrier MA, Bordelon BR, Laseter JL (1972) Adsorption and concentration of dissolved carbon-14 DDT by coloring colloids in surface water. Env Sci Technol 6:1033–1035

Sherwood TK, Pigford RL, Wilke CR (1975) Mass Transfer. Mc Graw-Hill, NewYork

Stumm W, Morgan IJ (1981) Aquatic chemistry. New York

Tanford C (1980) The hydrophobic effect. John Wiley & Sons, New York

Ziechmann W, Müller-Wegener U (1990) Bodenchemie. BI-Wissenschaftsverlag, Mannheim Wien Zürich

5 Sorption Properties of Biofilms

H.-C. Flemming, J. Schmitt and K.C. Marshall

5.1 Introduction

The distribution and fate of pollutants in soils, in sediments and in the water phase is highly influenced by sorption processes. Sorbing surfaces and materials can provide a sink for dissolved matter. If the conditions change, desorption may occur and, thus, turn the former sorbent into a new source of pollutants.

Many experiments have been conducted with mineral surfaces that can sorb dissolved ions and organic molecules. It is well known that data, gained with those mineralogical clean surfaces, cannot be simply applied and compared to natural conditions. *Organic matter* is masking the original mineral surfaces and thus, these surfaces change their sorption behaviour drastically.

To the authors, it seems unsatisfactory that the true nature of this *organic matter* is not yet well characterized. The material appears to represent a kind of black box. In Chapter 5 we do not attempt to define the architecture of this black box, but attempt to find a door into it.

"Organic matter" can be grossly divided into debris, humic substances and biomass. A quantitative estimation of the partition of each of these components within the entire body of organic matter cannot be given here. We will focus on the biomass as produced by microbial activity. This focus seems to be justified by the dynamic nature of biomass as opposed to abiotic organic matter found in nature.

Most biomass in natural environments consists of organisms that are more or less immobilized at surfaces in a matrix of biological origin. Immobilized biomass, often termed *slime*, can be considered in terms of *biofilms*, which probably represent more than 99.9% of all micro-organisms on earth (Costerton et al. 1987).

Biofilms are very diverse and one of their characteristics is their heterogeneity: they consist of various micro-organisms, develop on various surfaces and under

various conditions. In many cases under natural conditions, mineral surfaces will be at least partially covered by a biofilm having developed on the surface. Dissolved substances, thus, will sorb first on this biofilm before they reach the original mineral surface. The sorption properties of this *insulating layer* are therefore, of importance for the dissolved and the sorbed state of pollutants. Weber et al. (1991) pointed out that the behaviour, transport and ultimate fate of contaminants in subsurface environments may be affected significantly by their participation in sorption reactions and related phenomena. The degree to which the resulting effects can be quantified and predicted depends upon the extent to which certain fundamental aspects of sorption are understood, and upon the accuracy with which these phenomena can be characterized and modelled in complex subsurface systems. However, this is only possible if the dynamic nature of biofilms is taken into account.

Information about the nature and properties of biofilms is available, but must be gleaned from different fields. For example, publications on the technical attempts to remove heavy metals and organic pollutants from water streams by immobilized bioreactors contain a large body of data about the aspects of biofilms as sorbents. In this chapter, we have compiled and organized some of these data in the framework of the following questions: i) what do biofilms sorb, ii) which are the sorption sites, iii) what is the sorption capacity and iv) which is the remobilization potential? Of course these questions cannot be answered comprehensively, but are addressed and provide a basis for further study.

5.2 Composition and Properties of Biofilms

Biofilms result from the adhesion and growth of micro-organisms at interfaces. The tendency of suspended micro-organisms to adhere seems to be part of a general microbial strategy for survival, especially in low-nutrient environments (Marshall 1988). Thus, biofilms are almost ubiquitous and can be found even in extreme environments (Flemming 1991).

After attachment, the micro-organisms excrete extracellular polymer substances (EPS) as a slime matrix and are embedded in it. The cell material usually accounts only to a minor part of the total organic mass of the biofilm. Thus, the properties of the EPS matrix are very important for the sorption properties of biofilms.

The photograph in Fig. 5.1 shows a transmission electron micrograph (TEM) of bacteria attached to a plastic disc after exposure to river-water for a few weeks (courtesy of G.G. Geesey).

The loose structure of the EPS which surround the microbial cells is clearly visible. However, the picture shows only the dewatered structure because for TEM preparation the sample has to be dried. Thus, the original structure is not

Fig. 5.1. Transmission electron micrograph of bacteria attached to a plastic disc in river-water. (courtesy of G.G. Geesey)

completely represented by electron microscopy. Scanning confocal laser microscopy (SCLM) can be used with fully hydrated biological specimens and has been applied to biofilms (Lawrence et al. 1991). This technique revealed significant channeling and porosity of the investigated biofilm. Thus, nutrient flux throughout the biofilm will be facilitated.

The biofilm is a microenvironment which is entirely different from that which prevails in the bulk water phase. It represents a high local concentration of readily biodegradable organic carbon and a local accumulation of microbes between 10^9 and 10^{11} cells/ml in the biofilm mass (Characklis 1990). The cells respond physiologically and genetically to the contact with a solid surface (Dagostino et al. 1991). The biofilm mode of growth provides better nutrient conditions to microorganisms in flowing water systems, because adherent bacteria utilize surface-bound organics (Marshall 1988). In addition, biofilms sequester nutrients from the water stream (Costerton et al. 1987; Decho 1990). The processes in an epilithic biofilm in a river sediment (Costerton and Boivin 1991) are summarized in Fig. 5.2.

Principally, natural biofilms act in the way which is exploited in biofilm reactors: they sorb nutrients from the water stream, convert a substantial part into

Fig. 5.2. River epilithon: exchange processes between biofilm and water phase (after Costerton and Boivin 1991, modified)

biomass and immobilize it locally. In biofilms, the cells are fixed in juxtaposition to each other for long periods of time, compared to suspended micro-organisms. This allows the development of synergistic microconsortia which can orchestrate multi-staged degradation processes. These microconsortia can harbour slow growing specialists and break down less readily degradable organic compounds. In addition, gene exchange is enhanced and therefore, more frequent in biofilms. Enhanced plasmid transfer in the epilithion is described by various authors (see Fry and Day 1990). The phenomenon is attributed to the spatial proximity of the cells and the long retention time (Wuertz et al., in prep.).

One of the common features of biofilms is their heterogeneity, and therefore, uniform development, properties and behaviour of biofilms should not be anticipated. Figure 5.3 depicts an idealized biofilm structure with microchannels, pores, crevices, regions of EPS with different density etc., according to the results of microscopical studies by SCLM (Lawrence et al. 1991).

In spite of their heterogeneity, some common aspects of biofilms are clearly conceivable (Characklis and Marshall 1990). In general, they consist of:

* Water (85–95% wet weight)
* Extracellular polymer substances (EPS, 85–95% dry weight)
* Micro-organisms
* Entrapped particles, both organic and inorganic
* Sorbed, soluble substances

Thus, chemical analysis has to be interpreted with respect to the chemical nature of biofilms. Otherwise one can be can mislead because under certain circumstances, abiotic material collected in the biofilm may represent the major portion of the sample, although the biofilm caused the accumulation of this material, acting as a "glue" which agglomerates the particles.

Fig. 5.3. Idealized biofilm with microchannels, pores, EPS of different density, and mobile and immobilized micro-organisms

5.2.1 Role of Extracellular Polymer Substances (EPS)

The EPS are of paramount importance for biofilms. They are the material which first makes contact between a cell and a surface and which attaches the cell for irreversible adhesion. These EPS have adhesive properties and may be different from EPS in later stages of biofilm formation, as studies on the "footprints" of primarily adhering cells revealed (Neu and Marshall 1991). After adhesion, living cells excrete EPS which forms a matrix embedding the cells and provides the mechanical stability of the biofilm. This matrix represents a barrier for convective transport. The cells only allow diffusive transport of dissolved matter, and do so,

not homogeneously, but according to their varying density within the biofilm. It has been reported that the structure of biofilms can include pores, channels and very dense areas (Lawrence et al. 1991). In these, convective transport has to be expected to a certain extent. The chemical nature of the EPS supports ion exchange mechanisms and, thus, sorption of dissolved substances. In this context, it is necessary to provide additional details about EPS.

The EPS are excreted by and surround the bacterial cell. They are bound to the cell surface with differing levels of cohesion (Whitfield 1988). In gram-negative cells, even the outer part of the membrane and the lipopolysaccharides (LPS) can be considered as a part of the EPS. Some strains will carry a capsule on top of a LPS-containing membrane. Geesey and Jang (1989) pointed out that capsules are invariably observed anchoring sessile micro-organisms to each other, to other types of living cells and to inanimate objects. Capsules are elaborated by both gram-negative and gram-positive eubacteria, cyanobacteria, archebacteria, and unicellular algae. Capsular exopolymers are firmly retained by the cell and can withstand large shear forces (Pedersen et al. 1990). This association enables the cell to establish a stable orientation with respect to its neighbours and environment. Capsular EPS may extend from 0.1 to 2 micrometers or more from the cell surface into the surrounding environment, creating a buffer zone between the outer cell wall and the external environments. The widespread appearance of capsules among natural populations of micro-organisms indicates that they are essential to survival (Geesey and Jang 1989). Capsules and slime are differentiated from one another in that the former are extracellular polymers tightly associated with the cell surface and can exclude a negative stain such as Indian ink when viewed by light microscopy, whereas the latter is more loosely associated with the cell surface, often sloughs off the cell into the external milieu and does not exclude Indian ink. This behaviour indicates a different density of capsule and slime. All extracellular bacterial polymers, including capsules and slime, but excluding lipopolysaccharides, teichoic acid, lipoteichoic acid and teichuronic acid, have been broadly grouped together and called the glycocalyx (Costerton and Irvin 1981). In chemical terms, EPS have to be further differentiated. Additionally, in natural biofilms, fibrillar glycocalyx is formed which is not formed in laboratory cultures. In contrast to the rigid cell wall, the bacterial capsule is a loosely arranged structure containing over 90% water. Because capsules may contain anionic groups such as carboxyl groups and, occasionally phosphate and sulphate groups (Smiley and Wilkinson 1983; Sutherland 1984; Altman et al. 1986), the EPS represent an anion exchange potential. The amino groups of amino sugars and amino sugar acids provide a limited cationic exchange capacity.

Although most capsule and slime EPS described to date are composed of polysaccharide, considerable amounts of protein are frequently recovered in crude EPS preparations. Whether the proteins perform any function – apart from acting as exoenzymes – in these exocellular locations remains to be determined. Investigations on *Pseudomonas diminuta*, using biotin as a marker, may indicate

that proteins are an integral part of the EPS of this strain (Schaule 1992). This might be true for other organisms as well and could support the assumption that proteins contribute to the mass of EPS. Extracellular enzymes may also be trapped and concentrated in EPS. Other cell products such as DNA and RNA are also recovered with the capsule and slime EPS. It is not unlikely that some or all of this material represents excretory or lysis products that have become trapped by the polysaccharide molecules. The extent to which they contribute to the overall interactions between the capsular polysaccharide component and other substances in the environment is a question that has not been widely addressed. Extracellular proteins have to be taken into account as possible sorbents, without regard to their physiological origin or function.

The polysaccharide component of microbial exopolymers usually possesses a repeating sequence of between two and six sugar subunits. Some polymers, such as alginic acid, however, possess blocks of repeating sugars in one segment of the polymer that are different from those in other segments. Many bacterial and algal capsular polymers are composed of acidic polysaccharides. The acidic properties are contributed primarily by free carboxyl groups of uronic acid subunits or pyruvylated sugars. In some instances, the EPS exhibit branching where two or five sugar subunits extend away from the polymer backbone. When both capsule and slime are produced by the same strain, the chemical composition of their polysaccharides is usually similar (Cohen and Johnstone 1964).

It seems that the composition of the EPS excreted by single species and by micro-organisms in biofilms is not constant, but influenced by growth conditions and environmental stress (Schmitt et al. 1992). Many strains form more than one strain-specific EPS, and their composition may change during the life cycle (Uhlinger and White 1983). Some EPS exhibit surface activity, such as emulsan (Rosenberg and Kaplan 1986), in particular during growth on hydrophobic nutrients such as oil, grease and fat (Ramsay et al. 1988; Krynitsky and McLaren 1962; Zuckerberg et al. 1979; Bannerjee et al. 1983; Sar and Rosenberg 1983; Cameron et al. 1988). Surface activity in return will change the sorption properties of a biofilm towards traces of dissolved hydrophobic substances, such as some pesticides. Surface active EPS are in use for the tertiary recovery of oil since long (LaRivière 1955) and for the bioremediation of soils because they increase the solubility and bioavailability of sorbed pollutants.

Biofilm cells are dynamic and respond to environmental conditions by excreting more and/or different EPS. This has been confirmed by Fourier transform infrared spectroscopy (FTIR) with an internal reflection element (IRE; Nichols et al. 1985; Schmitt et al. 1994) in a flow cell (Fig. 5.4).

MONITORING MICROBIAL BIOFILMS USING ATR/FT-IR

Fig. 5.4. Schematical depiction of the FTIR-IRE system (Schmitt et al. 1994)

This system was used to monitor attachment and subsequent formation of a biofilm in presence of 5 ppm toluene, compared to a toluene-free assay. The IR spectra have been obtained from a bioluminescent reporter strain *Pseudomonas putida* RB1401 of toluene degradation, containing a fusion of the TOL plasmid pWWO and the *lux* genes of *Vibrio fischeri*. Light emission, detected through a light conductor connected to a photomultiplier, indicated the degradation activity of the biofilm in the IRE. Increases in the absorption bands of C-O-C and C-O-P stretching vibrations derived from oligo- and polysaccharidic molecules were attributed to increased EPS production as a stress response to toluene (Schmitt et al. in prep; Fig. 5.5). If the concentration of the toluene is further increased, carboxyl bands appear, indicating the formation of charged groups. Such groups will increase the ion exchange capacity of the biofilm and, thus, the binding rate of metal ions and charged groups from the solution. These results illustrate the dynamic and complex nature of the biofilm, which can respond to an organic substance by increasing the ion exchange capacity which, in turn, influences the sorption of metal ions.

EPS are in general hydrophilic due to the free OH-groups with the potential for H-bridging, carboxyl groups as H-acceptors and NH_2 groups in proteins with H-donor properties. But EPS also can be sorbed to hydrophobic surfaces (Costerton

Fig. 5.5. FTIR spectrum of a biofilm on a germanium crystal after 24 h of growth, *1* without toluene, *2* with toluene (5 ppm), *3* with toluene (15 ppm). *Arrows* indicate increase of polysaccharide (*2*) and carboxylic groups (*3*) in the biofilm

et al. 1985; Fletcher 1985), converting these into hydrophilic surfaces. In nature this is well known from the colonization of leaves or the decomposition of oils and fats. The "stickiness" of biofilms favours the adhesion of particles of all kind which will be integrated into the matrix – e.g. clay, gypsum, humic substances (Characklis et al. 1990). Organic portions of this particulate material may be utilized as a substrate and, in so doing, further change the properties of the biofilm.

It is assumed that biofilms tend to form more EPS when they develop under N- and/or P-limitation (Williams and Schwartzhoff 1978; Zevenhuizen 1981; Eighmy et al. 1983). The stronger or lesser formation of EPS in natural biofilms, however, seems to be due to more complex interactions.

5.2.2 Biofilms on Particulate Matter

Principally, the development of biofilms is largely independent of the size of the particle. This means that bacteria can attach to suspended detritus and abiotic particles and grow on them. An example is the calcium carbonate particles of coral reef sediments which carry magnificent biofilms around each particle (Moriarty

Fig. 5.6. Particulate substance (anthracite) embedded in a bacterial floc (courtesy of F. Opalla)

and Hayward 1982; Bicknell 1986). In most cases, bacterial flocs will evolve including one or more abiotic particles and enhancing particle stability, as demonstrated in Fig. 5.6, until the floc will sedimentate.

Suspended matter plays an important role in the transport of pollutants. It provides a transport vehicle and is responsible for the distribution of hydrophobic substances and metal ions. This illustration depicts the interaction: in many cases, it is not the free surface of the particle that a dissolved molecule or ion interacts with, but it is the porous EPS matrix of the bacterial floc. Thus, the sorption properties of the biofilm are obviously controlling a large proportion of the overall transport characteristics, including desorption.

5.3 Biofilms and Sorption

Acting as a gel matrix and exhibiting functionally-charged and non-charged groups, biofilms provide sorption capacities for the following:

* Water
* Inorganic dissolved substances
* Organic dissolved substances
* Particles and precipitates

The biofilm can bind substances by different mechanisms, not all being considered as sorption. However, other binding mechanisms, such as precipitation, should be taken into account when considering the fate of pollutants in the environment. The binding mechanism can be active or passive.

Active Binding. Through the excretion of binding, chelating or precipitating cell products in response to the presence of the dissolved substance, active binding occurs. These mechanisms are known from studies on detoxification, mainly investigated in terms of heavy metal resistance (Silver 1981). In addition, active transport systems may allow the uptake of e.g. metal ions into the cytoplasm. This might be due to the inability of the cell to discriminate between essential metal ions such as calcium and toxic metal ions such as cadmium. Methylation and demethylation as well as oxidation and reduction are active processes altering the sorbed species and contributing to the overall sorption capacity of a biofilm (Brierley 1990).

Passive Binding. Metal binding by bacterial surfaces is considered largely as a passive phenomenon within the process of electrostatic interaction between cationic metals and anionic cell surface groups. Consequently, it is not necessary that the organisms be viable, only that their surfaces remain intact (McLean and Beveridge 1990). Passive binding mechanisms occur through ion exchange, mainly of the EPS and the cell wall, or through precipitation and/or complexation by biofilm components *already present* and not formed in response to the bound species. Passive binding can be performed by dead biomass as well as by living biomass by partitioning and non-metabolic processes. In some instances, the sorption capacity of the dead biomass can be greater, which is utilized in metal recovery from waste water streams (Brierley 1990).

5.3.1 Sorption of Water

Biofilms exhibit a high binding capacity for water, which they sorb from the environment. In air exposed habitats, this is considered a mechanism against desiccation (Potts 1994). This property is mainly attributed to the EPS (Christensen 1989). The concentration of EPS in a biofilm ranges mostly between 1 and 2% (w/w; Christensen and Characklis 1990). In this concentration, EPS bind water, forming gels similar to agar or gelatine. Gel formation can be synergistic: if different EPS are combined, it can occur at concentrations as low as 0.8% (w/w), as demonstrated with the EPS galactomannan and xanthan (Christensen and Characklis 1990). From a standpoint of microbial ecology, this

is an advantage, because the cells can form a biofilm with very low concentrations of EPS.

In alginate and some other bacterial EPS, gel formation is mainly influenced by divalent cations, which bridge the poly-L-guluronic acid sequences. Those matrices can be weakened by chelating agents (Turakhia et al. 1983). Bruus et al. (1992) have shown that ethyleneglycol-bis-(2-aminoethylether)-tetra acetic acid (EGTA) competes with the EPS of activated sludge flocs and leads to an increase in the number of small particles, as a result of the depletion of EPS-bridging, gel-forming Ca^{2+} ions. In agarose, gelation is achieved by the formation of double helices.

The water binding capacity of the EPS allows for the existence of biofilms in extremely dry environments (Palmer and Friedman 1990; Roberson and Firestone 1992). There are also indications that the EPS bind hygroscopic salts (Schaule 1992). The salt content might then contribute to the water binding capacity of biofims.

The immobilization of water by biofilms has significant economical and ecological implications. For example, the water content of sewage sludge is primarily bound by the EPS of the biomass (Gulas and Bond 1979). If the EPS structures could be changed in such a way that they release the water, this would represent substantial progress in sewage sludge dewatering. Attempts were made to destroy the hydrated structure by enzymatic attack (Carlson 1979; Friedrich and Holesovsky 1987). Such processes have been patented (Wiatr 1990; Jungschaffer et al. 1988), but they have not yet proven in actual practice to be applicable. Water plays a key role in corrosion (Kaesche 1980). Probably, even the local accumulation of water at interfaces provides an important prerequisite for the dissolution processes occuring during microbially-influenced corrosion and weathering. These then, are influenced by microbial products dissolved in the EPS matrix and interacting with the interfaces.

5.3.2 Sorption of Inorganic Solutes

Numerous bacteria in aquatic sediments encounter and bind a wide variety of metals in their environment. As the sediments accumulate, they and their bacterial components become subject to geological forces that eventually result in rock formation. During this time, chemical and physical changes occur within the sediments as diagenetic processes. The surfaces of bacteria present in these sediments make suitable biological templates for the concentration of metals and the nucleation of crystals (Beveridge et al. 1983; Mann 1988) and can often greatly influence the initial mineralization process (Krumbein et al. 1994).

Bacteria and their remains in metal-rich waters (Ferris et al. 1987a) and sediments around geothermal sites (Ferris et al. 1986, 1987b) have been shown by transmission electron microscopy (TEM), analytical electron microscopy, and electron diffraction to act as nucleation sites for authigenic minerals. Iron-silica

precipitating species were associated with bacteria in geothermal sediments (Ferris et al. 1986), whereas the bacteria in metal-rich waters were associated with Fe-Al silicate (clay) polymorphs, millerite (NiS), and mackinawite (FeS_{1-x}). In the latter environment, bacteria are thought to initally bind and concentrate Fe and Ni. These concentrated metals then react with bacterially-produced sulphide from the sediments or soluble silica to produce the corresponding mineral. Similarly, todokorite (MnO_6; Ferris et al. 1987b) and Mg-calcite (marine peloids) deposits have also been shown to be nucleated by bacteria (Chafetz 1986; McLean and Beveridge 1990). A common example of bacterial metal binding is the slimy, rust-colored coating of Fe(III)-oxyhydrates in water pipes and in ponds and creeks. This is due to the binding of iron by the microbial biofilms which grow on these surfaces (Costerton et al. 1987). Manganese(IV)oxide is found in similar environments, precipitated by biofilms (Tyler and Marshall 1967; Marshall 1979).

The external environment and internal metabolism of living bacteria often exert a profound influence on the chemistry of bound metals. These influences include changes in oxidation state, formation of organometallic compounds and formation of precipitates due to detoxification or energy-yielding mechanisms of the bacterial protoplast. Alternatively, the metals may be affected indirectly by the production of metabolic end-products such as SO_4^{2-} and S^{2-} or an alteration in the local pH and/or E_h. The results of microbial activity may ultimately lead to metal immobilization, remobilization and/or the formation of metal aggregates (McLean and Beveridge 1990).

There are many examples where biotechnology uses the sorption capacity of biofilms. Chromium removal from synthetic waste streams by biofilms on disks of a laboratory scale rotating biodisc contactor was about 98% effective during steady state addition of Cr(VI) at concentrations of 100–200 mg l^{-1}. Analysis of the major soluble chromium ions demonstrated that Cr(VI) was first reduced to Cr(III) which then appeared to be adsorbed by the biofilm. Chromium levels associated with the biofilm were found by chemical analysis to be 6.5–7.0% of its dried weight (Coleman and Paran 1991). Deans and Dixon (1992) report the biotechnological use of the uptake of Pb^{2+} and Cu^{2+} by EPS.

The mechanisms of active and passive metal immobilization have been extensively reviewed (Kelly et al. 1979; Brierley et al. 1985a; Hutchins et al. 1986; Lakshmanan et al. 1986).

Precipitation. Precipitation results when living bacteria produce and excrete a substance that chemically reacts with metals present in solution to produce an insoluble metal compound. The generally-cited example of precipitation of metals is the production of hydrogen sulphide by sulphate-reducing bacteria and the subsequent formation of insoluble metal sulphides. Sulfate-reducing bacteria, of which the most prevalent genera are *Desulfovibrio* and *Desulfotomaculum*, live in anaerobic environments such as lake sediments, swamps, and anoxic soils, but also in aerobic systems which contain biofilms above 50–200 μm of thickness. In biofilms that thick, the base region is anerobic and provides a habitat for anaero-

bic micro-organisms such as sulphate-reducing bacteria (SRB), according to Costerton and Boivin (1987). Recent investigations, however, combining micro-electrode measurements and scanning laser confocal microscopy reveal that there may exist anaerobic clusters of microconsortia rather than complete anerobic layers, while channels penetrate the biofilms in which aerobic conditions still may prevail (Lewandowski et al. 1993). The SRB oxidize organic matter and use sulphate as electron acceptor, reducing it to sulphide:

$$SO_4^{2-} + 10\,(H) - H_2S + 4\,H_2O$$

The hydrogen sulphide reacts with metals to form water-insoluble metal sulphide compounds. Precipitation of metals by the activity of SRB has been documented to occur in the natural environment leading to the removal of metals from streams and lakes (Jackson 1978). This naturally occurring activity has been reproduced in purpose-built impoundments to remove soluble metals emanating from mining operations (Brierley and Brierley 1980; Brierley et al. 1980). From calculations based on primary productivity and sulphate reduction rates in microbial mat communities, it has been estimated that sufficient sulphide could be produced in order for ore-grade iron and heavy metal sulphides to form (Skyring 1981; Skyring and Bauld 1990).

Metals can also be precipitated from solution by HPO_4^{2-} derived from a phosphatase located on the bacterial cell surface. The phosphatase enzyme cleaves glycerol 2-phosphate liberating HPO_4^{2-}. Macaskie and Dean (1987a,b, 1988), working with *Citrobacter* sp., have demonstrated the cell-bound precipitation of Cd^{2+}, Pb^{2+}, and U^{4+} ions by HPO_4^{2-}. *Citrobacter* sp., immobilized as a biofilm on reticulated foam, was found effective in removal of 90% of the uranium (1 mM uranylacetate, pH 6.9) provided at a flow rate of 1.25 ml min^{-1}. After passage of 40 l of uranyl nitrate, the amount of uranium precipitated by the phosphatase activity was 9 g uranium per gram bacterial dry weight (Macaskie and Dean 1987b). Polyacrylamide gel-immobilized cells were comparatively effective in uranium removal (Macaskie and Dean 1988).

At a concentration of 10 mg l^{-1}, precipitation of Cd, Co and Mn have been the major mechanism of metal removal. The more soluble metals generally displayed the lowest removals. Concentrations of extracellular polymers and soluble chelating agents may be important in controlling removal of metals which are largely soluble in activated sludge (Brown and Lester 1982 a).

Oxidation and Reduction. Although oxidation and reduction of metal require the function of living organisms in most cases, some metal ion reductions can occur passively by chemically reactive sites on the cell walls of micro-organisms (Beveridge and Murray 1976, 1980; Gee and Dudeney 1988). Depending on the metal species, both oxidation and reduction result in metal immobilization.

Microgradients of O_2 and CO_2 produced by respiration play an important role in the oxidation and reduction potential within biofilms.

The oxidation of manganese is performed by a limited number of bacterial species (see detailed review by Marshall 1979; also Nealson and Ford 1982; Boogerd and de Vrind 1987; Adams and Ghiorse 1987; Boegard and de Vrind 1987) and mature spores, like a marine *Bacillus* spores (Mann et al. 1988). The oxidation by bacterial cells and *Bacillus* spores results in the formation of insoluble manganese oxide:

$$Mn^{2+} + 1/2 \ O_2 + H_2O - MnO_2 + 2 \ H^+$$

Microbially mediated oxidation of Mn^{2+} has been demonstrated to actively occur on a geological scale. Even under arctic conditions, rates were found to be sufficient to oxidize all of the Mn^{2+} in a lake under study (Johnston and Kipphut 1988).

A partial reduction to a metal ion of a smaller net charge can occur by certain actively metabolizing bacteria; an example is the conversion of ferric iron to ferrous iron by species of the bacteria *Bacillus*, *Thiobacillus*, *Micrococcus*, *Rhodopseudomonas*, and *Pseudomonas* (Brock and Gustafson 1976; Ehrlich 1981; Lundgren et al. 1983). This reduction reaction is the result of electron transport reactions (Silverman and Ehrlich 1964).

Under anaerobic conditions, the acidothermophilic archebacterium, *Sulfolobolus* sp., has been demonstrated to reduce Mo^{6+} to Mo^{5+}. This metal-ion reduction requires actively metabolizing bacteria because the reaction appears coupled to the oxidation of S° to SO_4^{2-} (Brierley and Brierley 1980).

Methylation and Demethylation. Methylation and demethylation reactions are mechanisms that living bacteria can use to transform metals in their environment (Pettibone and Cooney 1988). Methylation reactions of tin for example are environmentally signficant because methylation increases the volatility and the toxicity of tin, as well as altering its mobility in the environment. The same is true for mercury (Hintelmann et al. 1993).

Butylmethyltins have been reported in several sediments of harbours. Since antifouling paints from ship hulls were the principle source of tin in those harbours, it was assumed that butyltins could be partially degraded and then methylated in situ. Although 95% of the tributyltin in the water column is associated with particles, environmental concentrations are about three orders of magnitude greater in the microlayer of sediments than in the water column (Cooney et al. 1981).

Brinckman et al. (1983) suggested that adsorption and concentration of tin compounds and micro-organisms in surface slicks and on suspended particles could provide sites for biomethylation in aquatic systems.

Dimethyltin and trimethyltin were detected by GC-MS when sediments of Chesepeake Bay were incubated with $SnCl_4$. Methylated tins were not detected in poisoned or in sterile controls (Cooney et al. 1981). The production of monomethyltin was correlated positively with the number of sulphate reducers and of sulphur oxidizers. *Desulfovibrio* spp. isolated from the sediments were able

to carry out the methylation (Gilmour et. al. 1985). Potential mechanisms for methylation were reviewed by Thayer and Brinckman (1982). Methylation experiments in mixed or co-cultures with a *Bacillus* spp. do not seem to be a simple matter of cross-feeding, because organisms did not produce methylated products when transferred to the partially spent medium in which other organism had grown. Therefore, it does not appear that the contribution of a co-factor or extracellular enzyme is required by the other organism to carry out the conversion. Demethylation reactions yield an inorganic metal; these metal ions can subsequently be involved in other microbe-mediated transformations (Brierley 1990).

5.3.3 Sorption of Organic Solutes

The sorption of organic solutes to biofilms is a well known phenomenon. Biofilms sequester them as nutrients from the water stream as symbolized in Fig. 5.2 and, thus, manage to maintain a complex habitat (Decho 1990). The sorption is a prerequisite for biodegradation and as such, is part of any technology using biofilms for water purification. In some instances, substances are not degraded in the biofilm and accumulated only. This is true for some pesticides and aromatic chlorinated hydrocarbons which occur in high concentrations in digested sewage sludge (Laschka et al. 1989). The unwanted sorption of organic pollutants by biofilms is vastly documented, especially in sewage treatment plants (Mac Rae 1986; Tsezos and Seto 1986; Bell and Tsezos 1987; Gantzer et al. 1989; Kennedy et al. 1992).

The potential for enhanced mobility of hydrophobic pollutants by co-transport with bacteria in saturated soils was evaluated from measurements of biosorption of [14]C-labeled hexachlorobenzene and dichlordiphenyltrichlorethane (DDT) to five strains of soil and sewage bacteria (Lindqvist and Enfield 1992). The sorption process could be described by a linear partition equation and appeared to be reversible, but desorption kinetics were slow and/or partially irreversible. The DDT partition coefficients varied with equilibrium time, possibly reflecting DDT-induced changes in the physiology of the bacteria. The partition coefficients, normalized in the masses of bacteria, ranged from 250 to 14 000 for live cells, but the largest coefficients were associated with autoclaved cells of a *Pseudomonas* sp. This is an example for passive sorption. In a further experiment on the sorption capacity of the bacterial biomass, the sorption capacity was greater for DDT than for hexachlorobenzene, but was not correlated to overall bacterial hydrophobicity, measured by hydrophobic interaction chromatography. The 1.2×10^9 cells of a *Bacillus* sp. strain per ml enhanced DDT transport by about 8-fold, whereas an advective-dispersive-sorptive equilibrium model for two mobile phases, waste, and free living bacteria, suggested a 14-fold enhancement, based on the DDT partition coefficient. Model calculations based on literature data covering a wide range of organisms and compounds suggested that 10^6 cells/ml would increase the

mobility of very hydrophobic compounds (Octanol-water partition coefficient $k_{ow} > 6$), whereas higher densities of bacteria (10^8 cells/ml) would have a significant impact on compounds with a log $k_{ow} > 4$ (Lindqvist and Enfield 1992).

Investigations on the uptake of nutrients support considerations on the sorption properties of biofilms. Stream sediment with a natural biofilm generally adsorbed more glycine, aspartic acid, and lysine than sterile sediment (Armstrong and Bärlocher 1989a). The adsorption of glycine, aspartic acid and lysine to glass-beads was measured at three pH values. The positively charged lysine molecules were generally adsorbed more readily than the others. Beads coated with sterile leaf leachate as opposed to sterile glass beads adsorbed more lysine and aspartic acid, but no glycine. The same pattern was found when beads coated with leachate and colonized by bacteria were compared with beads coated with sterile leachate. Killing the bacteria on the beads with glutaraldehye did not affect adsorption rates. Increasing Ca^{2+} concentration from 5 to 50 mg l^{-1} decreased adsorption of lysine and increased adsorption of aspartic acid (Armstrong and Bärlocher 1989b). In natural habitats, low molecular weight substrates are more likely to be used as substrates for bacteria suspended in the water phase. Macromolecules, however, are more likely to be sorbed at surfaces and in biofilms (Marshall 1988).

Sorption data gained in sterile systems may differ profoundly from those gained in presence of micro-organisms and colonized surfaces. Klaas and Flemming (in prep.) have found, that pesticides such as chlortolurone are not sorbed on clean clay surfaces and not degraded by a suspended mixed soil microbiota. However, in the presence of clay, significant sorption was observed and the microbiota was able to degrade the pesticide, as indicated by the occurence of metabolites. Similar effects of clay surfaces on microbial activity have been reported earlier (Söhngen 1915; Filip 1975; Stotzky 1980). Higher concentrations of montmorillonite (10 g l^{-1}), however, might inhibit bacterial growth, as results from Marshman and Marshall (1981) indicate. The effect of surfaces on bacterial activity has certainly to be taken into account. Thus, organic substances in water, which may have appeared to be non-biodegradable in suspended biodegradation tests, may under certain instances be degraded in the presence of surface-rich material. The most likely explanation for this effect is that the cells settle on the surfaces and then, exhibit different physiological properties (Fletcher 1991). The occurence of this has been proven by experiments of Dagostino et al. (1991).

5.4 Sorption Sites in Biofilms

A characteristic feature of biofilms is heterogeneity. Thus, they offer a variety of very different sorption sites including:

- *Extracellular polymer substances* (EPS)
 - Charged groups
 e.g.: -COO⁻, -SH⁻, -SO₄, -H₂PO₄, -NH⁺, -NRH²⁺, etc.
- *Apolar groups*
 e.g.: aromatics, aliphatics such as found in proteins
- *Cell walls*
 - Outer membrane of gram-negative cells (lipids)
 - Murein or teichoic acid layer
 - Cytoplasmatic membrane (lipids)
- *Cytoplasm*

In Fig. 5.7 these compartments are schematically depicted.

gram negative gram positive

Fig. 5.7. Different compartments in a biofilm, including a gram-negative (*left*) and a gram-positive (*right*) organism. CY = cytoplasm; CM = cytoplasmic membrane; M = murein; OM = outer membrane; LPS = lipopolysaccharide; C = capsule; LTA = lipoteichoic acid

5.4.1 Extracellular Polymer Substances (EPS)

Bacterial EPS appear to play an important role in flocculation; divalent cations may also be involved in this process by bridging polysaccharide strands and, thus, forming gels. Metals present in the ionic form may be removed from solution by adsorption to sites on bacterial EPS and accumulation in the cytoplasm of the cell,

or sorbed on the cell wall (Brown and Lester 1979). In activated sludge EPS concentrations increased considerably with sludge age, showing a direct relationship with mixed liquor suspended solids. The EPS can act as chelating agents, such as siderophores (Hughes and Poole 1989; Plessner et al. 1993) or metal binding EPS. The production of EPS requires an actively metabolizing cell or enzymatic production of the polymer outside the cell. Once produced, however, the complexing agents immobilize metals passively. These complexing agents can in some cases be harvested and used repeatedly for metal removal.

Binding of Cations. Extensively studied for their metal-binding action are the polymers produced by bacteria in sewage treatment facilities (Dugan 1975; Brown and Lester 1979 1982 a, b; Norberg 1983; Norberg and Persson 1984; Scott et al. 1986 a, b; Sterrit and Lester 1986). These polymers, which bind soluble metal ions and flocculate insoluble metals by physical entrapment, are instrumental in removing toxic metals from industrial wastewaters discharged to sewage treatment facilities. Microbial production of polymers has been explored for treating industrial wastewaters.

EPS bind a wide variety of metals such as Pb^{2+}, Sr^{2+}, Zn^{2+}, Cd^{2+}, Co^{2+}, Cu^{2+}, Mn^{2+}, Mg^{2+}, Fe^{2+}, Ag^+ and Ni^{2+}. Theoretical predictions of EPS-binding capacities, based on estimated numbers of available carboxyl and hydroxyl groups, suggest a very high capacity, especially for acidic polysaccharides. This means that a small amount of EPS could theoretically bind a large amount of a given metal. Since EPS are sometimes composed of a matrix of fibrils with high-molecular weight, binding capacities of EPS cannot easily be expressed in terms of surface area, but rather must be expressed in terms of weight-specific binding (Harvey and Luoma 1985; Geesey and Jang 1989).

Most empirical measurements so far have indicated high weight-specific binding capacities of EPS for many metals. The affinities of EPS vary depending on the specific metal and micro-organisms involved (Harvey and Luoma 1985; Geesey and Jang 1989). Harvey (1981) found a binding capacity for lead to be 0.13 μmol of Pb per mg EPS. He calculated that if EPS represented only a very small portion (ca. 3%) of the organic matter in sediments, the EPS could still complex all available Pb^{2+} in the surface-layer sediments of a Palo Alto salt marsh. Adsorption densities as high as 22 ng metal per μg EPS have been found for Cu^{2+} (Kaplan et al. 1987). Scott and Palmer (1988) found that Cr^{2+} adsorbs 3.28 ng metal per μg EPS, while Brown and Lester (1982 a, b) observed that Co^{2+} adsorption was 5 ng metal per μg EPS. Black et al. (1988) investigated the role of metal-binding EPS from iron- and manganese-depositing bacteria in terms of microbially influenced corrosion. The stability constants for Ni^{2+}-, Cu^{2+}-, Pb^{2+}-, Cd^{2+}- and Zn^{2+}-complexes with EPS have been reported to lie between 10^5 and 10^9 (Neufeld and Hermann 1975; Rudd et al. 1984; Kaplan et al. 1987; Geesey et al. 1988). As there exists a competition between H^+ and metal ions, the stability constants strongly depend on the pH (Gold and Genetelli 1978; O'Shea and Mancy 1979). Some EPS have a high affinity to certain metal ions (Mittelman and

Geesey 1985; Scott et al. 1986). The stability constant of Cu^{2+}-complexes with the EPS of bacterial isolates from river sediments has been reported to be 10^9; the stability of Cu^{2+}-complexes with humic substances is in the same range (Geesey et al. 1988).

Metal binding to EPS is strongly influenced by pH. This further suggests the involvement of H^+ ions in this binding (Chen et al. 1988; Rudd et al. 1983). The pH dependence of the sorption holds a high potential for remobilization under changing pH conditions. It is both interesting and highly relevant to note that the greatest binding affinities occur at or near the pH of ambient sea water, at approximately pH 8.0 (which is also the normal pH of sewage treatment processes). At ambient sea water pH, Harvey and Luoma (1985) found >99% removal of Zn^{2+} and Ag^+ by EPS. Brown and Lester (1982a, b) found 96% removal of Co, whereas less than 1% Mn removal via EPS was observed. In studies of freshwater lakes, microbial biofilms under near-neutral pH scavenged metals up to 12 orders of magnitude higher than biofilms under lower pH (acidic) conditions (Ferris et al. 1989). This is because at low pH values the availability of negatively charged sites such as carboxylates is greatly reduced, so fewer metal cations can be adsorbed. At higher pH values (i.e. values closer to those of surface waters) metal binding is enhanced by a proportional increase in the number of ionized acidic groups which are free to bind ions (Ferris 1989). Not all association of metals with EPS may, however, be due to adsorption. Because the solubility of the metal itself depends on pH, it is probable that the metals associated with EPS under natural conditions are the result of both ion-exchange binding processes and the precipitation and later physical entrapment of the metal by the polymer (Brown and Lester 1979).

EPS-binding processes can be important in the downward transport of metals in ocean environments. Close associations between bacterial EPS and a variety of metals have been found in both sediments (Nealson 1983) and the water column (Cowen and Silver 1984). Open-ocean bacterial aggregates called "bacterial snow" below 100 m often have been found with EPS capsules containing metal precipitates. Their frequency increased with depth and implies the downward accumulation of metals on EPS aggregates. Similar associations have been found in freshwater lakes (Mittelman and Geesey 1985).

Conditional stability constants for complexes have been found between EPS of *Klebsiella aerogenes* and $Cu^{2+} = 7.69$, $Cd^{2+} = 5.16$, $Co^{2+} = 5.48$, $Ni^{2+} = 5.49$ (Rudd et al. 1984). The overall specific metal uptake by polymers extracted from activated sludge was approximately 10 times higher than by *Klebsiella aerogenes* polymer.

Table 5.1 shows the metal binding by bacterial capsules that has been investigated by McLean and Beveridge (1988). The binding capacity was measured by suspending lyophilized capsule (1 mg l^{-1}) in a trace metal solution containing 1 mM of the chloride salts of the above metals.

Surface active polymers are produced by primary film forming marine bacteria. Such polymers formed insoluble precipitates with cationic detergents and soluble

Table 5.1. Metal binding by bacterial capsules (McLean and Beveridge 1988); data expressed as μmol metal bound per mg capsule (dry weight)

Metal	Bacillus licheniformis	Klebsiella pneumoniae
Na^+	0.259	0.018
K^+	0.223	0.001
Mg^{2+}	0.073	0.246
Ca^{2+}	1.044	3.500
Cr^{3+}	0.940	0.083
Mn^{2+}	0.071	0.001
Fe^{3+}	1.340	2.340
Co^{2+}	0.100	0.001
Ni^{2+}	0.080	0.001
Cu^{2+}	0.890	0.003
Zn^{2+}	0.149	0.011

salts of iron, copper and lead. The neutral polymer solutions also formed non-diffusible complexes with Co^{2+}, Ni^{2+}, Zn^{2+}, Ca^{2+} and Mg^{2+}. While the bacteria were rather tolerant of toxic metal cations, there is no evidence that they ever produced larger quantities of polymers when grown in the presence of added metal than in the absence (Corpe 1975).

It must be emphasized that the determination of the interactions between micro-organisms and metals is complicated by the inability to determine actual effective metal concentration at the site of complex formation (Duxbury 1985). Determination of thermodynamic properties, such as stability constants and binding site density, depend on accurate estimations of the concentrations or activites of interacting chemical species at the site of complex formation. In the case of capsular polymers, a variety of factors complicates efforts to obtain accurate estimates of the amount of metal that is bound (Geesey and Jang 1989). In addition, the configuration of EPS such as used in binding studies may differ from that of EPS in natural biofilms because of the influence of EPS isolation procedures.

Binding of Anions. The slightly negative overall electrostatic charge of the EPS and the cell walls leads to the expectation that mainly cations will be attracted. However, if those cations are multivalent, there is a possibility of over-compensation of the charge which will attract anions. Amino groups in sugars, sugar acids and proteins provide additional groups capable to bind anions.

Additionally, there is evidence for the binding of anions from analyses of biofilms and EPS. For example, the biofilm which caused irreversible biofouling of a reverse osmosis membrane contained about one order of magnitude higher SO_4^{2-} concentration as compared to the feed water (Flemming and Schaule 1989). In another study, the EPS of *Pseudomonas diminuta*, an organism which rapidly

colonizes reverse osmosis membranes, significantly accumulated anions such as sulphate, phosphate and nitrate from the nutrient broth. This finding was obtained after the usual method to isolate the EPS had been modified. The usual preparation includes the precipitation of EPS by ethanol or acetone, but it was felt that this step denatures the original structure of the EPS. In this modified method, the EPS were dialyzed for several days against distilled water in order to remove small molecules. Analysis after this procedure revealed that the product held a much higher concentration of anions than the precipitated product (Schaule et al. 1992). This indicates that there must be several binding mechanisms for anions in the EPS which have not been further explored and reported at this time. The result demonstrates that biofilms may act as a sink for anions as well as for cations, albeit to a lesser degree.

Binding of Organic Molecules. As explained earlier, the EPS can contain significant amounts of proteins. These may well provide apolar sections which interact and sorb apolar organic molecules. However, the polysaccharide moiety, although hydrophilic by nature, must interact with hydrophobic molecules as well. This consideration is supported by the colonization of hydrophobic surfaces by means of the EPS of hydrophilic bacteria and by the performance of biofilm reactors which degrade hydrophobic organics. The degradation of 22 phenols, N-heterocyclic compounds and aromatic amines was performed by biofilms on fluidized particles of sand or activated carbon; the substances were first sorbed, then degraded. It was assumed that biodegradation was superimposed by adsorption of organic matter up to the saturation of the carbon. Only a minor part of the biomass was immobilized on the coal particles, while the major part was aggregated as flocs and pellets. The low metabolic activity of the activated carbon system is ascribed to inhibitory effects of by-products (Koch et al. 1991).

As demonstrated, even hydrophobic solutes can be sorbed in biofilms. However, there is not yet much work published which deals with the binding mechanisms and sites of sorption.

Comparatively few studies have directly examined the binding of organic compounds to EPS. Decho (1990) describes the situation in an extensive and excellent review. The binding mechanism of dissolved organic matter (DOM) and metals to EPS are thought to be similar, because DOM competes with trace metals for binding sites on organic ligands (Luoma and Davis 1983). EPS are often in close association with microbial cells so they can also potentially bind exudation and cell lysis products of algae and bacteria.

It might be initially predicted that not only the EPS of suspended bacteria but also the EPS in biofilms should preferentially bind only compounds with high molecular weight (MW) such as proteins and large peptides (which will be later hydrolysed by exoenzymes and taken up by the cell). Components with lower MW such as amino acids, simple sugars etc., would not readily bind to EPS because they can be directly taken up by the cell without substantial enzymatic modification prior to crossing the cell membrane. Surface chemistry studies also pre-

dict that, in general, high MW compounds adsorb to a given surface with greater abundance than compounds of lower MW. Examinations of adsorption to marine detritus give partial support to such predictions (Marshall 1988).

The limited evidence available from studies specifically examining EPS, however, does not support the predictions that EPS only bind DOM of large MW. In fact, small-MW compounds such as the amino acids alanine and arginine (Joyce and Dugan 1970) and a variety of phosphates (Dudman 1977) readily bind to EPS. Also, simple sugars such as glucose are significantly retarded (from 2 to 100% of the equivalent diffusion coefficient in water) when diffusing through an EPS matrix (Mateson and Characklis 1976; La Motta 1976). These studies suggest that EPS readily bind at least some low-MW compounds.

As microbial cells change their physiological state, the composition of their secreted EPS (and hence their ability to bind DOM) may also change. Vasse et al. (1984) observed that the lectin-binding ability of *Rhizobium* EPS decreased with age of the culture. The implications of these changes in the adsorptive properties of EPS, however, are not known.

Adsorption of DOM by EPS can be especially important to microbial communities in nutrient-poor environments. For example, in freshwater streams, more than 90% of the bacteria are enclosed within epilithic biofilms (Geesey et al. 1978), which are followed by the subsequent development of diverse grazer communities.

Because EPS are highly hydrated molecules (ca. 99% water), they act effectively as a "sponge". Most binding occurs through cation exchange processes with water, resulting in a loss of "bound water" for organics or metals (Rees 1976; Rendelman 1978a,b). The physical structure of EPS resembles a highly dispersed matrix of fibrils. This implies that EPS possess a large surface area for exchange reactions to occur (Sutherland 1977).

The polysaccharide moieties of EPS are especially important in binding reactions because they possess abundant carboxyl and hydroxyl groups which act as binding ligands for dissolved organic compounds (Aspinall 1982). A survey of a wide variety of marine and freshwater bacteria by Kennedy and Sutherland (1987) has shown that bacterial EPS typically contained 20–50% of their polysaccharides as uronic acids, which represent a high binding potential for cationic solutes. Recently, Smith and Geesey (1989) have developed HPLC methods to quantify ketal-linked pyruvates present on EPS. These pyruvates contain carboxyl groups which are free to react with positive-charged molecules and certain metals (Decho 1990).

5.4.2 Cell Surface

Cell surfaces of biofilm bacteria are able to interact with and bind metals in their environment due to their anionic nature. The localized high concentrations of metals which may develop on bacterial surfaces are subject to mineral formation

and diagenesis, which in some cases may be modified by bacterial metabolism (McLean and Beveridge 1990).

Metals are first bound passively to the cell surfaces by physico-chemical processes such as electrochemical attraction. Next the metals are accumulated inside the cell usually by an energy-dependent transport system. The inability of the bacterial cell's transport system to differentiate between metabolically essential and non-metabolic metals of the same charge can result in substantial intracellular accumulation of toxic heavy metals. This phenomenon has been reviewed by Brierley et al. (1985a). There are at least three mechanisms for the binding of metals on cell walls:

1. Ion exchange reactions
2. Precipitation
3. Complexation

While most bacteria possess a cell wall structure to bind metals, the gram-positive bacteria, principially the members of the genus *Bacillus*, have enhanced capacity for metal binding because of their thick network of peptidoglycan with attached macromolecules (Beveridge 1984).

In contrast, gram-negative bacteria, which have more limited metal-binding capacity, have cell walls that are chemically and structurally very different from the gram-positive cell walls. Gram-negative cell walls have a thin peptidoglycan layer between two membrane bilayers (Beveridge and Koval 1981). Gram-positive cell walls and surfaces have a negative charge density owing to the presence of teichoic acid and teichuronic acid attached to the peptidoglycan network. The phosphodiesters of the teichoic acid and the carboxylate groups of the teichuronic acids contribute to ion exchange capacity to the cell wall (Hancock 1986). This is the primary mechanism for metal binding to the cell wall.

Bacterial metal-binding capacities are frequently higher than can be accounted for by ion-exchange processes alone. The increased binding capacity may come from a nucleation reaction, i.e. biofilm components serve as nuclei for crystallization of inorganic and organic salts (Doyle et al. 1980; Hancock 1986). Nucleation, however, seems to be not yet well understood.

A third mechanism of metal binding that has been identified in bacterial cell walls is complexation. It is proposed that nitrogen and oxygen ligands in the wall contribute to complexation of transition-metal ions (Hancock 1986).

Gram-positive Cell Walls. A segment of the cell wall of *B. subtilis* is given in Fig. 5.8. The structure reveals both positively and negatively charged sites. The teichoic acid proportion may contain a D-alanine residue in place of the α-D-glucose as shown. The "X" represents the linkage between a muramic acid residue and a teichoic acid polymer. Anionic sites include carboxylate (from peptidoglycan) and phosphate (from teichoic acid). Positively charged sites are exclusively ammonium, from D-alanine (teichoic acid), amino sugar (glycan) and diaminopimelic acid (peptide portion of peptidoglycan).

Fig. 5.8. Segment of the cell wall of *Bacillus subtilis* 168 (after Doyle 1989)

Metal binding by *Bacillus subtilis* walls is considered to be a two-stage process. Initially metal ions are bound to the anionic sites in the peptidoglycan. These bound metals then act as nucleation sites whereby additional metals might be bound (often forming aggregates visible by TEM). In contrast, most metal binding by *Bacillus licheniformis* walls appears to be a single-step process which is associated with teichoic acid (TA) and teichuronic acid (TUA) in that metal aggregates are not usually detected by TEM (Beveridge 1981). Thus, it would appear that the wall chemistry of bacteria does not only affect its metal-binding profile but also influences the nucleation of metallic aggregates and crystals (McLean and Beveridge 1990). Table 5.2 illustrates the different binding capacities of cell walls of *B. subtilis* and *B. licheniformis* (McLean and Beveridge 1990).

Isolated *Escherichia coli* K-12 envelopes or *Bacillus subtilis* 168 cell walls were reacted with smectite or kaolinite clay in distilled deionized water (pH 6.0); unbound envelopes or walls were separated by a sucrose density gradient centrifugation, and the extent of adsorption was calculated. At saturation, both clays adsorbed ca. 1.0 mg (dr. wt.) of envelopes or walls per mg (dr. wt.) of clay. Clays showed a preference for "edge-on" orientation with both walls and envelopes (Marshall 1968, 1969a,b), which was indicative of an aluminum polynuclear-bridging mechanism between the wall or envelope surface and the clay edge. The addition of heavy metals increased the incidence of planar surface orientations, which suggested that multivalent metal cation bridging was coming into play and was of increasing importance. The metal-binding capacity of isolated envelopes, walls, clays, and envelope-clay or wall-clay mixtures were determined by atomic absorption spectroscopy after exposure to aqueous 5.0 mM Ag+, Cu^{2+}, Cd^{2+}, Ni^{2+},

Pb^{2+}, Zn^{2+}, and Cr^{3+} nitrate salt solutions at pH values determined by the buffering capacity of wall, envelope, clay, or composite system. The order of metal uptake was walls>envelopes>smectite clay, for the individual components, and walls plus smectite>walls plus kaolinite>envelopes plus smectite>envelopes plus kaolonite, for the mixtures.

Table 5.2. Metal binding of gram-positive cell walls of *Bacillus subtilis* and *Bacillus licheniformis* expressed as μmol of metal bound per mg (dr. wt.) of wall fragments (compiled by McLean and Beveridge 1990)

Metal	Bacillus subtilis	Bacillus licheniformis
Na^+	2.697	0.910
K^+	1.944	0.560
Mg^{2+}	8.226	0.400
Ca^{2+}	0.399	0.590
Mn^{2+}	0.801	0.662
Fe^{3+}	3.581	0.760
Ni^{2+}	0.107	0.520
Cu^{2+}	2.990	0.490
Au^{3+}	0.363	0.031

On a dry-weight basis, the envelope-clay, and wall-clay mixtures bound 20–90% less metal than did equal amounts of the individual components. This reduction in metal-binding capacity indicates that the adsorption of the wall or envelope on clay has masked or neutralized chemically reactive adsorption sites normally available to metal ions. TEM and energy-dispersive X-ray spectroscopy confirmed these results and gave a high spatial resolution to the sites of high binding density. For example, small electron-dense precipitates were seen associated with Ag^{2+}-, Pb^{2+}-, and Cr^{2+}-exposed *B. subtilis* walls, but not with any of the composites or with *E. coli* envelopes (Walker et al. 1989).

Gram-negative Bacterial Envelopes. The envelopes of gram-negative cells diplay a lower overall charge density and are chemically and structurally very different from gram-positive cell walls (Beveridge 1981). As in the case of gram-positive bacteria such as *B. subtilis*, purified peptidoglycan (Pg) from gram-negative *E. coli* binds metals quite well. The overall contribution of peptidoglycan to metal binding in gram-negative organisms is quite small due to the limited quantity of this substance in the cell envelope and its shielding from the environment by the overlying outer membrane. The binding capacity of gram-negative cell walls is summarized in Table 5.3.

Table 5.3. Metal binding of gram-negative cell wall envelopes of *Escherichia coli* AB264 and of purified peptidoglycan expressed as μmol of metal bound per mg (dr. wt.) of wall fragments (compiled by McLean and Beveridge 1990)

Metal	Wall envelopes	Purified peptidoglycan
Na^+	0.042	0.290
K^+	0.082	0.058
Mg^{2+}	0.256	0.035
Ca^{2+}	0.035	0.038
Mn^{2+}	0.140	0.052
Fe^{3+}	0.200	0.100
Ni^{2+}	0.002	0.019
Cu^{2+}	0.090	not done
Au^{3+}	0.056	not done

Organic Molecules. It is especially the lipid moiety of the cell walls, such as the outer and inner membrane, which provide sorption sites for hydrophobic pollutants. However, the sorption capacity of these must be considered as very limited. The amount of a hydrophobic compound which can be maximally bound to

Table 5.4. Bacterial sorption expressed as the biosorption factor (k_B) and participation coefficient (K_{ow}) of organic compounds (Baughman and Paris 1981)

Compound	k_B	k_{ow}
Benzo(a)anthracene	3.6×10^4	4.1×10^5
Benzo(a)pyrene	3.4×10^5	2.2×10^6
Benzo(f)chinoline	1.5×10^2	1.6×10^3
Benzo(b)thiophene	1.3×10^2	1.3×10^3
Dibenzocarbazol	8.7×10^4	5.6×10^5
Dibenzothiophene	1.5×10^3	1.3×10^4
Methylparathione	4.9×10^2	6.5×10^3
Methoxychlor	2.5×10^4	1.2×10^5
Mirex	4.0×10^5	7.8×10^6
Phenanthrene	6.3×10^3	3.7×10^4
Pyrene	2.7×10^4	1.5×10^5
Toxaphen	4.3×10^3	3.3×10^3
PCBs		
Aroclor 1016	6.1×10^3	2.4×10^4
Aroclor 1242	5.2×10^3	1.3×10^4

the lipid membranes of biofilm cells can be estimated from bioconcentration and k_{ow} factors. The biosorption factor k_B is defined as follows:

$$k_B = C_B/C_M$$

with k_B = biosorption factor
 C_B = concentration in biomass
 C_M = concentration in medium

Baughman and Paris (1981) compared the partition coefficients of different organic compounds in activated sludge with the k_{ow}, i.e. the partition coefficient between octanol and water (Table 5.4 on page 141).

Especially for some pesticides (Paris et al. 1977) and PCB (Urey et al. 1976), bioaccumulation in micro-organisms in activated sludge has been reported. Accumulation rates are similar for dead and living cells, indicating a sorption process independent of physiological activity. Bacterial sorption was characterized by the partition coefficient between biomass and medium (Baughman and Paris 1981). This approach, however, does not differentiate between different sorption sites. It has to be pointed out that many other authors could not confirm the correlation between k_{ow} and k_B and vast differences in the bioaccumulation by different bacteria have been observed in the meantime (Kennedy et al. 1992; Kim and Pirbazari 1989; MacRae 1986; Mackay 1982). Thus, the selection of the strains or mixed cultures and the experimental conditions may lead to very different results in sorption studies.

5.4.3 Lipid Membranes

No studies could be found about the sorption of metal ions in the lipid membranes, which is not very surprising. The lipid membranes are the most probable site of accumulation of lipohilic substances. It is surprising that practically nothing in the literature could be found, in which the accumulation of apolar molecules in lipid membranes was investigated. Assuming that 10% of the dry mass of an average microbial cell consist of lipids and that 20% of a biofilm consist of bacteria, 2% of the overall biofilm dry mass provide a potential capacity for hydrophobic matter. If the biofilm contained 90% water, 0.2% of the wet weight is made up of lipids. In sewage sludge, chlorinated organic compounds were found in concentrations between 15 and 150 mg kg^{-1} (Friege et al. 1989). Laschka and Trumpp (1991) report 500 mg kg^{-1} AOX in sewer biofilms. These values are far below the lipid content. Thus, such amounts may well be taken up by the lipid membranes. This assumption is plausible but not experimentally verified. The work of Wolfaardt et al. (1995) reveals that the EPS seem to be another potential sorbent for apolar molecules. Thus, further research is required in order to obtain deeper insight into the function of lipid membranes as sorption sites for lipophilic molecules.

5.4.4 Cytoplasm – Intracellular Accumulation

The interior of biofilm bacteria is also a possible location for binding and accumulation of dissolved substances (Ford and Mitchell 1993). Significant intracellular accumulation of metals by bacteria was demonstrated by Charley and Bull (1979), using a mixed culture of bacteria consisting of *Pseudomonas maltophila*, *Staphylococcus aureus* and a *coryneform* organism, which accumulated over 300 mg Ag^+ per gram dr. wt. biomass. Accumulation was observed at a rate of 21 mg Ag^+ per gram biomass. Cadmium is accumulated by a large number of organisms, as reported by Ford and Mitchell (1993). Research by Macaskie et al. (1987c) on *Citrobacter* suggests that it is accumulated as cell-bound cadmium phosphate. This is presumably a detoxification mechanism and is similar to the accumulation of lead as $PbHPO_4$ by a different *Citrobacter* species suggested by Aickin et al. (1979). Intracellular formation of a metal sulphide was observed when growing cultures of *Klebsiella aerogenes* accumulated Cd^{2+} ions and inorganic sulphide in a molar ration close to unity. A subsequent intracellular formation of CdS was noted (Aiking et al. 1984). Uranium, radium and cesium accumulate intracellularly in *Pseudomonas aeruginosa* (Strandberg et al. 1981). In this case, however, metabolism was not required. A chemoautotrophic bacterium of genus *Thiobacillus* reportedly accumulated silver, forming precipitates of Ag_2S on its cell surface (Pooley 1982). This metal sulphide compound represented up to 25% of the dry weight of the cell. A number of metals are enzymatically altered intracellularly. This does not usually result in accumulation. Rather, transformation occurs to a less toxic and, in certain cases, a volatile form. This has been discussed earlier. The ability of certain bacterial cells to accumulate metals intracellularly has been stressed as an explanation for the removal of metals from water streams. However, it is doubtful that intracellular accumulation is the main mechanism in all cases of biotechnological sequestering of metals from liquids; it is quite likely that the role of the EPS has been underestimated.

Metal-loading Capacity. The principal factors affecting the metal loading are: (1) the number and nature of the exchange, chelation, and nucleation sites on the bacterial material, (2) the initial metal concentration in solution, and (3) the presence of chelating or complexing agents (Brierley 1990). For natural environments, until now none of these factors has been addressed and quantified. It seems, however, very useful to focus further investigations on this field in order to obtain meaningful data for modelling of metal distribution processes in the aquatic environment – an aspect which has been neglected in otherwise highly valuable studies (e.g. Weber et al. 1991). Guezennec and Therène (1988) could demonstrate an accumulation of iron, manganese and copper ions from sea water in biofilms (see Table 5.5 on page 144).

Table 5.5. Sorption and accumulation of heavy metals from sea water in biofilms of different age; concentrations given in ppm (after Guezennec and Therène 1988)

Metal	Concentration in sea water	Concentration in biofilm (5 days)	Concentration in biofilm (30 days)
Iron	5–10	7–100	> 800
Manganese	50–60	50– 60	500–700
Copper	30–40	30– 40	60–100

In some studies, cell floc of EPS was shown to accumulate up to 25% of its weight as metal ions (Dugan and Pickrum 1972). Microalgal EPS similarly show high complexing capacities for certain metals, such as copper. EPS can effectively remove metal ions from solution depending on their speciation, even at relatively high or low metal concentrations (Brierley et al. 1988). For example, more than 95% of most metals are removed from solution by bacterial EPS, even at high (1 mg l^{-1}) metal concentrations. This reflects their tremendous potential to sequester and concentrate ions (Jang et al. 1990a). Alginate has been proposed in biotechnological applications for absorbing dissolved copper from aqueous media (Jang et al. 1990b).

Tomioka et al. (1992) report that the maximum values of cesium removed by a cesium-accumulating bacterium in a liquid culture containing 10 μmol l^{-1} of cesium was between 47 and 90%. The maximum contents in the strains was between 18.8 and 52.0 μmol g^{-1} (dry wt.). The maximum cesium concentration factors ranged between 3.6 x 10^3 and 3.5 x 10^4. Charley and Bull (1979) demonstrated the bioaccumulation of silver in biomass was as high as 300 mg g^{-1} (wet weight) while the silver was accumulated in aggregates at the cell wall (Goddard and Bull 1988). Gadd (1988) has compiled reported metal uptake for various micro-organisms. Table 5.6 summarizes metal accumulation by bacteria, algae and fungi – reporting surprisingly high values. However, the quantitative assessment of the sorption capacity of natural biofilms is still an open question.

A practical example may illustrate the importance of the sorption potential of biofilms, as it is exploited for the identification of emitters of pollutants into sewage lining systems. For this purpose, the metal content of the biofilms on the surfaces of the sewers are analyzed systematically upstream (Gutekunst and Hahn 1985; Gutekunst 1988). As soon as the concentration of heavy metals decreases, the site of emission has been intersected and crossed. Thus, the emitter must be between the sampling points with the last high and the first low metal concentration. The "memory effect" of this biofilm is based on the accumulation of heavy metals from the wastewater stream (Flemming and Ruck 1990). The system, based directly on the sorption properties of biofilms, has improved the effectivity of the monitoring authorities and reputedly helped to maintain lower pollutant concentrations in sewage sludge (Boes and Caspary 1987).

Table 5.6. Examples for metal accumulation by some bacterial, algal and fungal strains (selected from a compilation of Gadd 1988)

Organism	Element	Uptake (% dr. wt.)
Bacillus cereus	Cadmium	3.9–8.9
Zooglea ramigera	Cadmium	40
	Copper	17–40
Thiobac. ferrooxidans	Silver	25
Bacteria (3 Strains)	Uranium	8–19
S. viridochromogenes	Uranium	30
S. longwoodensis	Uranium	44
Citrobacter sp.	Lead	34–40
	Copper	13.5
Pseudomonas fluorescens	Lead	32–35
Pseudomonas aeruginosa	Uranium	15
Chlorella vulgaris	Gold	10
Chlorella regularis	Uranium	15
Rhizopus arrhizus	Uranium	19.5
Phoma sp.	Silver	2

Organic Molecules. The intracellular accumulation of apolar organic pollutants such as PCB or DDT seems possible but is not yet verified. Again, no studies could be found indicating that the accumulation of organic molecules occurred intracellularly.

5.5 Remobilization

Bacterial surfaces and biofilms are not inert chemical structures. They represent a dynamic system in which the various components are synthesized, assembled, modified and finally broken down by autolysins and sloughed off into the environment. (McLean and Beveridge 1990). Thus, they may contribute to the remobilization of the sorbed substances. McLean and Beveridge (1990) point out: "When the effects of pH, E_h, interfacial nature, and chemical attributes of the metal (heat of hydration, charge density, electronic shell etc.) and the presence of competing ions and other biological surfaces are also taken into account, we can begin to grasp the true complexity of bacterial metal binding".

By nature, the immobilization of metal ions in biomass cannot be irreversible. The biological binding sites sooner or later will be degraded. Fate and transport of the metal is directly related to the fate and transport of the bacterial cell. When the cell dies the metal is released. In some instances, this process will lead to mineral formation (Ferris et al. 1987a,b; Ehrlich 1990) and is responsible for the deposition of large ores. However, in other cases the sorbed metal ions will return in their more soluble form and be remobilized. Experiments with cell walls of *Bacillus subtilis* and *E. coli* envelopes adsorbed to kaolinite and smectite clays and with the corresponding organic material-clay aggregates showed the complexicity of remobilization processes (Flemming et al. 1990). Bound to these substances were Ag(I), Cu(II) and Cr(III). The sorbed metals were then leached with HNO_3, $Ca(NO_3)_2$, EDTA, fulvic acids and lysozyme at several concentrations. The findings on remobilization of the sorbed metals, in general showed the order $Cr^{2+} < Ag^+ < Cu^{2+}$. In the wall, clay and composite systems, Cr^{3+} was very stable; at pH 3, 500 μM EDTA, 120 ppm fulvic acid and 160 ppm Ca^{2+} released less than 32% (wet weight) of the sorbed chromium. Ag (45–87%) and Cu (up to 100%) were readily removed by these agents. The organic chelators were in general less effective at mobilizing certain metals than elevated Ca^{2+} or low (acidic) pH values. Lysozyme digestion of *Bacillus* walls remobilized Cu^{2+} from walls and Cu-wall-kaolinite composites. Ag^+ and Cr^{3+} smectite inhibited enzyme activity to some extent, and the metals remained insoluble.

The concentration dependency of the leaching agents was not always given; e.g. the Ag^+ mobility decreased with some clays and composites treated with high fulvic acid, EDTA, and lysozyme concentrations. It was assumed that the particle size of the deposited metal may account for some stability changes. The large and compact aggregates of Cr^{3+} and Ag^+ as seen in TEM were less likely to be remobilized.

Interesting is the comparison of the stability of metal complexes of low molecular weight complexing agents with the stability of metal-EPS-complexes. The EDTA complexes have stability constants of up to 10^{18}, the complex of Cu^{2+} with 1-hydroxyethane-1, 1-diphosphonic acid (HEDP) is about 10^{19} (Hartinger 1975). If these complexing agents occur in the water phase, it should be expected that eventually sorbed metal ions are rapidly complexed. However, the remobilization in this case is unexpectedly low. Gutekunst (1989) carried out elution experiments with EDTA and HEDP which showed only marginal remobilization of biofilm-bound metal ions. These observations indicate that other mechanisms than complex formation only must be effective in metal binding by biofilms.

Considering organic pollutants bound in lipid membranes, a rapid release after death of a cell is probable as the phospholipids belong to the first cell components to degrade when a cell dies (Tunlid and White 1990).

5.6 Concluding Remarks

In the complex system of sorbing species in soil and sediments, biofilms are a part of the *organic matter*. In this "lumped" parameter, the biofilm organisms and their polymer matrix represent a dynamic sink and source for sorbed substances. The biofilm forms a more or less dense gel phase between the solid mineral surface and the liquid phase. Sorption and desorption occur passively and actively. Both biotic and abiotic processes can be involved in the sorption process. The sorption capacity of biofilms has not been assessed yet. A prerequisite will be that biomass can be discriminated from other soil components, in particular humic acids. It is possible, that the overall sorption capacity of biofilms is not impressively high compared to abiotic sorbents. However, as the biofilm can influence all parameters involved in the sorption process – such as pH, E_h, oxygen content, ionic strength etc. – abiotic sorption will be influenced by biofilm processes as well. This is the indirect effect of the physiological activity of biofilm organisms, and this effect depends not on the amount of biomass but rather on its activity at specific sites. This aspect will have to be addressed in further research.

Although it is possible that the sorption capacity of biofilms does not exceed that of abiotic organic components such as humic acids, the dynamic nature of biofilms gives them a particular role in the sorption process.This is why they have to be considered in the assessment of the distribution of pollutants in surface and groundwater, if models of the distribution of pollutants in the aquatic or soil environment are to be realistic. The existing data on adsorption must be questioned if these are obtained under microbiologically undefined conditions. This chapter reveals a substantial research demand in many aspects. Usually, biomass is still considered as a black box; the particular binding sites for different sorbents are evaluated only in very few cases. The binding mechanisms for lipophilic molecules by EPS in particular are not well understood. The overall binding capacity of biofilms remains unclear, as well as the remobilization potential.

A practical example might illustrate the complexicity and the problems. It was common practice to trickle sewage water on trickling fields in many countries. The self purification potential of the soil matrix, i.e. the soil biofilms, degraded the sewage in the water more or less. Thus, an adapted micropopulation evolved, which not only was degrading but sorbing heavy metals and recalcitrant organics. Rising environmental concern curbed this practice of sewage water treatment, because the groundwater can be and often was polluted this way. Now the question is: what will happen to the soil microbiocoenosis that used to be supplied with nutrients by the sewage water? Of course it will decrease, because no more sewage water is coming. This will lead to the decay of the majority of the cells. At least part of the sorbed matter will be remobilized. The fate of the remobilized

substances depends on the soil components – the pollutants will be more mobile in sandy soils while they will tend to be adsorbed in soils high in montmorillonite and/or organic matter. To date there is no information about quantities of remobilizable matter and its fate in the environment. However, there is at least a significant potential for the formerly sorbed substances to be transported into the groundwater. Thus, a measure which seems environmentally reasonable – stopping the trickling of sewage water – may lead to environmental damage.

The questions posed in the introduction are not sufficiently answered by the body of data collected in this chapter. It may be incomplete, however, it reveals that the aspect of sorption by biofilms is important and deserves further attention. In particular, these questions seem not yet to be covered at all:

1. What is the influence of biofilm effects (pH, E_h, ionic strength etc.) on abiotically sorbed matter and how do biotic and abiotic sorption interact?

2. How can these data be included into model calculations about the distribution of pollutants in aquatic environments?

References

Adams LF, Ghiorse WC (1987) Characterization of extracellular Mn^{2+}-oxidizing activity and isolation of a Mn^{2+} oxidizing protein from Leptothrix discophora SS-1. J Bacteriol 169:1279–1285

Aickin RM, Dean ACR, Cheetham AK, Skarnulis AJ (1979) Electron microscope studies on the uptake of lead by a Citrobacter sp. Microbios Lett 7:7–14

Aiking H, Stijnman A, van Garderen C, van Heerikhuizen H, van't Riet J (1984) Inorganic phosphate accumulation and cadmium detoxification in Klebsiella aerogenes NCTC 418 growing in continuous culture. Appl Environ Microbiol 47:374–377

Altman E, Brisson J-R, Perry MB (1986) Structural studies of the capsular polysaccharide from Haemophilus pleumoniae serotype. Biochem Cell Biol 64:707–716

Armstrong SM, Bärlocher F (1989a) Adsorption and release of amino acids from epilithic biofilms in streams. Freshwater Biol 22:153–159

Armstrong SM, Bärlocher F (1989b) Adsorption of three amino acids to biofilms on glass beads. Arch Hydrobiol 115:391–399

Aspinall GO (1982) General introduction. In: Aspinall GO (ed): The polysaccharides. Acad Press, New York, pp 1–18

Bannerjee S, Duttagupta S, Chakrabarty AM (1983) Production of emulsifying agent during growth of Pseudomonas cepacia with 2,4,5 trichlorophenoxyacetic acid. Arch Microbiol 135:110–114

Baughman GL, Paris DF (1981) Microbioal bioconcentration of organic pollutants from aqueous systems. Crit Rev Microbiol 8:205–228

Bell JP, Teszos M (1987) Removal of hazardous organic pollutants by biomass adsorption. J Wat Poll Contr Fed 59:191–198

Beveridge TJ (1981) Ultrastructure, chemistry and function of the bacterial wall. Int Rev Cytol 72:229–317

Beveridge TJ (1984) Bioconversion of inorganic materials: mechanisms of the binding of metallic ions to bacterial walls and the possible impact on microbial ecology. In: Klug MJ, Reddy CA (eds) Current perspectives in Microbial Ecology. Am Soc Microb, Washington, DC, pp 601–607

Beveridge TJ, Koval SF (1981) Binding of metals to cell envelopes of Escherichia coli K-12. Appl Environ Microbiol 42:325–335

Beveridge TJ, Murray RGE (1976) Uptake and retention of metals by cell walls of Bacillus subtilis. J Bact 127:1502–1518

Beveridge TJ, Murray RGE (1980) Sites of metal deposition in the cell wall of Bacillus subtilis. J Bact 141:876–887

Beveridge TJ, Meloche JD, Fyfe WS, Murray RGE (1983) Diagenesis of metals chemically complexd to bacteria: Laboratory formation of metal phosphates, sulfides, and organic condensates in artificial sediments. Appl Envir Microbiol 45:1094–1108

Bicknell B (1986) Interaction between soluble and surface-active substrates,bacteria and particles in a coral reef lagoon. PhD Thesis, The University of New South Wales, Kensington, NSW, Australia

Black JP, Ford TE, Mitchell R (1988) Corrosion behaviour of metal-binding exopolymers from iron- and manganese-depositing bacteria. Corrosion '88, March 21–25, St Lous, Missouri; paper No. 94

Boes M, Caspary H (1987) Sielhautuntersuchungen – eine erfolgversprechende Methode zum Auffinden von Schwermetallemittenten im Kanalnetz. Korr Abw 34, 123–128

Boogerd FC, de Vrin JPM (1987) Manganese oxidation by Leptothrix discophora. J Bact 169:498–494

Brierley CL (1990) Metal immobilization using bacteria. In: Ehrlich HC, Brierley CL (eds) Microbial mineral recovery. McGraw-Hill, New York; pp 303–323

Brierley CL, Kelly DP, Seal KJ, Best DJ (1985 a) Biotechnology principles and applications. Blackwell Scient, Oxford, pp 163–212

Brierley JA, Brierley CL (1980) Biological methods to remove selected inorganic pollutants from uranium mine wastewater. In: Trudinger PA, Walter MR, Ralph BJ (eds) Biogeochemistry of Ancient and Modern Environments.Austral Acad Sci, Canberra; pp 661–667

Brierley JA, Brierley CL, Dreher KT (1980) Removal of selected inorganic pollutants from uranium mine waste water by biological methods. In: Brawner CO (ed) First Intern Conf on Uranium Mine Waste Disposal. Soc Mining Engg Am Inst Mining, Metallurg and Petrol Eng, New York; pp 365–376

Brierley JA, Brierley CL, Decker RF, Goyak GM (1988) Metal recovery. US Patent 4,789,481

Brinckman FE, Jackson JA, Blair WR, Olson GJ, Iverson WP (1983) Ultratrace speciation and biogenesis of methyltin transport species in estuarine waters. In:Wong CS, Boyle E,Bruland KW,Burton JD, Goldberg ED (eds) Trace metals in sea water (NATO Conference Series 4:9) Plenum Press, NewYork, pp 39–72

Brock TD, Gustafson J (1976) Ferric iron reduc tion by sulfur- and iron-oxidizing bacteria. Appl Environ Microbiol 32:567–571

Brown MJ, Lester JN (1979) Metal removal in activated sludge: the role of bacterial extracellular polymers. Wat Res 13:817–837

Brown MJ, Lester JN (1982 a) Role of bacterial extracellular polymers in metal uptake in pure bacterial culture and activated sludge - I. Effects of metal concentrations. Wat Res 16:1539–1548

Brown MJ, Lester JN (1982 b) Role of bacterial extracellular polymers in metal uptake in pure bacterial culture and activated sludge - II. Effects of mean cell retention time. Wat Res 16:1549–1560

Bruus JH, Nielsen PH, Keiding K (1992) On the stability of activated sludge flocs with implications to dewatering. Wat Res 26:1597–1604

Cameron JA, Bunch CL, Huang SJ (1988) Microbial degradation of synthetic polymers. In: Houghton DR, Smith RN, Eggins HOW (eds) Biodeterioration 7, Elsevier Appl Sci, New York; pp 553–561

Carlson CG (1979) Improved filtration of biosludges by enzyme treatment. Filtr Sep Jan/Feb 1979, pp 82–84

Chafetz HS (1986) Marine peloids: a product of bacterially induced precipitation of calcite. J Sediment Petrol 56:812–817

Characklis WG (1990) Biofilm processes. In: Characklis WG and Marshall KC (eds) Biofilms. John Wiley, New York, pp 195–232

Characklis WG, Marshall KC (eds)(1990) Biofilms. John Wiley, New York

Characklis WG, Turakhia MH, Zelver N (1990) Transport and interfacial transfer phenomena. In: Characklis WG, Marshall KC (eds) Biofilms. John Wiley, New York; pp 341–394

Charley RC, Bull AT (1979) Bioaccumulation of silver by a multispecies community of bacteria. Arch Microbiol 123:239–244

Chen XH, Vedry B, Rogalla F, Lesty Y, Lesavre J (1988) Copper fixation by biofilms in waste water treatment process. In: Astruc M, Lester JN (eds) Heavy metal hydrological cycle. Selper Ltd, London; pp 563–570

Christensen BE (1989) The role of extracellular polysaccharides in biofilms. J Biotec 10, pp 181–196

Christensen BE, Characklis WG (1990) Physical and chemical properties of biofilms. In: Characklis WG, Marshall KC (eds) Biofilms. John Wiley, New York; pp 93–130

Cohen GH, Johnstone DB (1964) Extracellular polysaccharides of Azotobacter vinlandii. J Bact 88, pp 329–335

Coleman RN, Paran JH (1991) Biofilm concentration of chromium. EnvironTechnol 12:1079–1094

Cooney JJ, Hallas LE, Means JC (1981) Tin and microbes in the Chesepeake Bay, USA. In: Proc of the 3rd Intern Conf on heavy metals in the environment, Amsterdam. The Netherlands, Sept 14–18, pp 413–482, CEP Consultants Ltd, Edinburgh

Corpe WA (1975) Metal-binding properties of surface materials from marine bacteria. Dev Ind Microbiol 16:249–255

Costerton JW, Boivin J (1987) Microbially influenced corrosion. In: Mittelman MW, Geesey GG (eds) Biological fouling of industrial water systems. A problem solving approach. Wat Micro Associates, San Diego, pp 56–76

Costerton JW, Boivin J (1991) Biofilms and Corrosion. In: Flemming H-C, Geesey GG (eds) Biofouling and biocorrosion in industrial water systems. Springer, Berlin, Heidelberg New York, pp 195–204

Costerton JW, Irvin RT (1981) The bacterial glycocalyx in nature anddisease. Ann Rev Microbiol 35:299–324

Costerton JW, Cheng K-J, Geesey GG, Ladd TI, Nickel JC, Dasgupta M, Marrie TJ (1987) Bacterial biofilms in nature and disease. Ann Rev Microbiol 41:435–464

Costerton JW, Marrie TJ, Cheng K-J (1985) Phenomena of bacterial adhesion. In: Savage DC, Fletcher MM (eds) Bacterial adhesion. Plenum Press, New York 1985; pp 3–44

Cowen JP, Silver MW (1984) The association of iron and manganese with bacteria on marine macro particulate material. Science 224:1340–1342

Dagostino L, Goodman A, Marshall KC (1991) Physiological responses induced in bacteria adhering to surfaces. Biofouling 4:113–119

Deans JR, Dixon BG (1992) Uptake of Pb^{2+} and Cu^{2+} by novel biopolymers. Wat Res 26:469–472

Decho AW (1990) Microbial exopolymer secretions in ocean environments: their role(s) in food webs and marine processes. Oceanogr Mar Biol Ann Rev 28:73–153

Doyle RJ (1989) How cell walls of Gram-positive bacteria interact with metal ions. In: Beveridge TJ, Doyle RJ (eds) Metal ions and bacteria. John Wiley, New York; pp 27–293

Doyle RJ, Matthews TH, Streips UN (1980) Chemical basis for selectivity of metal ions by the Bacillus subtilis cell wall. J Bact 143l:471–480

Dudman WF (1977) The role of surface polysaccharides in natural environments. In: Sutherland IW (ed) Surface carbohydrates of the procaryote cell. Acad. Press, New York, pp 537–414

Dugan PR (1975) Bioflocculation and the accumulation of chemicals by floc-forming organisms. EPA-600/2-75-032, September 1975, Nat Tech Inform Service, Springfield, VA

Dugan PR, Pickrum HM (1972) Removal of mineral ions from water by microbially produced polymers. Proc 27th Ind Waste Conf, Purdue Univ, Lafayette, IN; pp1019–1032

Duxbury T (1985) Ecological aspects of heavy metal responses in microorganisms. Adv Microb Ecol 8:185–235

Ehrlich HL (1981) The geomicrobiology of iron. In: Geomicrobiology. Marcel Dekker, New York; pp 165–201

Ehrlich HL (1990) Geomicrobiology. Marcel Dekker, New York, Basel; pp 557–602

Eighmy TT, Maratea D, Bishop PL (1983) Electron microscopic examination of wastewater biofilm formation and structural components. Appl Environ Microbiol 45:1921–1931

Ferris FG (1989) Metallic interactions with the outer membrane of Gram-negative bacteria. In: Beveridge TJ, Doyle RJ (eds) Metal ions and bacteria. John Wiley, New York; pp 295–323

Ferris FG, Beveridge TJ, Fyfe WS (1986) Iron-silica crystallite nucleation by bacteria in a geothermal sediment. Nature (London) 320:609–611

Ferris FG, Fyfe WS, Beveridge TJ (1987a) Bacteria as nucleation sites for authigenic minerals in a metal-contaminated lake sediment. Chem Geol 63:225–232

Ferris FG, Fyge WS, Beveridge TJ (1987b) Manganese oxide deposition in a hot spring microbial mat. Geomicrobiol J 5:33–42

Ferris FG, Schultze S, Witten TC, Fyfe WS, Beveridge TJ (1989) Metal interactions with microbial biofilms in acidic and neutral pH environments. Appl Envir Microbiol 55:1249–1257

Filip Z (1975) Wechselbeziehungen zwischen Mikroorganismen und Tonmineralien und ihre Auswirkung auf die Bodendynamik. Habilitationsschrift, Universität Gießen

Flemming HC (1991) Biofilms as a particular form of microbial life. In: Flemming HC, Geesey GG (eds) Biofouling and Biocorrosion in Industrial Wat Systems. Springer, Berlin Heidelberg New York, pp 1–7

Flemming HC, Ruck W (1990) Lokalisierung von Schadstoffeinleitern vom Kanalnetz aus. In: Wagner R (Hrsg) Wasserkalender. Verlag Erich Schmidt, Berlin, pp 115–133

Flemming H-C, Schaule G (1989) Biofouling auf Umkehrosmose- und Ultrafiltrationsmembranen. Teil II: Analyse und Entfernung des Belages. Vom Wasser 73:287–301

Flemming CA, Ferris FG, Beveridge TJ, Bailey GW (1990) Remobilisation of toxic heavy metals adsorbed to bacterial wall-clay composites. Appl Envir Microbiol 56:3191–203

Fletcher MM (1985) Effect of solid surfaces on the activity of attached bacteria. In: Savage DD, Fletcher MM (eds) Bacterial adhesion. Plenum Press, New York, pp 339–362

Fletcher MM (1991) The physiological activity of bacteria attached to solid surfaces. Adv Microb Physiol 32:53–85

Ford T, Mitchell R (1993) Microbial transport of toxic metals. In: Mitchell R (ed) Environmental Microbiology. John Wiley, New York, pp 83–101

Friedrich E, Holesovsky U (1987) Enzymatische Schlamm-Stabilisierung. Entsorgungs Praxis 10/87:474–480

Fry JC, Day MJ (1990) Plasmid transfer in the epilithon. In: Fry JC, Day MJ (eds) Bacterial genetics in natural environments. Chapman and Hall, London, pp 55–80

Gadd GM (1988) Accumulation of metals by microorganisms and algae. In: Rehm HJ (ed) Biotechnology, Vol 6 B: Special microbial processes. VCH Verlags, Weinheim, Germany, pp 401–433

Gantzer CJ et al. (1989) Group report: Exchange processes at the fluid-biofilm interface. In: Characklis WG, Wilderer P (eds) Structure and function of biofilms. John Wiley, New York, pp 73–89

Gee AR, Dudeney AWL (1988) Adsorption and crystallisation of gold at biological surfaces. In: Norris PR, Kelly DP (eds) Biohydrometallurgy. Science and Technol Letters, Kew Surrey, UK, pp 437–451

Geesey GG, Jang L (1989) Inteactions between metal ions and capsular polymers. In: Beveridge TJ, Doyle RJ (eds) Metal ions and bacteria. John Wiley, New York, pp 325–357

Geesey G, Jang L (1990) Extracellular polymers for metal binding. In: Ehrlich HC, Brierley CL (eds) Microbial mineral recovery. McGraw-Hill, New York, pp 223–247

Geesey GG, Mucht R, Costerton JW, Green RB (1978) Sessile bacteria: an important component of themicrobial population in small mountain streams. Limnol Oceanogr 23:1214–1223

Geesey GG, Jang L, Jolley JG, Hankins MR, Iwaoka T, Griffiths PR (1988) Binding of metal ions by extracellular polymers of biofilm bacteria. Proc Int Conf Wat Wastewater Microbiol, Newport Beach, Ca; 8–11 Feb 1988; Vol. I, 26.1–26.11

Gilmour CC, Tuttle JH, Means JC (1985) Tin methylation in sulfide bearing sediments. In: Sieglo AC, Hattori A (eds) Marine and estuarine biogeochemistry. Lewis Publishers Inc Chelsea, MI, pp 239–258

Goddard PA, Bull AT (1988a) Accumulation of silver by growing and non-growing populations of Citrobacter intermedius B 6. Eur J Appl Microbiol 31:314–319

Gold MS, Genetelli EJ (1978) Heavy metal complexation behavior in anaerobically digested sludges. Wat Res 12:505–512

Guezennec J, Therène M (1988) A study of the influence of cathodic protection on the growth of SRB and corrosion in marine sediments by electrochemical techniques. In: Sequeira CAC, Tiller AK (eds) Microbial corrosion 1. Elsevier Appl Sci, London, New York, pp 256–264

Gulas V, Bond M (1979) Use of exocellular polymers for thickening and dewatering activated sludge. J Wat Poll Contr Fed 51:798–807

Gutekunst B (1988) Sielhautuntersuchungen zur Einkreisung schwermetallhaltiger Einleitungen. Schriftenreihe des Instituts für Siedlungswasserwirtschaft, UniversitätKarlsruhe, Band 49

Gutekunst B (1989) Wechselwirkung zwischen Schwermetallen und Sielhaut. GWF Wasser Abwasser 130, 456–462

Gutekunst B, Hahn HH (1985) Schwermetallgehalte in Sielhäuten – eine Möglichkeit zum Nachweis von Einleitungen schwermetallhaltigen Abwassers in die Kanalisation. Vom Wasser 65:127–137

Hancock IC (1986) The use of gram-positive bacteria for the removal of metals from aqueous solution. In: Thompson R (ed) Trace Metal Removal from Aqueous Solutions. The Royal Society of Chemistry, London; pp 25–43

Hartinger, L (1975) Komplexchemie in der Abwassertechnik. Vom Wasser 44, 69–117

Harvey RW (1981) Lead-bacterial interactions in an estuarine salt marsch microlayer. PhD thesis, Stanford University, Stanford, 161 pp

Harvey RW, Luoma SN (1985) Effect of adherent bacteria and bacterial extracellular polymers upon assimilation by Macoma balthica of sediment-bound Cd, Zn and Ag. Mar Ecol 22:281–289

Hintelmann H, Ebinghaus R, Wilken R-D (1993) Accumulation of mercury(II) and methylmercury by microbial biofilms. Wat. Res. 27, pp 237–242

Hughes MN, Poole RK (1989) Metals and micro-organisms. Chapman and Hall, London, pp 93–140

Hutchins SR, Davidson MS, Brierley JA, Brierley CL (1986) Microorganisms in reclamation of metals. Ann Rev Microbiol 40:311–336

Jackson TA (1978) The biogeochemistry of heavy metals in polluted lakes and streams at Flin Flon, Canada, a proposed method for limiting heavy-metal pollution of natural rivers. Environ Geol 2:173–189

Jang LK, Brand W, Resong M, Mainieri W, Geesey GG (1990a) Feasibility of using alginate to absorb dissolved copper from aqueous media. Environ Prog 9:269–281

Jang LK, Geesey GG, Lopez SL, Eastman SL, Wichlacz PL (1990b) Use of a gel-forming biopolymer directly dispensed into a loop fluidized bed reactor to recover dissolved copper. Wat Res 24:889–897

Johnston CG, Kipphut GW (1988) Microbially mediated Mn(II) oxidation in an oligotrophic Arctic lake. Appl Environ Microbiol 54:1140–1145

Joyce GH, Dugan PR (1970) The role of floc-forming bacteria in BOD removal from waste water. Dev Ind Microbiol 11:377–386

Jungschaffer G, Reiner R, Sprössler B, Scorialo A (1988) Verfahren zum Verbessern der Entwässerbarkeit von biologischem Klärschlamm. Eur Pat 0 291 665 B 1 v. 25.3.88

Kaesche H (1990) Die Korrosion der Metalle. 3. Auflage. Springer, Berlin Heidelberg New York

Kaplan D, Christiaen D, Shoshana A (1987) Chelating properties of extracellular polysaccharides from Chlorella spp. Appl Environ Microbiol 53:2953–2956

Kelly DP, Norris PR, Brierley CL (1979) Microbiological methods for the extraction and recovery of metals. In: Bull AT, Ellwood DC, Ratledge C (eds) Microbial Technology: Current Status, Future Prospects. Cambridge Univ Press, Cambride, pp 263–308

Kennedy AFD, Sutherland IW (1987) Analysis of bacterial exopolysaccharides. Biotechnol Appl Biochem 9:12–19

Kennedy KJ, Lu J, Mohn WW (1992) Biosorption of chlorophenols to anaerobic granular sludge. Wat Res 26:1085–1092

Kim SH, Pirbazari M (1989) Bioactive adsorber model for industrial wastewater treatment. J Environ Eng 115:1235–1256

Koch B, Ostermann M, Hoke H, Hempel D-C (1991) Sand and activated carbon as biofilm carriers for microbial degradation of phenols and nitrogen containing aromatic compounds. Wat Res 25:1–8

Krumbein WE, Paterson DM, Stal LJ (eds) Biostabilization of sediments. Univ. Oldenburg, P.O.Box 2541, 26015 Oldenburg, Germany, ISBN 3-8142-0483-2

Krynitsky JA, McLaren GW (1962) Some effects of microbial growths on surfactant properties of fuels. Biotechnol Bioeng IV, pp 357–367

Lakshmanan VI, Chrisison J, Knapp RA, Scharer JM, Snmugasunderm V (1986) A review of bioadsorption techniques to recover heavy metals from mineral-processing streams. In: Proc Second Ann Gen Meet Biominet, CANMET Special Publication SP85-6, Can Gov Publ Centre, Ottawa, pp 75–96

LaMotta EJ (1976) Internal diffusion and reaction in biological films. Environ Sci Technol 10:765–769

LaRivière JWM (1955) The productivity of surface active compounds by micro-organisms and its possible significance in oil recovery. Ant Leuvenhoek 21:1–27

Laschka D and Trumpp M (1991) Sielhautuntersuchungen zur Lokalisierung von AOX-Eminttenten im Kanalnetz. Korr. Abw. 38:495–496

Laschka D, Braun F, Kalbfus W, Metzner G (1989) Schadstoffe im Klärschlamm. Korr Abw 36:706–713

Lawrence JR, Korber DR, Hoyle BD, Costerton JW, Caldwell DE (1991) Optical sectioning of microbial biofilms. J Bact 173:6558–6567

Lewandowski Z, Stoodley P, Altobelli S, Fukushima E (1993) Hydrodynamic and kinetics in biofilm systems – recent advances and new problems. 2nd Int Specialized Conf on Biofilm Reactors, 30.9.–1.10., Paris, pp 313–319

Lindqvist R, Enfield CG (1992) Biosorption of Dichlorodiphenyltrichlormethane and hexachlorobenzene in groundwater and its implications for facilitated transport. Appl Environ Microbiol 58:2211–2218

Lundgren DG (1989) Biotic and abiotic release of inorganic substances exploited by bacteria. In: Poindexter JS, Leadbetter ER (eds) Bacteria in nature. Plenum Press, New York, pp 293–335

Lundgren DG, Boucheron J, Mahoney W (1983) Geomicrobiology of iron: mechanism of ferric iron reduction. In: Rossi G, Torma AE (eds) Recent Progress in Biohydrometallurgy. Assoc Mineraria Sarda, Iglesias, Italy, pp 55–70

Luoma SN, Davis JA (1983) Requirements for modeling trace metal partitioning in oxidized estuarine sediments. Mar Chem 12:159–181

Macaskie LE, Dean ACR (1987a) Use of immobilized biofilm of Citrobacter sp. for the removal of uranium and lead from aqueous flow. Enz Microbiol Technol 9:2–4

Macaskie LE, Dean ACR (1987b) Uranium accumulation by a Citrobacter sp. immobilized as biofilm onvarious support materials. In: Neijssel OM, van der Meer RR, Luyben KChAM (eds) Proc 4th Eur Cong Biotehcnol, Elsevier, Amsterdam, pp 37–40

Macaskie LE, Dean ACR (1988) Uranium accumulation by immobilized biofilms on a Citrobacter sp. In: Norris PR, Kelly DP (eds) Biohydrometallurgy. ScienceTechnol Lett, Kew Surrey, UK, pp 556–557

Macaskie LE, Dean ACR, Cheetham AK, Jakeman RJB, Skarnulis AJ (1987c) Cadmium accumulation by a Citrobacter species: the chemical nature of the accumulated metal precipitate and its location on the bacterial cells. J Gen Microbiol 133:539–546

Mackay D (1982) Correlation of bioconcentration factors. Environ Sci Technol 16:274–278

MacNicol RD, Beckett PHT (1989) The distribution of heavy metals between the principal components of digested sewage sludge. Wat Res 23:199–206

MacRae IC (1986) Removal of chlorinated hydrocarbons from water and wastewater by bacterial cells adsorbed to magnetite. Wat Res 20:1149–1152

Mann S (1988) Molecular recognition in biomineralization. Nature (London) 332:119–124

Mann S, Sparks NHC, Scott GHE, de Vrind-de Jong EW (1988) Oxidation of manganese and formation of Mn3O4 (hausmannite) by spore coats of a marine Bacillus sp. Appl Environ Microbiol 54:2140–2143

Marshall KC (1969a) Orientation of clay particles sorbed on bacteria possessing different ionogenic surfaces. Biochim Biophys Acta 193:472–474 (?)

Marshall KC (1969b) Studies by microelectrophoretic and microscopic techniques of the sorption of illite and montmorillonite to Rhizobia. J Gen Microbiol 56:301–306

Marshall KC (1979) Biogeochemistry of manganese minerals. In: Trudinger PA, Swaine DJ (eds) Biogeochemical cycling of mineral-forming elements. Elsevier, Amsterdam pp 253–292

Marshall KC (1988) Adhesion and growth of bacteria at surfaces in oligotrophic habitats. Can J Microb 34:503–506

Marshman NA, Marshall KC (1981) Some effects of montmorillonite on the growth of mixed microbial cultures. Soil Biol Biochem 13:135–141

Mateson JV, Characklis WG (1976) Diffusion into microbial aggregates. Wat Res 10:877–881

McLean RJC, Beveridge TJ (1988) Influence of metal ion charge on their binding capacity to bacterial capsules. Abst 88th Ann Meet Am Soc Microbiol Q 146, pp 307

McLean RJC, Beveridge TJ (1990) Metal-binding capacity of bacterial surfaces and their ability to form mineralized aggregates. In: Ehrlich HC, Brierley CL (eds) Microbial mineral recovery. McGraw-Hill, New York, pp 185–222

Mittelman MW, Geesey GG (1985) Copper-binding characteristics of exopolymers from a freshwater-sediment bacterium. Appl Environ Microbiol 49:846–851

Moriarty DJW, Hayward AC (1982) Ultrastructure of bacteria and the proportion of Gram-positive bacteria in marine sediments. Microb Ecol 8:1–14

Nealson KA (1983) The microbial manganese cycle. In: Krumbein WE (ed) Microbial Geochemistry. Blackwell Sci Publ, Oxford, pp 98–155

Nealson KA, Ford J (1982) Surface enhancement of bacterial manganese oxidation: implications for aquatic environments. Geomicrobiology J 2:88–91

Neu TR, Marshall KC (1991) Microbial "footprints" – a new approach to adhesive polymers. Biofouling 3:101–112

Neufeld RD, Hermann ER (1975) Heavy metal removal by acclimated activated sludge. J Wat Poll Contr Fed 47:310–329

Nichols PD, Henson JM, Guckert JB, Nivens DE, White DC (1985) Fourier transform-infrared spectroscopic methods for microbial ecology: analysis of bacteria, bacteria-polymer mixtures and biofilms. J Microbiol Meth 4:79–94

Nichols WW, Dorrington SM, Slack MPE, Walmsley HL (1988) Inhibition of tobramycin diffusion by binding to alginate. Antimicrob Agents Chemother 32:518–523

Norberg A (1983) Production of extracellular polysaccharide by Zoogloea ramigera and its use as an adsorbing agent for heavy metals. PhD dissertation, University of Lund, Lund, Sweden 1983

Norberg AB, Persson H (1984) Accumulation of heavy-metal ions by Zoogloea ramigera. Biotech Bioeng 26:239–246

O'Shea TA, Mancy KH (1979) The effect of pH and hardness metal ions on the competitive interaction between trace metal ions and inorganic and organic complexing agents found in natural waters. Wat Res 12:703–711

Palmer RJ, Friedman EI (1990) Wat. relations and photosynthesis in the cryptoendolithic microbial habitats of hot and cold deserts. Microb Ecol 19:111–118

Paris DF, Lewis DL, Barnett JT (1977) Bioconcentration of Toxaphene by microorganisms. Bull Envir Cont Toxicol 17:564–572

Pedersen K (1990) Biofilm development on stainless steel and PVC surfaces in drinking water. Wat Res 24:239–243

Pettibone GW, Cooney JJ (1988) Toxicity of methyltins to microbial populations in estuarine sediments. J Ind Microbiol 2:373–378

Plessner O, Klapatch T, Guerinot ML (1993) Siderophore utilization by Bradorhi zobium japonicum. Appl Environ Microbiol 59:1688–1690

Pooley FD (1982) Bacteria accumulate silver during leaching of sulphide ore minerals. Nature 296:642–643

Potts M (1994) Desiccation tolerance of procaryotes. Microb Rev 58:755–805

Ramsay B, McCarthy J, Guerra-Santos L, Kaeppeli O, Fiechter A (1988) Biosurfactant production and diauxic growth of Rhodococcus aurantiacus when using n-alkanes as the carbon source. Can J Microbiol 34:1209–1212

Rees DA (1976) Stereochemistry and binding behavior of carbohydrate chains. In: Whelan WJ (ed) Biochemistry of carbohydrates, Vol. 5. University Park Press, Baltimore, pp 1–42

Rendelman JA (1978a) Metal-polysaccharide complexes. Part I. Food Chem 3:47–79

Rendelman JA (1978 b) Metal-polysaccharide complexes. Part II. Food Chem 3:127–162

Roberson EB, Firestone MK (1992) Relationship between desiccation and exopolysaccharide production in a soil Pseudomonas sp. Appl Environ Microbiol 58:1284–1291

Rosenberg E, Kaplan N (1986) Surface-active properties of Acinetobacter exopolysac-charides. In: Inouye M (ed) Bacterial outer membranes as a model system. Interscience Publ, New York, pp 311–342

Rudd T, Sterritt RM, Lester JN (1983) Stability constants and complexation capacities of complexes formed between heavy metals and extracellular polymers from activated sludge. J Chem Tech Biotechnol 33A:374–380

Rudd T, Sterrittt RM, Lester JN (1984) Formation and conditional stability constants of complexes formed between heavy metals and bacterial extracellular polymers. Wat Res 18:379–384

Sar N, Rosenberg E (1983) Emulsifier production by Acinetobacter calcoaceticus strain. Curr Microbiol 9:309–314

Schaule G (1992) Primäradhäsion von Pseudomonas diminuta an Filtermembranen. Dissertation, Universität Tübingen

Schaule G, Flemming HC, Poralla K (1992) Forces involved in primary adhesion of Pseudomonas diminuta to filtration membranes. 5th Int Conf on Microb Ecol, Barcelona, Sept 8–12 1992

Schmitt J, Nivens D, Flemming HC, Symader W, White DC (1992) In-situ-Monitoring der Entwicklung von Biofilmen mit Hilfe der FTIR-ART-Spektroskopie und die Rolle der EPS für die Sorptionseigenschaften von Biofilmen. Jahrestagung der Fachgruppe Wasserchemie in der GdCh, Dresden

Scott JA, Palmer SJ (1988) Cadmium bio-sorption by bacterial exopolysaccharide. Biotechnol Lett 10:21–24

Scott JA, Palmer SJ, Ingham J (1986b) Microbial metal adsorption enhancement by naturally excreted polysaccharide coatings. In: Eccles H, Hunt S (eds) Immobilization of ions by bio-sorption. Ellis Horwood, Chichester, UK, pp 81–88

Scott JA, Sage GK, Palmer SJ, Powell DS (1986a) Cadmium adsorption by bacterial capsular polysaccharide coatings. Biotechnol Lett 8:711–714

Silver S (1981) Mechanism of bacterial resistances to toxic heavy metals: arsenic, antimony, silver, cadmium and mercury. NBS Special Publ, pp 301–324

Silverman MP, Ehrlich H (1964) Microbial formation and degradation of minerals. Adv Appl Microbiol 6:153–206

Skyring GW (1981) Sulphate reduction in modern sediments and implications for ore formation. BMR J Aust Geol Geophys 6: 335

Skyring GW, Bauld J (1990) Microbial mats in Australian coastal environments. Adv Microb Ecol 11:461–498

Smiley DW, Wilkinson BJ (1983) Survey of taurine uptake and metabolism in Staphylococcus aureus. J Gen Microbiol 129:2421–2428

Smith JJ, Geesey GG (1989) Detection and quantification of polymeric pyruvate in bacterial exopolymers and aquatic sediments. Abstr Am Soc Microbiol, New Orleans, pp 287

Söhngen NL (1915) Einfluß von Kolloiden auf mikrobiologische Prozesse. Centralbl. f. Bakt. Abt. II Bd. 26:621–647

Sterrit RM, Lester JN (1986) Heavy metal immobilization by bacterial extracellular polymers. In: Eccles H, Hunt S (eds) Immobilization of ions by bio-sorption. Ellis Horwood, Chichester, UK, pp 121–134

Stotzky G (1980) Surface interactions between clay minerals and microbes, viruses and soluble organics, and the probable importance of these interactions to the ecology of microbes in soil. In: Berkeley RCW, Lynch JM, Rutter PR, Vincent B (eds) Microbial adhesion to surfaces. Ellis Horwood, Chichester, pp 231–247

Strandberg GW, Shumate SE, Parrott JR (1981) Microbioal cells as biosorbents for heavy metals: accumulation of uranium by Saccharomyces cerevisiae and Pseudomonas aeruginosa. Appl Environ Microbiol 41:237–245

Sutherland IW (1977) Microbial exopolysaccharide synthesis. In: Stanford PA, Laskin A (eds) Extracellular microbial polysaccharides. Am Chem Soc, Washington DC, pp 40–57

Sutherland IW (1984) Microbial exopolysaccharides – their role in microbial adhesion in aqueous systems. CRC Crit Rev Microbiol 10:173–201

Thayer JS, Brinckman FE (1982) Denitrification and removal of heavy metals from waste water by immobilized microorganisms. Appl Biochem Biotechnol 6:3–13

Tomioka N, Uchiyama H, Yagi O (1992) Isolation and characterization of cesium-accumulating bacteria. Appl Environ Microbiol 58:1019–1023

Tsezos M, Seto W (1986) The adsorption of chloroethanes by microbial biomass. Wat Res 20:851–858

Turakhia MH, Cooksey KE, Characklis WG (1983) Influence of a calcium-specific chelant on biofilm removal. Appl Environ Microbiol 46:1236–1238

Tyler PA, Marshall KC (1967) Microbial oxidation of manganese in hydroelectric pipelines. Antonie van Leuwenhoek 33:171–183

Uhlinger DJ, White DC (1983) Relationship between physiological status and formation of extracellular polysaccharide glycocalyx in P. atlantica. Appl Environ Microbiol 45:64–70

Urey JC, Kricher JC, Boylan JM (1976) Bioconcentration of four pure PCB isomers by Chlorella pyrenoidosa. Bull Envir Contam Toxicol 16:81–85

Vasse JM, Dazzo FB, Truchet GL (1984) Re-examination of capsule development and lectin-binding sites on Rhizobium japonicum 311 B 110 by the glutaraldehyde/ruthenium red/uranyl acetate staining method. J Gen Microbiol 130:3037–3047

Walker SG, Flemming CA, Ferris FG, Beveridge TJ, Gailey GW (1989) Physicochemical interaction of Escherichia coli cell envelopes and Bacillus subtilis cell walls with two clays and ability of the composite to immobilize heavy metals from solution. Appl Environ Microbiol 55:2976–2984

Weber WJ, McGinley PM, Katz LE (1991) Sorption phenomena in subsurfacesystems: concepts, models and effects on contaminant fate and transport. Wat Res 25:499–528

Whitfield C (1988) Bacterial extracellular polysaccharides. Can J Microbiol 34:415–420

Wiatr CL (1990) Controlling industrial slime. Eur Pat 0388 115 v. 12.3.90

Williams FD, Schwarzhoff RH (1978) Nature of the swarming phenomenon in Proteus. Ann Rev Microbiol 32:101–122

Zevenhuizen LPTM (1981) Cellular glycogen, ß-1,2-glucan, poly-ß-hydroxybutyric acid and extracellular polysaccharides in fast-growing species of Rhizobium. Ant v Leuwenhoek J Microbiol Serol 47:481–497

Zuckerberg A, Diver A, Perri Z, Gutnick DL, Rosenberg E (1979) Emulsifier of Arthobacter RAG-1; chemical and physical properties. Appl Environ Microbiol 37:414–420

6 Ecotoxicity of Certain Heavy Metals Affecting Bacteria-Mediated Biogeochemical Pathways in Sediments

W. Reichardt

6.1 Introduction

Biogeochemical pathways of carbon turnover in sediments are mainly mediated by bacteria. From the early stages of their evolution on, these procaryotes have had to cope with elevated concentrations of non-essential and toxic heavy metals. As a result, effective mechanisms of resistance to heavy metal ions have evolved (Beveridge 1989). Ecotoxicity of heavy metals via the main biogeochemical pathways of assimilation and dissimilation of carbon in aquatic sediments, however, cannot simply be extrapolated from metal tolerance observed in physiological experiments using cultures. To obtain a realistic basis for ecological prognoses and ecotoxicological risk assessment, there is a need for integrated methods in microbiological analysis of sediment communities. These will have to be developed further and combined with ecophysiological experiments.

Bacteria in sediments differ from their counterparts in pure cultures not only because of their specific physico-chemical environment. Also, the physiological state of bacterial populations in sediments is often subject to extreme variations depending on rapidly changing environmental conditions that control bacterial growth in situ. Among these, particularly organic carbon supply and redox conditions have not yet received adequate attention in conventional ecotoxicity assessment (e.g. Luoma and Carter 1993). The following is an attempt to summarize our current knowledge about ecotoxicity of cadmium, mercury and copper ions via bacteria-mediated carbon turnover in aquatic sediments.

6.2 Heavy Metal Resistance of Bacteria

In general, toxicity of heavy metals is characterized by their interaction with biomolecules. Toxic metals can either block essential functional groups, displace essential metal ions, or induce conformational changes (Vallee and Ulmer 1972; Collins and Stotzky 1989). For a relatively small number of easily culturable bacterial isolates, physiological mechanisms of heavy metal resistance and their genetical background have been elucidated. Examples for responses of bacteria to heavy metals are summarized in Table 6.1.

Table 6.1. Main responses of cultured bacteria to heavy metals

IMMOBILIZATION

by binding of the metal to inducible biomolecules

- binding proteins
- metallothionein
- inducible exopolymers
- intracellular compounds
- sulphide

by non-specific interactions with the cell surface

- ionic exchange at bacterial cell walls and exopolymers

EFFLUX-SYSTEMS (energy-dependent, often acquired)

- plasmid controlled
- chromosome-controlled

VOLATILIZATION

- inducible mercury reductase

DERIVATIZATION (METHYLATION)

- cobalamin-mediated

6.2.1 Immobilization

Several bacterial mechanisms of detoxification can be summarized as immobilization. This comprises sequestration and trapping of toxic metal ions by intracellular or extra-cellular biomolecules. Adsorption of soft or "class B" metal ions ("weak acids" as characterized by high charge densities and unstable electron configurations) such as Cd and Cu (Nieboer and Richardson 1980) to microbial cells is dominated by covalent rather than ionic bonding (Avery and Tobin 1993). Both Gram-positive and Gram-negative cell walls provide highly anionic binding sites

for metal ions. Maximum binding capacities are ensured by the peptidoglycanes and teichoic acids of Gram-positive cell walls.

Often quantities of metal ions being bound exceed the number of available binding sites. This means that initially bound metals can act as nucleation sites in a multiple-step deposition process. Thus, cell walls become nuclei for mineral formation (Beveridge 1984; Mullen et al. 1989). In living cells which are energized by proton pumps, metal ions have to face competition with protons for anionic binding sites on the bacterial membranes. Therefore, living cells (with energized membranes) tend to bind less metal ions than metabolically inactive bacterial cells (Urrutia-Mera et al. 1992).

Ion trapping rather than efflux or exclusion systems is considered as the main cause of bacterial resistance to cadmium (Higham et al. 1985). There are specific as well as non-specific metal ions traps. Doyle et al. (1975) note extra-cellular as well as intracellular accumulation of cadmium. Like copper (Cooksey and Azad 1992), cadmium can be sequestered by inducible binding proteins (Khazaeli and Mitra 1981; Higham et al. 1984; Henneke 1990; Cunningham and Lundie 1993) or by inorganic phosphates (Aiking et al. 1984). It can be precipitated as CdS (Flatau et al. 1991) or bind in a non-specific way, again like copper (Aiking et al. 1982; Fleming et al. 1990), to cell walls (Flatau et al. 1986; Hambuckers-Berhin and Remacle 1990) and to extra-cellular polysaccharides (Bitton and Friehofer 1978; Mittelman and Geesey 1985).

Sorption of metal ions to bacterial cell walls (Beveridge and Murray 1980; Mullen et al. 1989; Beveridge 1989) may be prevented by strong chelators such as EDTA or NTA (Schreiber et al. 1985; Shuttleworth and Unz 1993). Aquatic bacteria themselves are known as potent producers of complexing agents. These are mainly exopolymers with high affinity binding sites containing ionized carboxyl groups (Geesey et al. 1988). Besides widespread metalloproteins of the metallothionein type (Hamer 1986), further very specific metal-binding proteins can be induced by metal-stressed bacteria (Olafson et al. 1979; Remacle and Vercheval 1992). Micromolar, temporarily growth-inhibiting concentrations of Cu ions induce the *de novo* synthesis and excretion of Cu-complexing proteins by the marine bacterium *Vibrio alginolyticus* (Harwood-Sears and Gordon 1990; Schreiber et al. 1990). This kind of bacterial conditioning of metal speciation is achieved only by one quarter of the Cu-stressed population, indicating a co-operative phenomenon (Gordon et al. 1993).

6.2.3 Efflux Systems

Efflux systems are highly specific pumps energized directly by ATP or by the membrane potential. They provide a most efficient mechanism to pump out toxic metal ions once these have reached the cytoplasm. These systems are either plasmid- or chromosome-controlled (Summers and Silver 1978; Perry and Silver 1982; Silver and Misra 1988; Silver et al. 1989).

6.2.4 Volatilization

Enzymatic reduction to the volatile elemental form, or enzymatic alkylation are the major mechanisms of bacterial resistance to mercury (Silver et al. 1989). Reduction of Hg^{2+} to the volatile and less toxic Hg^0 is based on inducible enzymes (mercury reductases) and often mediated by plasmid or transposon genes, *mer genes*, (Komura and Izaki 1971; Nelson and Colwell 1975; Clark et al. 1977; Weiss et al. 1977; Schottel 1978; Olson et al. 1981; Olson et al. 1982; Barkay et al. 1991). Volatile Hg^0 is also released after an enzymatic cleavage of methyl-mercury by organomercurial lyases (Robinson and Tuovinen 1984; Summers 1986). Substantial reduction and volatilization of mercury occurs only under aerobic conditions (Radosevich and Klein 1993).

Apparently, only in long-term mercury-polluted habitats such as Minamata Bay, organomercurial-volatilizing bacteria with organomercurial lyases of wide ranging substrate specificity have evolved (Nakamura et al. 1990). The expression of methyl-mercury resistance genes is positively correlated with environmental concentrations of mercury (Barkay and Olson 1986). But there are also great variations with respect to different species and their physiological states (Tsai and Olson 1990).

6.2.5 Biomethylation

Mercury and some other elements such as Pb, Sn, Pd, Au, Tl, As, Se and Te can undergo biomethylation. S-adenosylmethionine, N-methyl-tetrahydrofolate or methyl corrinoid (vitamin B_{12}) derivatives are envisaged as methylating agents in a process whose enzymatic nature has still to be clarified (Ridley et al. 1977; Craig 1980; Robinson and Tuovinen 1984). Methyl cobalamin appears to be the most likely methyl donor during biosynthesis of the more volatile, but highly neurotoxic methyl-mercury. This compound accumulates only above pH 5.0 (Ramlal et al. 1985) and preferably under anoxic conditions and in the presence of sulphate-reducing bacteria (Compeau and Bartha 1985,1987; Baldi et al. 1989; Berman et al. 1990). Mercury methylation by the sulphate-reducing bacterium, *Desulfovibrio desulfuricans*, is, in contrast to its Hg-tolerance, high during fermentative growth, but only low during sulphate respiration (Choi and Bartha 1993).

Irrespective of the induction of resistance mechanisms in bacteria, metal ions may cause various other ecophysiological effects as, for example, regulation of motility of the flagellae in *E. coli*. In this case transcription of the *flic gene* (encoding flagellin) can be induced by Al, Cu, Fe and Ni (Guzzo et al. 1991).

6.3 The Gap Between Microbial Physiology and Ecology

To understand ecotoxicity of heavy metals in natural microbial communities and, accordingly, in bacteria-mediated biogeochemical processes, our knowledge of metal tolerance of single species alone is completely insufficient. Firstly, only a tiny fraction of the bacterial populations being present in a sediment sample can be isolated by using indirect and more or less selective techniques. "Total" viable counts of heterotrophic bacteria in a sediment sample may differ from total direct cell counts by a factor of 1000 (e.g. Reichardt 1978). Furthermore, physiological data on metal tolerance are available only from an extremely limited number of usually "copiotrophic" isolates, viz. fast-growing heterotrophic bacteria with a high substrate turnover, which are easy to cultivate.

Secondly, most of the metal resistance data on bacteria refer to physiological conditions being unlike those prevailing in natural sediment microbiota. According to Novitsky (1987) the vast majority of sediment bacteria must be considered as starved and *"quiescent"*. Steep gradients of nutrients and particularly of oxygen tend to form a variety of tiny microhabitats for sediment bacteria.

The physiological diversity of microbial communities in sediments is unparalleled. Nevertheless, microbial ecologists have pursued their goal to develop assays for complex, but ecologically meaningful measures of microbial functions in sediments. This is illustrated by assays that measure the incorporation of tritiated thymidine into the DNA of bacterial communities. This technique was designed to assess the total productivity of heterotrophic prokaryotic communities during short-term incubation of sediment samples (Moriarty and Pollard 1990). Unfortunately, there are inherent problems to agree upon some kind of "universal" conversion factor related to the production of bacterial carbon. Furthermore, replication rates of certain bacterial groups such as sulphate-reducers go undetected with this method (Piker and Reichardt 1991).

Yet currently, thymidine incorporation into bacterial DNA seems to be almost irreplaceable as a criterium of the productivity of main segments of heterotrophic bacterial communities in sediments. Its sensitivity to heavy metal ions (often in contrast to other classes of toxicants) has made this technique a effective tool of bacteriological toxicity assessment, too (Martinez and Vives-Rego 1988; Tubbing et al. 1993; see following section).

6.4 Influence of Cd, Hg, and Cu on Some Key Measures of Microbiological Sediment Analysis

6.4.1 Community Structure

Biofilms. In intertidal sediments of the North Frisian Wadden Sea, polychaete burrow linings represent internal surfaces at which thymidine incorporation assays indicate maximum productivity of the heterotrophic microbiota (Reichardt 1988a). Further measures of microbial activity such as mineralization rates reveal similar trends. According to X-ray fluorescence analyses of paired samples from the reduced bulk sediment and the burrow linings, these "perizoic" biofilms, which by definition cover the surface of macrofaunal burrows (Reichardt 1988b), turned out to be the main accumulation sites for several metals including copper (Fig. 6.1). This coincidence can be explained by a double function of the complex "perizoic" biofilm serving as both matrix of bacterial growth and metal ion trap.

Apparently, mucus and bacterial extra-cellular polymers in such biofilm layers possess high adsorption capacities for metal ions. They are likely to play an important role in controlling toxic metal concentrations in sediments (Juniper et al. 1986; Kurek et al. 1982; Mittelman and Geesey 1985; Hsieh et al. 1985; Ferris et al. 1989; Schmiechen et al. 1992). Most probably, any association with bulk organic material will reduce the toxicity of heavy metals to aquatic bacteria (Litchfield et al. 1975; Melanovich and Wilson 1975; Fu et al. 1992). Whether the metal binding strengths of bacterial exopolymers are always sufficient to cause achange of metal partitioning in sediments, however, is not clear. (Kellems and Lion 1989).

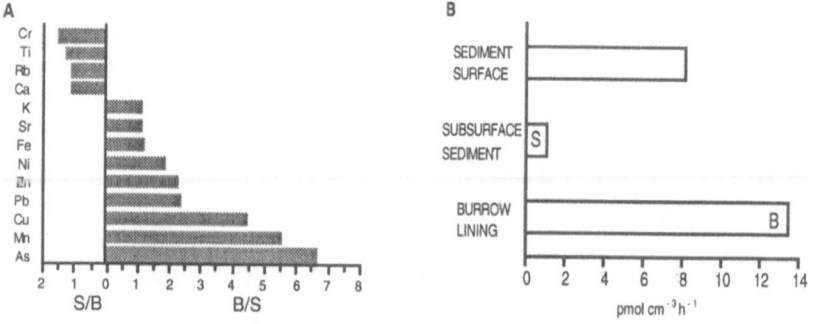

Fig. 6.1. A Metal distribution in perizoic biofilms (burrow linings) *B* of *Arenicola marina* burrows relative to surrounding sediment *S* in intertidal sediments of the North Frisian Wadden Sea at Westerhever Sand. Ratios *B/S* and *S/B* for median values based on n=10 samples each. **B** Rates of thymidine incorporation into bacterial DNA at different microsites of the same sampling station

As illustrated for the binding of ferric iron to bacterial capsule polymers, redox conditions and oxidation states of the metal ions are extremely important (McLean et al. 1992). Initial stages of the colonization of solid substrata by aquatic bacteria seem to indicate that essential metal ions are captured by metal-oxidizing chemoautotrophs (Kepkay et al. 1986). Moreover, the production of metal scavenging bacterial exopolymers can be induced during starvation survival. (Wrangstadh et al. 1990). This could finally explain why bacteria from the nutrient-poor environment of the open ocean are believed to reveal maximum metal resistance (Gonye and Jones 1973).

Metal Resistant Populations. Metal resistance of microbial communities is a widespread response to heavy metal pollution (Nelson and Colwell 1975; Timoney et al. 1978; Houba and Remacle 1980; Titus et al. 1980; Hines and Jones 1982; Olson and Thornton 1982; Duxbury and Bicknell 1983; Barkay et al. 1985; Hornor and Hilt 1985; Duxbury 1986). Yet, such a response of aquatic bacteria is not a coercive consequence of metal pollution, as lacking toxicity responses at elevated pH values reveal (Hahne and Kroontje1973; Dean-Ross and Mills 1989).

Sediment samples are enriched with metal ions to determine the thresholds of heavy metal resistance. This depends on heavy metal bioavailability in nature, which, in turn, is controlled by the metal ion speciation (Hughes and Poole 1991; Liebert et al. 1991; Farrell et al. 1993) and physico-chemical sequestration (Dyrssen and Kremling 1990; Elskens et al. 1991; Ankley et al. 1993). Acid volatile sulphide (AVS = sulphide removed by cold acid extraction) is a major precipitant. Its formation in anaerobic sediments of both marine and limnic waters (Dyrssen and Kremling 1990; Bagarinao 1992) has led to controversial views whether AVS can be regarded as an appropriate partitioning phase for predicting the bioavailability of divalent cations in sediments (Ankley et al. 1991, 1993; Howard and Evans 1993; Allen et al. 1993).

Elevated thresholds of bacterial metal resistance as found in reduced Elbe river sediment (Reichardt et al. 1994) may appear exaggerated. This sediment had to be treated with millimolar concentrations of cadmium, mercury, or copper salts to cause a decrease of the total viable counts of aerobic heterotrophic bacteria below the threshold value of untreated controls (Table 6.2; Reichardt et al. 1994).

Moreover, up to millimolar concentrations caused even an increase of the viable counts by several hundred percent with all three metal ions tested. Observations of this kind were not made for the first time (Nelson and Colwell 1975; Mills and Colwell 1977; Hines and Jones 1982; Barkay 1987). As shown for Cu-resistance of *Pseudomonas fluorescens*, metal resistance genes may confer enhanced competitive fitness, too (Yang et al. 1993). Yet, plate counts (indirect viable counts) suffer from an inherent lack of sensitivity as compared with direct viable counts (Liebert and Barkay 1988). Aquatic microbial communities that had been polluted with mercury, showed a decline of their Shannon-Weaver diversity index (Barkay 1987). It seems that during metal-induced changes of microbial community structures (Singleton and Guthrie 1977), only small and fast-growing

Table 6.2. Colony forming units (CFU) in surface sediment from Elbe River (Mühlenberger Loch) as determined 6 days after adding $3\,CdSO_4 \times 8\,H_2O$, $HgCl_2$ or $CuSO_4$, respectively. Plate counts on casein-peptone-starch agar (Reichardt 1978) after 10 days at 20°C

Condition	CFU ($cm^3 \times 10^6$)	CFU (in % of control)
Metal-free control	14	100
Cadmium		
2 μM	19	137
20 μM	40	289
200 μM	30	217
2 mM	5	36
Mercury		
18 μM	6	48
184 μM	13	93
2 mM	50	362
18 mM	1	7
Copper		
31 μM	19	137
313 μM	n.d.	–
3 mM	76	549
31 mM	1	7

segments of the heterotrophic microbiota will be stimulated. Similar phenomena have been noted for cadmium pollution on superimposed trophic levels (protozoans; Fernandez-Leborans and Novillo-Villajos 1993).

As a strategy of remediation avoiding bioaccumulation of mercury in limnic food-chains, it has even been suggested to suspend organic-poor sediment in heavy metal-polluted waters (Rudd and Turner 1983). Aquatic bacteria acclimatize predominantly to the Hg^{II} species of mercury (Liebert et al. 1991). Potent biomethylators such as sulphate-reducing bacteria, on the other hand, are resistant to high concentrations of methyl-mercury which is transformed to insoluble dimethyl-mercury (Baldi et al. 1993).

Removal of mercury via volatilization is controlled by the *mer* operon with its genes coding for mercuric reductase. These genes are considered as a model describing the acquirement of heavy metal resistance. Using DNA gene probes, Barkay and Olson (1986) have shown that microbial communities that had been pre-exposed to mercury contained more *mer* genes than unaffected communities. Yet, the molecular mechanisms by which bacterial communities adapt to mercury stress are much more complex (Barkay 1987). Hence, relationships between molecular mechanisms of heavy metal resistance and structures of metal-adapted

communities remain largely a matter of speculation.

Recent progress in genetical taxonomy allows for the development of DNA- and RNA-probes for direct application in environmental samples. Still, it remains unknown which fingerprints represent actively metabolizing or biogeochemically-competent bacteria. Difficulties to analyze the complete make-up of all biogeochemically-competent members of a given microbial community persist and thwart the calculation of comprehensive bacterial diversity indices. Consequently, current attempts to analyze toxicant-induced changes of bacterial community structures are incomplete and could lead to controversial judgements.

Yet, as a compromise, it can be useful to consider certain groups such as the copiotrophic heterotrophs (Poindexter and Leadbetter 1986; Reichardt et al. 1991). This group is easily accessible with established enrichment and enumeration methods (Mills and Colwell 1977; Houba and Remacle 1980; Hines and Jones 1982; Barkay 1987; Reichardt et al. 1994). Enrichment techniques imply relatively long incubation periods which favour the selection of metal-resistant organisms (Barnhart and Vestal 1983). Thus, due to adaptive changes within the considered community segment, gross underestimates of acute metal ecotoxicity can be expected.

6.4.2 Community Functions

Estimates of metabolic activity requiring extremely short incubation times can be useful as instantaneous manifestations of toxic effects (Barnhart and Vestal 1983; Dutka and Bitton 1986). In addition, direct linkages between microbial activity measures and basic biogeochemical functions augment the ecological significance of any toxicity assay (Dutka and Kwan 1988; Dutka et al. 1991; Lanza and Dougherty 1991; Reichardt et al. 1993).

ETS Activity. In sediment biogeochemistry, respiratory electron transport (ETS) activities are frequently measured parameters (Reichardt 1978, 1979; Graf 1986; Oren 1987). ETS activities based on artificial reduction indicators have also a long tradition in ecotoxicity testing (Bucksteeg 1959; Lenhard and Nourse 1964). Til now, such dehydrogenase activity assays are considered as a "suitable low-cost and sensitive tool" in ecotoxicity assessment (Koopman and Bitton 1987; Anderson et al. 1988; Rossel and Tarradellas 1991; Torslov 1993; Perez-Garcia et al. 1993).

Frequently, toxicity assays based on dehydrogenase activities are carried out in vivo (e.g. Blondin et al. 1989). Recent tests with Elbe river sediments, however, shed doubts on the significance of in vivo ETS activities as indicators of metal ecotoxicity (Reichardt et al. 1994). Exposure to cadmium and mercury (but not copper) for 8 days caused a doubling of succinate dehydrogenase activity. Viable counts of heterotrophic bacteria increased at the same time (Table 6.2). Also soil dehydrogenase activities show a positive correlation with microbial biomass

(Rossel and Tarradellas 1991). Given the metal-induced rise of total bacterial abundance (viable counts) in our example, ETS activity reflects adaptation of the microbial community to heavy metals rather than acute ecotoxicity of these heavy metals.

DNA Replication (Thymidine Incorporation) Rates. How significant are alternative indicators of ecotoxic effects on microbial community functions? Incorporation of thymidine into the DNA of heterotrophic bacteria requires extremely short incubation times (Moriarty and Pollard 1990; Piker and Reichardt 1991). The still insufficiently defined target of this assay is probably a much larger segment of the heterotrophic microflora than being accessible to viable count determinations. DNA replication rates obtained with this technique constitute an ideal link between structural and functional properties. Moreover, very early events of metal-induced changes of heterotrophic bacterial replication (or biomass production) are likely to be recognized by using this method. When applied for the first time (to water samples from metal-polluted Baltimore Harbour/Chesapeake Bay), thymidine incorporation proved a most sensitive indicator of acute metal ecotoxicity (Jonas et al. 1984).

Later investigations, however, do hardly confirm this view. Under mercury stress (250 μg/l of $HgCl_2$) water samples from different aquatic environments did not only respond with increasing total viable counts (and decreasing species diversity), but also enhanced rates of thymidine incorporation (Barkay 1987). Sediment samples from the Elbe River that had been dosed with graded concentrations of Hg, Cu, or Cd, revealed stimulatory effects of Cd in the millimolar range (Reichardt et al. 1993,1994).

Furthermore, strong metal-induced stimulation was noted in cell suspensions of *Vibrio anguillarum* as function of their physiological state (Reichardt et al. 1993; Heise 1995). In this investigation, bacteria that had been starved for their organic carbon and energy sources, showed enhanced thymidine incorporation rates at micromolar metal concentrations. This result corroborates similar – though less exorbitant – observations with sediments. It also suggests how potential starvation-induced losses of heavy metal resistance, a plasmid-mediated property (Silver and Misra 1988; Griffiths et al. 1990), could be compensated for. Both carbon starvation and cadmium stress are often overlapping signals in the functioning of metabolic regulatory networks. These control the induction of stress proteins which confer increased cadmium resistance upon the cells (van Bogelen et al. 1987; Nyström et al. 1990; Blom et al. 1992).

Dose responses of bacterial DNA replication (thymidine incorporation) rates to Cd, Hg, and Cu can vary by more than one order of magnitude, depending on whether the assay is run under aerobic or anaerobic conditions, and whether growing or starved bacteria are tested (Reichardt et al. 1993). This should be taken into account, when thymidine incorporation is used as a sensitive indicator of ecotoxicity in metal-polluted waters (Martinez and Vives-Rego 1988; Tubbing and Admiraal 1991).

Heavy metal resistance of sediment microbiota is different from that of plankton communities. Patterns of metal resistance are determined by the predominant speciation of a metal ion in its environment (e.g. Jonas 1989). Furthermore, the techniques being applied in planktonic investigations are usually simpler than the sequential extraction procedures being recommended for sediment analyses (Moriarty and Pollard 1990). Hence, thymidine incorporation data from sediment and water samples may even lack sufficient compatibility.

Mineralization Rates. Evolution of CO_2 from ^{14}C-labelled organic substrates is a widely used measure of carbon mineralization. When under stress, bacteria may respond with increased respiration rates (Griffiths et al. 1984). In Elbe river sediments and in bacterial cell suspensions, the initial rates of glucose mineralization were stimulated at micromolar concentrations of Cu, Cd and Hg (Reichardt et al. 1994). In anoxic sediments sulphate reduction (Capone et al. 1983) and growth of sulphate-reducing bacteria (Taylor and Oremland 1979; Reichardt et al. 1993) were particularly sensitive to heavy metal ions. These are also able to shift the predominant pathways of carbon flow from sulphate reduction to methanogenesis (Mountfort and Asher 1981; Capone et al. 1983). On the other hand, methanogenesis seems to remain largely unaffected by mercury in lake sediment (Pederson and Sayler 1981) or is even stimulated following an initial decline, in salt-marsh sediment (Capone et al. 1983).

As noted already for thymidine incorporation, the inhibitory effects of metal ions on mineralization of carbon tend to be much stronger in the water column than in the sediment. Clear trends, however, are missing. Glucose assimilation can be enhanced by copper ions (Vaccaro et al. 1977), but ≤ 0.1 ppm Cu^{2+}, Cd^{2+}, or Hg^{2+} show a non-competitive inhibition of glucose turnover in plankton (Albright et al. 1972).

As compared with the water column, much less information is available on sediments. On the average, the acute ecotoxicity of heavy metal ions in water is by two to three orders of magnitude higher than in sediments (Barnhart and Vestal 1983; Reichardt et al. 1993). Strong physico-chemical sequestration of metal ions by precipitate formation or chelation with clay particles and organic matter in sediments (Babich and Stotzky 1980) have already been cited. Enhanced levels of acquired metal resistance depending on the physiological state of individual sediment bacteria in situ are another cause of reduced metal ecotoxicity (Reichardt et al. 1993).

Single Enzyme Assays. Due to the complexity of microbial communities, their responses show a high degree of variability. Single enzyme assays provide a more direct indication of acute toxicity (Burton and Stemmer 1988; Dutton et al. 1988; Obst et al. 1988; Perez-Garcia et al. 1993; Torslov 1993). To cause inhibitory effects, the toxic metal is required to interfere with the enzyme by dislodging essential ions from its active site or by binding to a different site on the enzyme (Eichhorn 1975). Alkaline phosphatase (Sayler et al. 1982), ß-glucosidase

(Barnhart and Vestal 1983; Dutton et al. 1988), ß-galactosidase (Katayama-Hirayama 1986) and urease (Doelman and Haanstra 1986) are examples for simple assays of hydrolytic enzymes being used to detect metal ecotoxicity in sediments or soils.

As indicators of ecotoxicity, enzyme assays ought to discriminate between responses on the level of enzyme synthesis and enzyme activity. Certain rates of enzyme synthesis constitute more sensitive targets than enzyme activities (Dutton et al. 1988). Moreover, several hydrolytic enzymes such as alkaline phosphatases are known as metal enzymes. The activity of these enzymes depends on divalent cations such as Mg^{2+} or Zn^{2+}. These, however, can be replaced in part by other ions such as Cd^{2+}, Cu^{2+}, or Co^{2+} without considerable loss of activity (Applebury et al. 1970; Reichardt unpubl.). As yet, however, this aspect of metal enzymes in heavy metal toxicity testing has not received proper attention.

Most Sensitive Assays. As illustrated above, microbial responses can be rather ambiguous indicators of heavy metal ecotoxicity, particularly when the assays are carried out with sediment samples. Nevertheless, certain bacteria-mediated transformations are sufficiently sensitive to qualify unequivocally as sensitive measures of heavy metal toxicity to benthic microbial communities. Among these are chemoautotrophic CO_2-fixation by sulphide-oxidizing or nitrifying bacteria (Tsai and Tuovinen 1989; Reichardt et al. 1994) and non-symbiotic N_2-fixation in soils (Skujins and Oden 1986).

6.5 Conclusions

Techniques of microbiological ecosystem analysis have been considerably improved during the recent past (Reichardt 1978; Karl 1986; Moriarty 1986). Designed to understand microbe-mediated energy flow and nutrient cycling in natural microbial communities, these methods were thought to be capable of detecting ecotoxicity in aquatic sediments as well.

When applied to sediments, however, such integrated measures of heterotrophic bioactivity often fail to be sensitive indicators of metal ecotoxicity. This lacking manifestation of ecotoxic effects is not only caused by strong physico-chemical sequestration. Acquired bacterial metal resistance too plays an eminent role. Moreover, as a function of physiological status and growth conditions of bacteria in sediments alone, ecotoxic effects can vary over several orders of magnitude.

Usually, linear dose response models prove inadequate for the needs of microbiological ecotoxicity testing (Liu 1985). This implies also that the expected inhibitory effect of a toxicant may occasionally turn out to be stimulatory. The occurrence of such stimulatory responses at subtoxic doses has long been known as

hormesis (Stebbing 1982). Obviously, this phenomenon deserves further attention in ecotoxicological testing (Brain and Cousens 1989; van Ewijk and Hoekstra 1993). As revealed by metabolic *"overshoot"* responses, hormesis reflects the reactivity of physiological and biogeochemical regulatory patterns (Stebbing 1982; Reichardt et al. 1993). Yet, as potential indicators of *ecosystem resilience* and reactivity under heavy metal stress, hormesis effects remain largely unexplored. Instead, these effects continue being treated as an unwanted by-product of mainstream ecotoxicity testing.

References

Aiking H, Kok H, Heerikhuizen H van, vant Riet J (1982) Adaptation to cadmium by *Klebsiella aerogenes* growing in continuous culture proceeds mainly via formation of cadmium sulfide. Appl Environ Microbiol 44:938–944

Aiking H, Stijnman C, Garderen C van, Keerikhuizen van, vant Riet J (1984) Inorganic phosphate accumulation and cadmium detoxification in *Klebsiella aerogenes* NCTC 418 growing in continuous culture. Appl Environ Microbiol 47:374–377

Albright LJ, Wentworth JW, Wilson EM (1972) Technique for measuring metallic salt effects upon the indigenous heterotrophic microflora of a natural water. Wat Res 6:1589–1596

Allen HE, Fu G, Deng B (1993) Analysis of acid-volatile sulfide (AVS) and simultaneously extracted metals (SEM) for the estimation of potential toxicity in aquatic sediments. Environ Toxicol Chem 12:1441–1453

Anderson K, Koopman B, Bitton G (1988) Evaluation of INT-dehydrogenase assay for heavy metal inhibition of activated sludge. Wat Res 22:349

Ankley GT, Mattson VR, Leonard EN, West CW, Bennett JL (1991) Predicting the acute toxicity of copper in freshwater sediments: evaluation of the role of acid-volatile sulfide. Environ Toxicol Chem 12:315–320

Ankley GT, Mattson VR, Leonard EN, West CW, Bennett JL (1993) Predicting the acute toxicity of copper in freshwater sediments: evaluation of the role of acid-volatile sulfide. Environ Toxicol Chem 12:315–320

Applebury ML, Johnson BP, Coleman JE (1970) Phosphate binding to alkaline phosphatase. Metal ion dependence. J Biol Chem 245:4968–4976

Avery SV, Tobin JM (1993) Mechanism of adsorption of hard and soft metal ions to *Saccharomyces cerevisiae* and influence of hard and soft anions. Appl Environ Microbiol 59:2851–2856

Babich H, Stotzky G (1980) Environmental factors that influence the toxicity of heavy metal and gaseous pollutants to microorganisms. CRC Critical Reviews in Microbiology 8:99–145

Bagarinao T (1992) Sulfide as an environmental factor and toxicant: Tolerance and adaptations in aquatic organisms. Aquatic Toxicol 24:21–62

Baldi F, Filipelli M, Olson GJ (1989) Biotransformation of mercury by bacteria isolated from a river collecting cinnabar mine waters. Microb Ecol 17:263–274

Baldi F, Pepi M, Filipelli M (1993) Methylmercury resistance in *Desulfovibrio desulfuricans* strains in relation to methylmercury degradation. Appl Environ Microbiol 59:2479–2485

Barkay T, Fouts DL, Olson BH (1985) Preparation of a DNA gene probe for detection of mercury resistance genes in gram-negative bacterial communities. Appl Environ Microbiol 49:686–692

Barkay T, Olson BH (1986) Phenotypic and genotypic adaptation of aerobic heterotrophic sediment bacterial communities to mercury stress. Appl Environ Microbiol 52:403–406

Barkay T (1987) Adaptation of aquatic microbial communities to Hg^{2+} stress. Appl Environ Microbiol 53:2725–2732

Barkay T, Turner RR, Brook A vanden, Liebert C (1991) The relationships of Hg(II) volatilization from a freshwater pond to the abundance of *mer* genes in the gene pool of the indigenous microbial community. Microb Ecol 21:151–161

Barnhart LH, Vestal JR (1983) Effects of environmental toxicants on metabolic activity of natural microbial communities. Appl Environ Microbiol 46:970–977

Berman M, Chase T, Bartha R (1990) Carbon flow in biomethylation by *Desulfovibrio desulfuricans*. Appl Environ Microbiol 56:298–300

Beveridge TJ, Murray RGE (1980) Sites of metal deposition in the cell wall of *Bacillus subtilis*. J Bact 141:876–887

Beveridge TJ (1984) Mechanisms of the binding of metallic ions to bacterial walls and the possible impact on microbial ecology. In: Klug MJ and Reddy CA (eds), Current Perspectives in Microbial Ecology. ASM, Washington, DC, pp 601–607

Beveridge TJ (1989) Role of cellular design in bacterial metal accumulation and mineralization. Annu Rev Microbiol 43:147–171

Bitton G, Friehofer V (1978) Influence of extracellular polysaccharides on the toxicity of copper and cadmium on *Klebsiella aerogenes*. Microb Ecol 4:119–123

Blom A, Harder W, Matin A (1992) Unique and overlapping pollutant stress proteins of *Escherichia coli*. Appl Environ Microbiol 58:331–334

Blondin GA, Knobeloch JM, Read HW, Harkin HW (1989) An in vitro submitochondrial bioassay for predicting acute toxictity in fish. In: Suter GW II, Lewis MA (eds) Aquatic Toxicology and Environmental Fate, vol 11, ASTM; STP 1007. American Society for Testing and Materials, Philadelphia, pp 551–563

Bogelen RA van, Kelly PM, Neidhardt FC (1987) Differential induction of heat shock, SOS, and oxidation stress regulons and accumulation of nucleotides in *Escherichia coli*. J Bacteriol 169:26–32

Brain P, Cousens R (1989) An equation to describe dose responses where there is stimulation of growth at low doses. Water Research 29:93–96

Burton jr GA, Stemmer BL (1988) Evaluation of surrogate tests in toxicant impact assessment. Toxicity Assessment 3:255–269

Bucksteeg W (1959) Die Beurteilung von Schlamm und Abwasser mittels TTC Gas Wasser Fach 36:916–920

Capone DG, Reese DR, Kiene RP (1983) Effects of metals on methanogenesis, sulfate reduction, carbon dioxide evolution and microbial biomass in anoxic salt marsh sediments. Appl Environ Microbiol 45:1586–1591

Choi S Ch, Bartha R (1993) Cobalamin-mediated mercury methylation by *Desulfovibrio desulfuricans* LS Appl Environ Microbiol 59:290–295

Clark DL, Weiss AA, Silver S (1977) Mercury and organomercurial resistances determined by plasmids in *Pseudomonas*. J Bacteriol 132:186–196

Collins YE, Stotzky G (1989) Factors affecting the toxicity of heavy metals to microbes. In: T J Beveridge and RJ Doyle (eds), Metal ions and bacteria. John Wiley Sons, New York, p 31–90

Compeau GC, Bartha R (1985) Sulfate-reducing bacteria: Principal methylators of mercury in anoxic estuarine sediment. Appl Environ Microbiol 50:498–502

Compeau GC, Bartha R (1987) Effect of salinity on mercury-methylating activity of sulfate-reducing bacteria in estuarine sediments. Appl Environ Microbiol 53:261–265

Cooksey DA, Azad HR (1992) Accumulation of copper and other metals by copper-resistant plant-pathogenic and saprophytic pseudomonads. Appl Environ Microbiol 58:274–278

Craig PJ (1980) Metal cycles and biological methylation. In: Hutzinger O (ed), Handbook of Environmental Chemistry 1A:169–227

Cunningham DP, Lundie jr LL (1993) Precipitation of cadmium by *Clostridium thermoaceticum.* Appl Environ Microbiol 59:7–14

Dean-Ross D, Mills AL (1989) Bacterial community structure and function along a heavy metal gradient. Appl Environ Microbiol 55:2002–2009

Doelman P, Haanstra L (1984) Short-term and long-term effects of cadmium, chromium, copper, nickel, lead, and zinc on soil microbial respiration in relation to abiotic soil factors. Plant and Soil 79:305–316

Doelman P, Haanstra L (1986) Short- and long-term effects of heavy metals on urease activity in soils. Biol Fertil Soils 2:213–218

Doyle JJ, Marshall TT, Pfander WH (1975) Effects of cadmium on the growth and uptake of cadmium by microorganisms. Appl Environ Microbiol 29:562–564

Dutka BJ, Bitton G (1986) Toxicity testing using microorganisms. Vol.2, CRC Press, Boca Raton, FL

Dutka BJ, Kwan KK (1988) Battery of screening tests approach applied to sediment extracts. Toxicity Assessment 3:303–314

Dutka BJ, Kwan KK, Rao SS, Jurkovic A, McInniss R (1991) Use of bioassays to evaluate river water and sediment quality. Environ Toxicol Wat Qual 6:309–327

Dutton RJ, Bitton G, Koopman B (1988) Enzyme biosynthesis versus enzyme activity as basis for microbial toxicity testing. Toxicity Assessment 3:245–253

Duxbury T, Bicknell B (1983) Metal-tolerant bacterial populations from natural and metal-polluted soils. Soil Biol Biochem 15:243–250

Duxbury T (1986) Ecological aspects of heavy metal responses in microorganisms. Adv Microb Ecol 8:185–235

Dyrssen D, Kremling K (1990) Increasing hydrogen sulfide concentration and trace metal behavior in the anoxic Baltic waters. Mar Chem 30:193–204

Eichhorn GL (1975) Active sites of biological macromolecules and their interactions with heavy metals. In: McIntire AD and Mills CF (eds), Effects of heavy metal and organohalogen compounds. Plenum Publ., New York, pp 123–142

Elskens M, Leermakers M, Panutrakul S, Montany F ,Baeyens W (1991) Microbial activity in sandy and muddy estuarine sediments. Geo-Marine Letters 11:194–198

Ewijk PH van, Hoekstra JA (1993) Calculation of the EC50 and its confidence interval when subtoxic stimulus is present. Ecotoxicol Environ Safety 25:25–32

Farrell RE, Germida JJ, Huang PM (1993) Effects of chemical speciation in growth media ˜on the toxicity of mercury (II). Appl Environ Microbiol 59:1507–1514

Fernandez-Leborans G, Novillo-Villajos A (1993) Changes in trophic structure of a protozoan community subjected to cadmium. Ecotoxicol and Environ Safety 25:271–279

Ferris FG, Schultze S, Witten TC, Fyfe WS, Beveridge TJ (1989) Metal interactions with microbial biofilms in acidic and neutral pH environments. Appl Environ Microbiol 55:1249–1257

Flatau GN, Clement RL, Gauthier MJ (1986) Influence of the cell envelope on the accumulation of cadmium by a marine pseudomonad. Microbios Lett 32:141–146

Flatau GN, Clement RL, Mahdyoun F, Gauthier MJ (1991) Resistance to cadmium in a marine bacterium. Kieler Meeresforsch Sonderh 8:343–347

Flemming CA, Ferris FG, Beveridge TJ, Bailey GW (1990) Remobilization of toxic heavy metals adsorbed to bacterial wall-clay composites. Appl Environ Microbiol 56:3191–3203

Fu G, Allen H, Cao Y (1992) The importance of humic acids to proton and cadmium binding in sediments. Environ Toxicol Chem 11:1363–1372

Geesey GG, Jang L, Jolley JG, Hankins MR, Iwaoka T, Griffith PR (1988) Binding of metal ions by extracellular polymers of biofilm bacteria. Wat Sci Tech 20:161–165

Gonye ER, Jones GE (1973) An ecological survey of open ocean and estuarine microbial populations II. The oligodynamic effect of nickel on marine bacteria. Marine Microbial Ecology, Columbia,SC, Univ. of South Carolina Press

Gordon AS, Harwood VJ, Sayyar S (1993) Growth, copper-tolerant cells, and extracellular protein production in copper-stressed chemostat cultures of *Vibrio alginolyticus*. Appl Environ Microbiol 59:60–66

Graf G (1986) Winter inversion of biomass and activity profile in a marine sediment. Mar Ecol Prog Ser 33:231–235

Griffiths RP, Caldwell BA, Morita RY (1984) Observations on microbial percent respiration values in Arctic and Subarctic marine waters and sediments. Microb Ecol 151–164

Griffiths RP, Moyer CL, Caldwell BA, Ye C, Morita RY (1990) Long-term starvation-induced loss of antibiotic resistance in bacteria. Microb Ecol 19:251–257

Guzzo A, Diorio C, DuBow MS (1991) Transcription of the *Escherichia coli* flic gene is regulated by metal ions. Appl Environ Microbiol 57:2255–2259

Hahne HCH, Kroontje W (1973) Significance of pH and chloride concentration on behavior of heavy metal pollutants: mercury (II), cadmium (II), zinc (II), and lead (II). J Environ Qual 2:444–450

Hambuckers-Berhin F, Remacle J (1990) Cadmium sequestration in cells of two strains of *Alcaligenes eutrophus*. FEMS Microb Ecol 73:309–316

Hamer DH (1986) Metallothionein. Annu Rev Biochem 55:913–951

Harwood-Sears V, Gordon AS (1990) Copper-induced production of copper-binding supernatant proteins by the marine bacterium *Vibrio alginolyticus*. Appl Environ Microbiol 56:1327–1332

Heise S (1995) Der Einfluß von Umweltfaktoren auf die Bildung von exopolymerer Substanz (EP) durch ein marines Bakterium. PhD thesis, University of Kiel

Henneke H (1990) Regulation of bacterial gene expression by metal-protein complexes. Mol Microbiol 4:1621–1628

Higham DP, Sadler PJ (1984) Cadmium resistant *Pseudomonas putida* synthesizes novel cadmium protein. Science 225:1043–1045

Higham DP, Sadler PJ, Scawen MD (1984) Cadmium-resistant *Pseudomonas putida* synthesizes novel cadmium protein. Science 225:1043–1046

Higham DP, Sadler PJ, Scawen MW (1985) Effect of the cadmium on the morphology, membrane integrity and permeability of *Pseudomonas putida*. J Gen Microbiol 131:2539–2544

Hines ME, Jones GE (1982) Microbial metal tolerance in Bermuda carbonate sediments. Appl Environ Microbiol 44:502–505

Hornor SG, Hilt BA (1985) Distribution of zinc-tolerant bacteria in stream sediments. Hydrobiologia 128:155–160

Houba C, Remacle J (1980) Composition of the saprophytic bacterial communities in freshwater systems contaminated by heavy metals. Microb Ecol 6:55–69

Howard DE, Evans RD (1993) Acid-volatile sulfide (AVS) in a seasonally anoxic mesotrophic lake: seasonal and spatial changes in sediment AVS. Environ Toxicol Chem 12:1051–1057

Hsieh KM, Lion LW, Shuler ML (1985) Bioreactor for the study of defined interactions of toxic metals and biofilms. Appl Environ Microbiol 50:1155–1161

Hughes MN, Poole RK (1991) Metal speciation and microbial growth – the hard (and soft) facts. J Gen Microbiol 137:125–734

Jonas RB, Gilmour CC, Stoner DL, Weir MM, Tuttle JH (1984) Comparison of methods to measure acute metal and organometal toxicity to natural aquatic microbial communities. Appl Environ Microbiol 47:1005–1011

Jonas RB (1989) Acute copper and cupric ion toxicity in an estuarine microbial community. Appl Environ Microbiol 55:43–49

Juniper SK, Thompson JAJ, Calvert SE (1986) Accumulation of minerals and trace elements in biogenic mucus at hydrothermal vents. Deep-Sea Res 33:339–347

Karl DM (1986) Determination of in situ microbial biomas, viability, metabolism and growth. In: Poindexter JS, Leadbetter ER (eds) Bacteria in Nature, Vol.2, Plenum, New York, pp 85–176

Katayama-Hirayama K (1986) Inhibition of the activities of ß-galactosidase and dehydrogenases of activated sludge by heavy metals. Wat Res 20:491

Kellems BL, Lion LW (1989) Effect of bacterial exopolymer on lead(II) adsorption by GAL2O3 in seawater. Estu Coast Shelf Sci 28:443–457

Kepkay PE, Schwinghamer P, Willar T, Bowen AJ (1986) Metabolism and metal binding by surface-colonizing bacteria: results of microgradient measurements. Appl Environ Microbiol 51:163–170

Khazaeli MB, Mitra RS (1981) Cadmium-binding component in *Escherichia coli* during accomodation to low levels of this ion. Appl Environ Microbiol 41:46–50

Komura I, Izaki K (1971) Mechanisms of mercuric chloride resistance in microorganisms. I. Vaporization of a mercury compound from mercuric chloride by multiple drug resistance strains of *Escherichia coli*. J Biochem 70:885–893

Koopman B, Bitton G (1987) INT-dehydrogenase assay for chemical toxicity in waste water systems. Toxicity Assessment 2:105

Kurek E, Czaban J, Bollag J (1982) Sorption of cadmium by microorganisms in competition with other soil constituents. Appl Environ Microbiol 43:1011–1015

Lanza GR, Dougherty JM (1991) Microbial enzyme activity and biomass relationships in soil ecotoxicology. Environ Toxicol Wat Qual 6:165–176

Lenhard G, Nourse LD (1964) The measurement of dehydrogenase activity of activated sludges. 2nd Int Conf Wat Pollut Res Section II, No.6

Liebert C, Barkay T (1988) A direct viable counting method for measuring tolerance of aquatic microbial communities to Hg^{2+}. Can J Microbiol 34:1090–1095

Liebert C, Barkay T, Turner RR (1991) Acclimation of aquatic microbial communities to Ag(II) and CH_3HgCl in polluted freshwater ponds. Microb Ecol 21:139–149

Litchfield CD, Rake JB, Zindulis J, Watanabe TT, Stein DJ (1975) Optimization of procedures for recovery of heterotrophic bacteria from marine sediments. Microb Ecol 1:219–233

Liu D (1985) Effect of bacterial cultures on microbial toxicity assessment. Bull Environ Contam Toxicol 34:331–339

Luoma SN, Carter JL (1993) Understanding the toxicity of contaminants in sediments: Beyond the bioassay-based paradigm. Environ Toxicol Chem 12:793–796

Martinez J, Vives-Rego J (1988) Thymidine incorporation and bacterial mortality as ecotoxicological assessment in aquatic habitats. Toxicity Assessment 3:33–40

McLean RJC, Beauchemin D, Beveridge TJ (1992) Influence of oxidation state on iron binding by *Bacillus licheniformis* capsule. Appl Environ Microbiol 58:405–408

Melanovich FP, Wilson DW (1975) Detoxifying effect of yellow substance on *Escherichia coli* in media containing copper. Nature 253:460–461

Mills AL, Colwell RR (1977) Microbiological effects of metal ions in Chesapeake Bay water and sediment. Bull Environ Contam Toxicol 18:99–103

Mittelman MW, Geesey GG (1985) Copper-binding characteristics of exopolymers from a fresh water sediment bacterium. Appl Environ Microbiol 49:846–851

Moriarty DJW (1986) Measurement of bacterial growth rates in aquatic systems from rates of nucleic acid synthesis. Adv Microb Ecol 9:245–292

Moriarty DJW, Pollard PC (1990) Effects of radioactive labelling of macromolecules, disturbance of bacteria and adsorption of thymidine to sediment on the determination of bacterial growth rates in sediment with tritiated thymidine. J Microbiol Meth 11:127–139

Mountfort DO, Asher RA (1981) Role of sulfate reduction versus methanogenesis in terminal carbon flow in polluted intertidal sediments of Waimea Inlet, Nelson, New Zealand. Appl Environ Microbiol 42:252–253

Mullen MD, Wolf DC, Ferris FG, Beveridge TJ, Fleming CA, Bailey GW (1989) Bacterial sorption of heavy metals. Appl Environ Microbiol 55:3143–3149

Nakamura K, Sakamoto M, Uchiyama H, Yagi O (1990) Organomercurial-volatilizing bacteria in the mercury-polluted sediment of Minamata Bay, Japan. Appl Environ Microbiol 56:304–305

Nelson JD, Colwell RR (1975) The ecology of mercury-resistant bacteria in Chesapeake Bay. Microb Ecol 2:191–218

Nieboer E, Richardson DHS (1980) The replacement of the nondescript term "heavy metals" by a biologically and chemically significant classification of metal ions. Environ Pollut Ser. B1:3–26

Novitsky JA (1987) Microbial growth rates and biomass production in a marine sediment: Evidence for a very active but mostly nongrowing community. Appl Environ Microbiol 2363–2372

Nyström T, Albertson NH, Flärdh K, Kjelleberg S (1990) Physiological and molecular adaptation to starvation and recovery from starvation by the marine *Vibrio* sp. S14. FEMS Microbiol Ecol 74:129–140

Obst U, Holzapfel-Pschorn A, Wiegand-Rosinus M (1988) Application of enzyme assays for toxicological water testing. Toxicity Assessment 3:81–91

Olafson RW, Abel K, Sim RG (1979) Prokaryotic metallothionein: Preliminary characterization of a blue-green alga heavy metal-binding protein. Biochem Biophys Res Commun 89:36–43

Olson BH, Barkay T, Colwell RR (1979) Role of plasmids in mercury transformation by bacteria isolated from the aquatic environment. Appl Environ Microbiol 38:478–485

Olson GJ, Iverson WP, Brinckman FE (1981) Volatilization of mercury by *Thiobacillus ferrooxidans*. Curr Microbiol 5:115–118

Olson GI, Porter FD, Rubinstein J, Silver S (1982) Mercuric reductase enzyme from a mercury-volatilizing strain of *Thiobacillus ferrooxidans*. J Bacteriol 151:1230–1236

Olson BH, Thornton I (1982) The resistance pattern to metals of bacterial populations in contaminated land. J Soil Sci 33:271–277

Oren A (1987) On the use of tetrazolium salts for the measurement of microbial activity in sediments. FEMS Microbiol Ecol 45:127–133

Pederson D, Sayler G (1981) Methanogenesis in freshwater sediment: inherent variability and effects of environmental contaminants. Can J Microbiol 27:198–205

Perez-Garcia A, Codina JC, Cazorla FM, deVicente A (1993) Rapid respirometric toxicity test: Sensitivity to metals. Bull Environ Contam Toxicol 50:703–708

Perry RD, Silver S (1982) Cadmium and manganese transport in *Staphylococcus aureus* membrane vesicles. J Bacteriol 150:973–976

Piker L, Reichardt W (1991) Do sulfate-reducing bacteria respond to thymidine incorporation assays in marine sediments? Kieler Meeresforsch Sonderh 8:102–106

Poindexter JS, Leadbetter ER (1986) Bacteria in nature. Vol. 2, Plenum, New York

Radosevich M, Klein DA (1993) Bacterial enumeration and mercury volatilization in deep subsurface sediment samples. Bull Environ Contam Toxicol 51:226–233

Ramamoorty S, Kushner DJ (1975) Binding of mercuric and other heavy metal ions by microbial growth media. Microb Ecol 2:162–176

Ramlal PS, Rudd JWM, Furutami A, Xun L (1985) The effect of pH on methylmercury production and decomposition in lake sediments. Can J Fish Aquat Sci 42:685–692

Reichardt W (1978) Einführung in die Methoden der Gewässer-Mikrobiologie. G Fischer-Verlag Stuttgart, New York, 250 pp

Reichardt W (1979) Significance of different methods to determine dehydrogenase (ETS)-activities in aquatic environments. Arch Hydrobiol Beih 12:105–114

Reichardt W (1988a) Impact of bioturbation by *Arenicola marina* on microbiological parameters in intertidal sediments. Mar Ecol Prog Ser 44:149–158

Reichardt W (1988b) Impact of Antarctic benthic fauna on the enrichment of biopolymer degrading psychrophilic bacteria. Microb Ecol 15:211–271

Reichardt W, Piker L, Juterzenka K von, Heise S, Grossmann S, Bussmann I (1991) Burrowing macrozoobenthos as major determinant of bacteria in sediments. Kieler Meeresforsch Sonderh 8:86–91

Reichardt W, Heise S, Piker L (1993) Ecotoxicity testing of heavy metals using methods of sediment microbiology. Environ Toxicol Wat Qual 8:299–311

Reichardt W, Heise S ,Piker L , Juterzenka K von (1994) Microbiological aspects of heavy metal exotoxicity in Elbe river sediments. Verh Int Verein Limnol 25: 2043–2046

Remacle J, Vercheval C (1992) A zinc-binding protein in a metal-resistance strain, *Alcaligenes eutrophus* CH34. Can J Microbiol 37:875–877

Ridley WP, Dizikes LJ, Wood JM (1977) Biomethylation of toxic elements in the environment. Science 197:329–332

Robinson JB, Tuovinen OH (1984) Mechanisms of microbial resistance and detoxification of mercury and organomercury compounds: Physiological, biochemical, and genetic analyses. Microbiol Revs 48:95–124

Rossel D, Tarradellas J (1991) Dehydrogenase activity of soil microflora. Significance in ecological tests. Environ Toxicol Wat Qual 6:17–33

Rudd JWM, Turner MA (1983) The English-Wabigoon river system: II. Suppression of mercury and selenium bioacccumulation by suspended and bottom sediments. Can J Fish Aquat Sci 40:2218–2227

Sayler GS, Sherrill TW, Perkins RE, Mallory LM, Shiaris MP, Pedersen D (1982) Impact of coal-cooking effluent on sediment microbial communities: a multivariate approach. Appl Environ Microbiol 44:1118–1129

Schmiechen H, Wittig H, Martin S (1992) Entfernung von Schwermetallen aus Abwässern mittels Sorption am Biopolymer des Mikroorganismus *Ectothiorhodospira shaposhnikovii*. BioEngineering 1:38–40

Schottel JL (1978) The mercury and organomercurial detoxifying enzymes from a plasmid-bearing strain of *Escherichia coli*. J Biol Chem 253:4341–4349

Schreiber DR, Gordon AS, Millero FJ (1985) The toxicity of copper to the marine bacterium *Vibrio alginolyticus*. Can J Microbiol 31:83–87

Schreiber DR, Millero FJ, Gordon AS (1990) Production of an extracellular copper-binding compound by the heterotrophic marine bacterium *Vibrio alginolyticus*. Mar Chem 28:275–284

Shuttleworth KL, Unz RF (1993) Sorption of heavy metals to the filamentous bacterium *Thiothrix* strain A1. Appl Environ Microbiol 59:1274–1282

Silver S, Misra TK, Laddaga RA (1989) Bacterial resistance to toxic heavy metals. Metal Ions and Bacteria. New York, John Wiley & Sons

Silver S, Misra TK (1988) Plasmid-mediated heavy metal resistances. Ann Rev Microbiol 42:717–742

Singleton FL, Guthrie RK (1977) Aquatic bacteria and heavy metals. I. composition of aquatic bacteria in the presence of copper and mercury salts. Wat Res 11:639–642

Skujins J, Oden S (1986) Sensitivity of soil dehydrogenase to the presence of low concentrations of metal ion. Proc IV ISME, pp 664–667

Stebbing ARD (1982) Hormesis – the stimulation of growth by low levels of inhibitors. The Science of the Total Environment 22:213–234

Summers AO (1986) Organization, expression, and evolution of genes for mercury resistance. Annu Rev Microbiol 40:607–634

Summers AO, Silver S (1978) Microbial transformations of metals. Ann Rev Microbiol 32:637–672

Taylor B, Oremland R (1979) Depletion of ATP in *Desulfovibrio* by oxyanions of group VI elements. Curr Microbiol 3:101–103

Timoney JF, Port J, Giles J, Spanier J (1978) Heavy metal and antibiotic resistance in the bacterial flora of sediments of NewYork Bight. Appl Environ Microbiol 36:465–472

Titus JA, Parsons JE, Pfister RM (1980) Translocation of mercury and microbial adaptation in a model aquatic system. Bull Environ Toxicol 25:456–464

Torslov J (1993) Comparison of bacterial toxicity tests based on growth, dehydrogenase activity, and esterase activity of *Pseudomonas fluorescens*. Ecotoxicol and Environ Safety 25:33–40

Tsai Y-L, Olson BH (1990) Effects of Hg^{2+}, CH_3-Hg^+, and temperature on the expression of mercury resistance genes in environmental bacteria. Appl Environ Microbiol 56:3266–3272

Tsai Y, Tuovinen OH (1989) Influence of metals on oxygen uptake, carbon dioxide fixation, and cytochrome reduction in *Nitrobacter agilis*. Toxicity Assessment 4:185

Tubbing DM, Admiraal W (1991) Inhibition of bacterial and phytoplanktonic metabolic activity in the lower river Rhine by ditallowdimethyl-ammonium chloride. Appl Environ Microbiol 57:3616–3622

Tubbing DMJ, Santhagens LR, Admiraal W, vanBeelen P (1993) Biological and chemical aspects of differences in sensitivity of natural populations of aquatic bacterial communities exposed to copper. Environ Toxicol Wat Qual 8:191–205

Urrutia-Mera M, Kemper K, Doyle R, Beveridge TJ (1992) The membrane-induced proton motive force influences the metal-binding ability of *Bacillus subtilis* cell walls. Appl Environ Microbiol 58:3837–3844

Vaccaro RF, Azam F, Hodson RE (1977) Response of natural marine bacterial populations to copper: controlled ecosystem pollution experiment. Bull Mar Sci 27:17–22

Vallee BL, Ulmer DD (1972) Biochemical effects of mercury, cadmium, and lead. Ann Rev Biochem 41:91–128

Weiss AA, Murphy SD, Silver S (1977) Mercury and organomercurial resistances determined by plasmids in *Staphylococcus aureus*. J Bact 132:197–208

Wrangstadh M, Szewzyk U, Östling J, Kjelleberg S (1990) Starvation-specific formation of a peripheral exopolysaccharide by a marine *Pseudomonas* sp., strain S9. Appl Environ Microbiol 56:2065–2072

Yang CH, Menge JA, Cooksey DA (1993) Role of copper resistance in competitive survival of *Pseudomonas fluorescens* in soil. Appl Environ Microbiol 59:580–584

7 Bioassays on Sediment Toxicity

P.-D. Hansen

7.1 Introduction

In order to evaluate the impacts of environmental pollutants on biota, it is essential to consider assessment strategies which include bioassay sensors for both sediments and water ecosystems. The different approaches currently applied in such environmental impact assessment programs appear to vary. The presently practised approaches to conserve aquatic sediment quality need to be re-evaluated in order to establish a uniform and meaningful criteria applicable to sediments/water which fulfil the requirements for the setting of uniform standards. We need at present more information to improve our knowledge in this direction. Even the small physico-chemical ecosystem alteration could effect the next level of ecosystem compartment (biota) through adsorption, desorption and eventual diffusion of many metabolic by-products. These effects are normally monitored via the responses in the organisms such as MFO (MFO = mixed function oxygenases activity – detoxification), genotoxicity and immunosuppression responses. Such bioassay approaches provide us with information in the area of toxicity and sublethal toxicity. Sublethal toxicity assessment includes such bioassays as DNA unwinding, umu-assay and phagocytosis which improves information on the genotoxic potential. These measurements, when applied to sediment/water components of an ecosystem, could provide information on the potential adverse effects in sediments to biota. Also this approach could be a powerful tool for setting standards or criteria for sediment conservation and subsequent protection of ecosystem health.

Chapter 7 outlines the application of the currently available biomonitoring tools to collect information on early warning signals on sediment damage due to both man-made and natural pollution. Emphasis is made on the use of recently developed "on-line monitoring" approaches as potential biotechnology approaches to assess environmental damages and manage ecosystem health.

7.2 Ecological Requirements

In order to establish effective pollution control measures for sediment conservation, both emission and emission levels must be taken into consideration so as to better understand ecological responses to these environmental impacts. Knowledge about the small temporal dimension of the individual physico-chemical processes is essential for developing larger models.

Principles associated with the different scales of biochemical processes relating to ecosystem functions in sediments are given in Fig. 7.1 which shows the structural and functional hierarchies of biological interactions as they relate to ecosystem complexity. In order to understand these complex ecosystem interactions, it is necessary to break down these functional and structural components into their respective parts for studying as indicated by McIntosh (1981), unlike Odum's (1969) holistic methods. However, in order to fully comprehend the overall ecosystem function and ecosystem health, it is necessary to apply both concepts.

Figure 7.2 for example outlines how biological systems i.e. sediment organisms respond to sublethal levels of environmental pollutants at suborganismic, organismic, population and ecosystem levels with their end-points. The rapid reaction level is generally initiated at the suborganismic level (1–3 days), where symptoms relating to initial stages of detoxification and regulation processes are manifested. Examples of some impact monitoring tools to assess biochemical responses at this level include mixed function oxygenases (MFO), acetylcholinesterase-inhibition (AChe) and immunosuppression (phagocytosis). Overlapping with this molecular

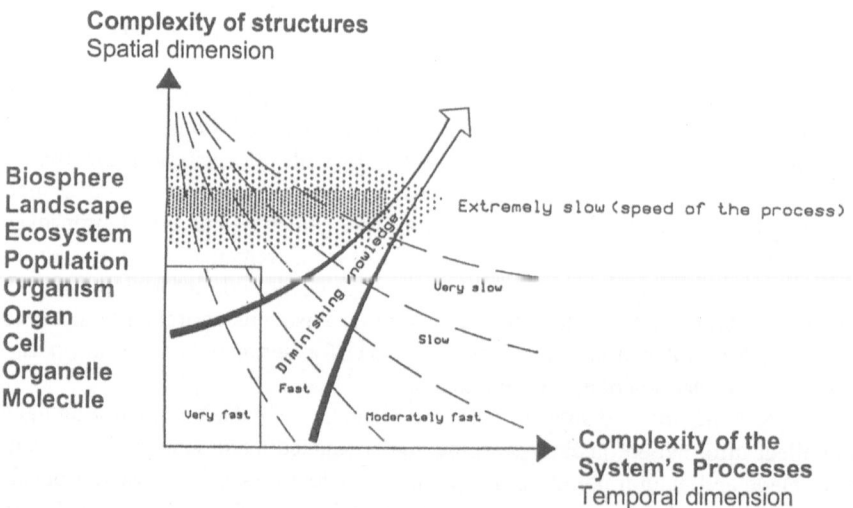

Fig. 7.1. Structural and functional hierarchies in biological systems (Hansen 1992b)

Responses in biological systems

(min)

		Ecosystem Level:
10^6	(> 2a)	alteration in ecosystems – redevelopment of the system's elements and structure
	(0.5–1a)	Population Level:
	(0.5–1a)	population dynamics – self organization – reorganization
10^5	(1–12 mo]	change in growth and adaptation of the system
	(20–120 d)	Organismic Level (exposure): "in situ Bioassays"
		- growth, reproduction, ELST (early life stage test), accumulation,
10^4		biotransformation: MFO, reaction with macromolecules, DNA-damage and repair, mutagenesis
		Organismic / Suborganismic Level:
	(1–3d)	Symptons in individuals: detoxification and regulation processes
10^3		-biochemical response ("MFO", AChE, phagocytosis)
		- change in behaviour
10^2		
10^1	(10 min)	Early Warning Systems - On Line Monitoring - "Biosensors"

10^0	Input of pollutants (sublethal level)

Fig. 7.2. Time-scale responses in biological systems (suborganismic to ecosystem levels) to sublethal pollutants

level of functional component, the effects at the organismic level can be perceived when organisms are exposed longer to these environmental pollutants. Effects at the organismic level would eventually result in contaminant genotoxic effects at macromolecular levels such as DNA damage, mutagenesis etc. Since environmental genotoxicity assessment approaches provide information on such effects, these would be the potential monitoring tools.

Table 7.1 gives information on the consequences of DNA damage at the organizational levels of biological systems of sediments.

Table 7.1. Consequences of DNA Damage on different oganizational levels of biological systems

Level of biological organization	Effects
DNA	– Mutations
Cell	– Cell death
	– Disordered proliferation and differentiation
	– Neoplastic transformation
Tissue / Organ	– Functional defects
	– Malformations
	– Tumours
Organism	– Reduced viability
	– Reduced fertility
Population	– Reduction of population size
	– Extinction
Ecosystem	– Reduction of species diversity

Extended exposure of organisms to environmental genotoxins would result in several physiological disorders such as reproductive impairments and other related abnormalities. Therefore, the response measurements to reproductive toxicity is essential for assessing the effects of anthropogenic sources.

For such assessment approaches incubation periods of 20 to 120 days would be required. This could be the beginning of the adaptation process in the organism which would reflect changes (feedback; see Fig. 7.1) in the ecosystem. These effects can also be monitored at population levels so that information on the overall functional potential of the ecosystem can be obtained. This assessment would include such processes as reorganization, redevelopment and structure of the sediment system.

In order to obtain information from the overall ecosystem assessment at least a two-year period is required, where only the initial signals on these events of reorganization and redevelopment are obtained. These signals could be critical and information on such signals in relation to time and space (site specific) would determine the ecosystem pulses. Such information would be highly helpful to the proper understanding and managing of the ecosystem health and restoration. Information on these initial signals can be obtained by the processes such as detoxification (MFO), immunosuppression and ability of the organism to reproduce. Presently, knowledge necessary to generate on-line signals is not available. Therefore, biosensor detection approaches for sediments would be useful for monitoring early warning signals (at the initial onset) of environmental deterioration process (Hansen 1992a). Biosensors are defined here as selective biological systems (enzymes, antibodies, organelles and cells) combined with a transducer (thermistor, potentiometric and amperiometric electrode, piezoelectric and optical receivers) which generate continuous information. A promising effective biosensors at this time is the immunoassay technique. Progress has already been made in this direction using enzymes and bacterial bioluminescence detection systems. Using biosensor approaches, once the information on the genotoxic potentials and the immunotoxic potentials in ecosystems is available, effective control measures can then be easily applied.

7.3 Organismic Responses in Sediments and Impacts of Environmental Stressors

Here certain examples of the effects of stressors from anthropogenic sources are outlined. A conceptual model of the response of an organism to xenobiotic pollutants is given in Fig.7.3. The initial exposure to xenobiotic pollutants and their uptake in biota precedes their distribution and accumulation in organisms. Several internal systems respond to these distributed and concentrated substances. The site-specific responses within the organism are controlled by these xenobiotic sub-

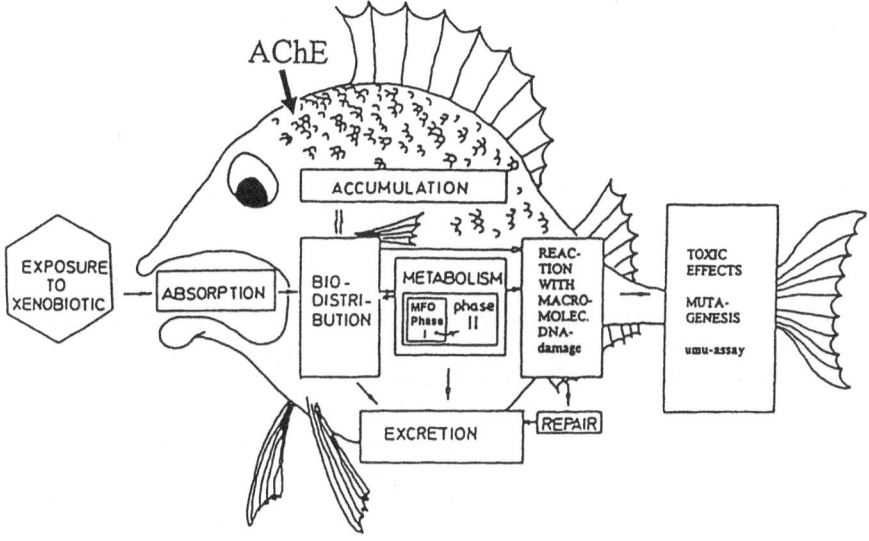

Fig. 7.3. Responses by the in biological system and the genotoxic damage end-points i.e. AChE = acetylcholinesterase inhibition, MFO = mixed function oxygenases activity (detoxification), DNA damage

stances. For example, the basic mechanism of toxic action of phosphorus-pesticides and carbamates is the inhibition of the acetylcholinesterase activity (AChE) in the nervous tissue. This results in the accumulation of the pollutants and eventual disruption of the nervous function.

Detoxification responses to these substances within organisms mainly occur in the liver tissues and kidneys. Measurement of hepatic mixed function oxygenase (MFO) activity is a sensitive indicator of an organism's ability to detoxify these pollutants.

These enzyme systems are located in the endoplasmatic reticulum (ER) of the hepatic cells. There are two phases in the metabolism of these xenobiotic substances.

Phase I includes hydroxylation, de-ethylation, de-alkalization and de-amination. Primary oxidation products from phase I are excreted or transformed into water-soluble products by the series of conjugating phase-II enzymes (Gelboin 1980). Phase II is the real detoxification stage where the polarity of these substances increase. However, the main process is the oxidation reaction of phase I products.

In addition to the detoxification systems, there are other systems such as phagocytosis which measures and quantifies the immunological defence activities.

The procedure allows us to establish the stress of the immunological defence mechanism of organisms exposed to chemical pollutants. Similarly, information on deoxyribo nucleic acid (DNA)-damaging activity provides information on genotoxic effects of these pollutants through macromolecular reactions (Fig. 7.3).

7.4 Enzymatic Inhibition (Cholinesterase) in Sediment Organisms

The basic mechanism of the toxic action of organophosphates and carbamates includes accumulation of these pollutants at the nerve endings with the eventual disruption of nerve functions. The detailed outline of this process is indicated in Fig.7.4. Essentially, recovery of acetylcholinesterase activity (AChE) in organism that survive acute effects is dependent on the spontaneous, but slow dephosphorylation of the inhibited site and the synthesis of new AChE. This monitoring

Fig. 7.4. Mode of action of inhibition of acetylcholinesterase (Peakall 1992)

approach is not only helpful in identifying the hot spots and pollution gradients near inputs, but it also characterizes the impact of pollutants (EEM = environmental effect monitoring). The effects of the substances as detected by cholinesterase inhibition are quantified as paraoxon equivalents and the inhibition constant (see Table 7.2).

The phosphoruspesticides and insectide carbamates inhibit the cholinesterases in different levels. The well known cholinesterases are acetylcholinesterase and butrylcholinesterase. The acetylcholinesterase was isolated from bovine and human erythrocytes and from electric eel (*Electrophorus electricus*); the butryl-cholinesterase was isolated from the serum from horses and humans. The strength of the inhibition effect is expressed as the inhibition constant. The inhibition constants of the most important phosporuspesticices and insectide carbamates have been published by Herzsprung et al. (1989).

A standard DIN protocol (DIN 38415-T1) is currently available (Beutler 1993) which outlines the details of a procedure applicable to water samples, with slight modifications for samples from organisms such as fish, worms and sediment feeders etc.

Table 7.2. Paraoxon equivalents and inhibition constant (Beutler 1993). Units given in 1 μg/l paraoxon-ethyl/l = 1 μg paraoxon equivalent/l

Substance	Paraoxon equivalents and inhibition constant
Paraoxonmethyl	2.0
Aldicarb (sulfoxid)	0.8
Azinphosethyl	0.5
Azinphosmethyl	0.6
Carbofuran	0.6
Carbaryl	25.0
Chlorfenvinphos	60.0
Demeton-S-methylsulfon	800.0
Diazinon	13.0
Methamidophos	360.0
Omethoat	505.0
Parathionethyl	1.1
Paraoxonethyl	1.0
Parathionmethyl	2.1
Primicarb	90.0
Propoxur	3.0

7.4.1 Enzymatic Induction (Detoxification Activity MFO) in Flatfish

This parameter MFO (MFO/EC-No. 1.14.14.1) which is widely used as a monitoring tool to detect impacts of pollutants on fish (Addison et al. 1991a,b; Hansen et al. 1993a,c) has also been used more recently in mussels (Michel et al. 1993). It is a sensitive indicator to detect the ability of organism to detoxify certain pollutants. MFO activity can be measured quantitatively to determine the extent of exposure to pollutants and is used as a bioassay to detect sublethal effects. As shown in Fig. 7.5 the most common substrates used for the determination of MFO induction are 7-Ethoxyresorufin (EROD) and 7-Ethoxycoumarin (ECOD). The end-product is determined as resorufin and umbelliferone.

Factors affecting MFO induction in fish for example include fish physiology (sexual maturity and/or reproductive status, nutritional status) and other environmental factors such as pH values, temperature and oxygen. The salinity apparently has no effect on the MFO induction (Pluta 1992). Application of the MFO-inducton system for monitoring environmental samples of the coastal North Sea waters is indicated in Fig. 7.6 (see p. 187) and Fig. 7.7 (p. 188).

Figure 7.6 shows the elevated MFO activity near input sources and declining with distance. However, at a distant site (station 9) an observable increase in MFO activity was seen due to coastal zone input. Such data supports the usefulness of

Fig. 7.5. Determination of mixed function oxygenases (MFO activity) using 7-Ethoxyresorufin (EROD) and 7-Ethoxycoumarin (ECOD). The measured products of the enzymatic reaction are resorufin and umbelliferone respectively

the parameter as a biomonitoring tool. Figure 7.6 also indicates that the sex of the fish is a contributing factor in MFO induction – only the male sex is useful. The adult females (see front-row columns with horizontal lines in Fig. 7.6) were spawning at that time and consequently the MFO levels were extremely low due to the elevated levels of their steroid hormones. Male fish are, therefore, prime candidates for monitoring MFO activity. Figure 7.7 shows the usefulness of MFO activity as a screening tool for detecting the areas where MFO induction in fish is due to anthropogenic impact. This approach is an ideal strategy for determining impact effects and managing ecosystem health. A standard protocol for MFO induction is currently available including data on inter-laboratory comparisons (Stagg and Addison 1994). Measurement parameter that would provide information about the effects of multiple environmental components is of extreme importance in environmental monitoring processes. Such parameters need to be standardized for uniform application.

Fig. 7.6. MFO activity (detoxification) in liver microsomes of fish (DAB, *Limanda limanda* L) exposed to the pollution gradient of the River Elbe estuary and sea until the Dogger Bank. The *rear* columns with *hatched lines* represent juvenile fish; the *dark centre* columns represent male fish; and the *front row* columns with *horizontal lines* represent female fish. EROD-activity = 7-Ethoxyresorufin O-de-Ethylase-activity.

■ **1.0 nmol Resorufin / g liver / min.**

Fig. 7.7. MFO activity – EROD – (detoxification) in flounder (*Platichthys flesus*) from sampling sites in the German Wadden Sea and its estuaries, August 1989

7.4.2 Immunosuppression of Sediment Organisms

The resilience of an organism is influenced by changes in the environment which are due to both natural and man-made pollution (Hansen 1993b, c). By recording immunological resistance (phagocytosis) in terms of quality and quantity, it is possible to detect pollution effects on organisms. Through phagocytic activity, foreign particles and attached pollutants are digested (Hansen et al. 1991; Hansen 1992c). A luminescent bacteria, *Photobacterium phosphoreum*, is used as a detection method. The hemocytes of the mussel feed on the luminescent bacteria which are treated as foreign particles and attached pollutants. The phagocytic activity is directly related to a decrease in luminescence (Fig. 7.8). Figure 7.8 is a schematic diagram of the collection and preparation of hemocytes and the determination of the phagocytic activity using yeast cells and *Photobacterium phosphoreum*.

Fig. 7.8. A Protocol for studying immunosuppression responses in mussels (*Mytilus edulis*).

B

Fig. 7.8. B Luminescence inhibition by phagocytic activity of the mussel-hemocytes

Hemocytes play a major part in the immunological defence system of many invertebrates. Measurement of phagocytic activity offers ample opportunities for detection of unknown biotoxins by their influence on mussel immunology. A good example for application of the phagocytic bioassay is the detection of algal toxins in marine and fresh-water environments. The number of blooms of toxic algae has

increased in recent time because of eutrophication of fresh-water and marine ecosystems. These blooms pose a threat not only to the natural ecosystems and fisheries, but also to human health. Since biotoxins excreted by algae have not been identified, a bioassay system such as phagocytosis could be extremely useful for detecting the biotoxic effects. Environmental-effects monitoring by this assay to record the integrated effects of biotoxins would be highly useful.

Experiments of feeding mussels with toxic algae (*Chrysochromulina polylepis*) and non-toxic algae (*Isochrysis galbana*) and the resulting phagocytosis analysis is reported by Krumbeck et al. (1994). For this experiment the phagocytosis was performed using microplate technique. Currently developments in this direction are underway using FITC-conjugated yeast cells and the final measurements of the phagocytosis index using a fluorescence microplate reader. This procedure has potential for becoming an on-line immunotoxicity biosensor system.

7.5 Genetic Damage

Monitoring of the effects of environmental toxic and genotoxic substances has gained increased importance in recent times. Assessment of environmental samples for the presence of genotoxins (carcinogens and mutagens) has therefore, become a valuable approach for evaluating genotoxic potentials of environmental pollutants. Several biomonitoring approaches currently exist. In Germany there are three protocols for environmental genotoxicity assessment. These are: DNA unwinding (Shugart 1988; Herbert 1990; Herbert and Zahn 1990; Herbert and Hansen 1992), AFE (alkaline filtration elution) technique (Waldmann 1993) and umu-assay (Oda et al. 1985). The German Institute for Standardization (DIN) has recently adopted umu-assay as an official protocol for monitoring environmental genotoxicity (DIN 38415-T3).

The DNA-unwinding assay is known to be a promising tool for detecting DNA damage due to environmental genotoxins in aquatic animals. Extensive work is currently in progress to develop a basis for uniform application of this bioassay for environmental samples. Recent work (Figs. 7.9 and 7.10 unpublished data) on the application of this protocol (Herbert 1990) has clearly indicated its usefulness as a genotoxic monitoring tool. Figure 7.9 shows DNA-damaging activity in mussel hemocytes (*Mytilus edulis*) after 96 h of static exposure to genotoxic chemicals. Figure 7.10 indicates that the DNA damage in *Mytilus edulis* is increased in the stressed area of the harbour as compared to an unpolluted reference station.

In the future, environmental monitoring approaches should include measurements of DNA damage in fish and mussels and umu-assay in other environmental samples. Finally, environmental mapping of hot spots for the genotoxic potential should be made mandatory for assessing and maintaining ecosystem health.

Fig. 7.9. Genotoxic effects: DNA fragmentation in *Mytilus edulis* (hemocytes) from laboratory studies (96 h static exposure). MNNG = N-Methyl-N'-nitro-N-nitrosoguanidine; DMS = Dimethylsulfate; NQO = 4-Nitroquinoline-1-Oxide

Fig. 7.10. Genotoxic effects: DNA fragmentation in *Mytilus edulis* (hemolymph and gills) from two field sites at the Island of Sylt (North Sea)

7.6 Ecological Relevance of Biological Responses in Sediments

Biological responses in sediment organisms to environmental stress could provide us with information that signals the potential damage. These responses in ecosystems, if perceived in the early stages, could prevent the eventual deterioration. On the other hand, once ecosystem damage has occurred, the remedial action process for its recovery could be expensive and pose certain logistical problems. Remedial action plan (RAP), however, has been used in certain instances to recover damaged water bodies. Prevention of ecosystem deterioration is always better than restoring of a damaged one. Ideally, "early warning signals" in ecosystems using biosensors (Hansen 1992a) would not only tell us the initial levels of damage, but these signals could provide answers to develop control strategies (precautionary measures).

Acute toxicity results in organism selection; genotoxicity results in mutagenicity and physiological impairments (genetic disease syndromes); and induction of MFO (biochemistry) tells us that fish species are induced with elevated detoxification levels and provides us with data on the effects of specific chemical species

(warning signals). However, these data do not have high ecological relevance, as do information from immunosuppression (phagocytosis), although information from immunosuppression assays does not provide signals of environmental damage similar to data from population levels. Genotoxic-damage end-points have high ecological significance as these relate to the ability of reproduction. Stress responses at population levels have direct ecological implications even though these exhibit low specificity (Fig. 7.11). There should be, therefore, a harmonized ecosystem assessment approach where the overall information (high specificity to low specificity) should be considered in parallel for proper ecosystem health management.

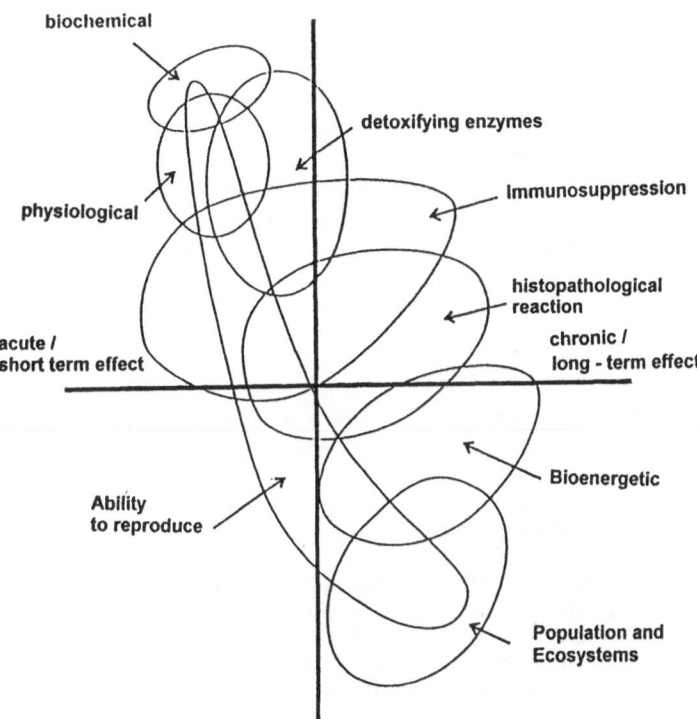

Fig. 7.11. Ecological response levels of biological markers (modified after Adams et al. 1989)

In considering the impact of either natural stress or man-made stress, we always encounter detoxification, disease defence, regulation and adaptation processes. This situation makes the assessment approach rather complicated. On the other hand, symptoms analysis, even though it includes functional (behaviour, activity and metabolism) and structural changes in organism (cellular, tissue and organs), it has ecological assessment potential.

7.7 Recommendations

From the practical standpoint, there are several concepts which can be used as powerful bioassays for management of sediments and deal with the day by day problems. There are principal grounds for determining the sediment quality requirements to protect sediments against dangerous substances. These sediment quality requirements have to be derived from scientific approaches, for example NOEC (no observed effect concentration) data. There is currently a battery of test approaches available for assessing the trophic levels of the food chain using bacteria, algae, crustacean and fish. In addition, we need to apply other assessment concepts which include effects parameters such as AChE-inhibition, MFO-induction, phagocytosis-index etc.

Among the existing genotoxicity assessment approaches, the umu-assay has become a DIN standard (DIN 38415-T3) and an ISO (International Standard Organization) standard (ISO/TC 147/SC5/WG9N10) and will become part of the Federal Regulatory Program based on the Federal Water Act (§7a WHG).

References

Adams SM, Shepard KL Greeley MS jr, Jimenez BD, Ryon MG, Shugart LR, McCarthy JS, Hinton DE (1989) The use of biomarkers for assessing the effects of pollutants stressed on fish. Mar Environ Res 28: 459–464

Addison RF, Hansen PD, Wright EC (1991 a) Heptic mono-oxygenase activities in American plaice (Hippoglossoides platessoides) from the Miramichi estuary. N.B. Can Tech Rep Fish Aquat Sci No. 1800

Addison RF, Pluta HJ, Hansen PD (1991 b) Absense of MFO-induction in several species of North Sea flatfish (Plytichthys flesus, Limanda limanda and Solea solea) following treatment with PCB substitute with Ugilec 141. Mar Environ Res 31: 137–144

Beutler HO (1993) Cholinesterase-Hemmtest. In: Fachgruppe Wasserchemie der GDCh (Hrsg) Biochemische methoden zur Schadstofferfassung im Wasser – Möglichkeiten und Grenzen. VCH Verlagsges, Weinheim

Gelboin HV (1980) Benzo (a) pyrene metabolism, activation, and carcinogenesis: role and regulation of mixed-function oxidases and related enzymes. Physiol Rev 60: 1107–1165

Hansen PD, Bock R, Brauer F (1991) Investigations of phagocytosis concerning the immunological defense mechanism of Mytilus edulis using a sublethal luminescent bacterial assay (Photobacterium phosphoreum). Comp Biochem Physiol 100C, 1/2: 129–132

Hansen PD (1992a) On-line monitoring mit Biosensoren am Gewässer zur ereignisgesteuerten Probenahme. Acta hydrochim hydrobiol 20,2: 92–95

Hansen PD (1992b) Suborganismische Testverfahren – Anwendungsbereich und Bewertungsfrage. In: Steinhäuser KG, Hansen PD (eds) Schriftenreihe Wasser-, Boden- und Lufthygiene, Bd 89. Gustav Fischer Verlag, Stuttgart

Hansen PD (1992c) Phygocytosis in Mytilus edulis, a system for understanding the sublethal effects of anthropogenic pollutants and the use of AOX as an integrating parameter for the study of equilibria between chlorinated organics in Dreissena polymorpha following long term exposures. In: Neumann, Jenner (eds) The zebra mussel Dreissena polymorpha. Limnologie aktuell, vol. 4. Gustav Fischer Verlag, Stuttgart

Hansen PD (1993a) Enzymatische Verfahren zur Erfassung der Biotransformation (Entgiftungsaktivität) in der Fischleber – ein Beitrag zum biologischen Effektmonitoring. In: Fachgruppe Wasserchemie der GDCh (Hrsg) Biochemische Methoden zur Schadstofferfassung im Wasser – Möglichkeiten und Grenzen.. VCH Verlagsges, Weinheim

Hansen, PD (1993b) Schadstoffwirkungen auf das Immunsystem. In: Fachgruppe Wasserchemie der GDCh (Hrsg) Biochemische Methoden zur Schadstofferfassung im Wasser – Möglichkeiten und Grenzen.. VCH Verlagsges, Weinheim

Hansen PD (1993c) Regulatory significance of toxicological monitoring by summarizing effect parameters. In: Richardson M (ed) Ecotoxicology Monitoring. VCH Publishers, New York

Herbert A, Zahn RK (1990) Monitoring DNA damage in Mytilus golloprovincialis and other aquatic animals, II. Pollution effects on DNA denaturation characteristics. Angew Zool 77, 1: 13–33

Herbert A (1990) Monitoring DNA damage in Mytilus golloprovincialis and other aquatic animals, III. A case study: DNA damage in fish from a Florida marsh. Angew Zool 77, 2: 143–150

Herbert A, Hansen PD (1992) Erfassung des erbgutverändernden Potentials von gewässern durch Messung von DNS-Schäden mittels alkalischer Denaturierungsverfahren. In: Steinhäuser KG, Hansen PD (eds) Schriftenreihe Wasser-, Boden- und Lufthygiene, Bd 89. Gustav Fischer Verlag Stuttgart

Herzsprung P, Weil L, Quentin KE (1989) Bestimmung von Phosphorpestiziden und insektiziden Carbamaten mittels Cholinesterasehemmung – 1: Hemmwirkung von Phosphorbestiziden und insektiziden Carbamaten auf immobilisierte Cholinesterase. Z Wasser Abwasser Forsch 22: 67–72

McIntosh RP (1981) Succession and ecological theory. In: West DC, Sugart HH, Botkin DB (eds) Forest succession: concepts and application. Springer, Berlin Heidelberg New York, pp 10–23

Michel XR, Cassand PM, Narbonne JF (1993) Activation of benzo(a)pyrene and 2-aminoanthracene to bacteria mutagens by mussel digestive gland postmitochondrial fraction. Mutation Res 301: 113–119

Oda Y, Nakamura SI, Oki I, Kato T, Shinagawa H (1985) Evaluation of the new system (umu-test) for the detection of environmental mutagens and carcinogens. Mutation Res 147: 219–229

Odum EP (1969) The strategy of ecosystem development. Science 164: 262–270

Peakall D (1992) Biomarkers of the nervous system. In: Animals as pollution indicators. Ecotoxicol Ser I. Chapman & Hall, pp 20–45;

Pluta HJ (1992) Joint monitoring in der Nordsee (North Sea Task Force (NSTF). MFO-Untersuchungen als Teil des Deutschen Beitrages zum Quality Status report (QSR). In: Steinhäuser KG, Hansen PD (eds) Schriftenreihe Wasser-, Boden- und Lufthygiene Bd 89. Gustav Fischer Verlag, Stuttgart

Shugart LR (1988) Quantification of chemically induced damage to DNA of aquatic organisms by alkaline unwinding assay. Aquatic Toxicol 13: 43–52

Stagg RM, Addison RF (1995) An inter-laboratory comparison of measurements of ethoxyresorufin O-de-ethylase activity in dab (Limanda limanda) liver. Mar Environ Res 40/1: 93–108

Waldmann P (1993) Die alkalische Filterelution als Methode zur Erfassung von gentoxischen Potentialen in Gewässern. PhD thesis (dissertation) University of Mainz, Biological Sciences

8 Trace Element Diagenesis
in Polluted Sediments of the River Elbe Estuary

K. Wallmann, W. Petersen, C. Reiners, and H. Gramm

8.1 Introduction

Due to large anthropogenic inputs, the Elbe river is heavily polluted with trace metals. These trace metals are predominantly bound onto suspended particulate matter which is transported into the sediment by flocculation and sedimentation (Irmer et al. 1985). During high water discharge and storm events trace metals are remobilized from the sediments into the river water by the resuspension of particles (Wilken et al. 1991). In addition to this exchange of particle bound trace metals, a permanent flux of dissolved elements across the sediment–water interface is induced by early diagenesis in surface sediments (Santschi 1988).

8.1.1 Early Diagenesis

Early diagenesis, i.e., the change in composition and characteristics of particles deposited as sediment in the aquatic environment, is mainly driven by the microbial mineralization of organic substances (Berner 1980). According to Froelich et al. (1979), sediments form oxic, suboxic, and anoxic environments. In oxic sediments the organic matter is mineralized by microorganisms which use dissolved oxygen as terminal electron acceptor. In suboxic sediments, nitrate and the oxide minerals of iron and manganese serve as oxidizing agents for the degradation of organic matter into inorganic carbonate species. In anoxic sediments the electrons of the organic matter are transferred to sulfate and bicarbonate which are reduced to sulfide and methane. The stoichiometry of these microbial redox reactions is shown in Table 8.1.

In the depth profile of an undisturbed sediment, the microbial reactions follow the sequence shown in Table 8.1. The oxic zone is found at the sediment surface

followed by suboxic and anoxic layers at greater depth. The products of manganese, iron and sulfate reduction form new authigenic minerals in the sediment. These are Mn(II) and Fe(II) carbonates in the suboxic layer and iron monosulfides and pyrite in the anoxic sediment (Berner 1981). The dissolved products of anaerobic mineralization reactions are transported via diffusion along concentration gradients in the porewater. Thus, Mn(II) and Fe(II) may reach the oxic zone and may be reoxidized by dissolved oxygen to form new oxide minerals at the sediment surface. The recent work of Fossing and Jorgensen (1990) and Jorgensen and Bak (1991) shows that sulfate reduction is not limited to the anoxic environment but may as well take place within the oxic zone of sediments which containing high amounts of degradable organic matter. The reduced sulfur species are reoxidized and do not accumulate in oxic environments.

Table 8.1. Mineralization of organic matter in oxic, suboxic, and anoxic sediments. $C(H_2O)$ is used as a model substance for organic matter

Oxic environment
$C(H_2O) + O_2(aq) = CO_2 + H_2O$
Suboxic environment
$C(H_2O) + 4/5 \ NO_3^- = 2/5 \ N_2 + 3/5 \ H_2O + 4/5 \ HCO_3^- + 1/5 \ CO_2$
$C(H_2O) + 2 \ MnO_2(s) + 3 \ H^+ = 2 \ Mn^{2+} + HCO_3^- + 2 \ H_2O$
$C(H_2O) + 4 \ FeOOH(s) + 7 \ H^+ = 4 \ Fe^{2+} + HCO_3^- + 6 \ H_2O$
Anoxic environment
$C(H_2O) + 1/2 \ SO_4^{2-} = 1/2 \ HS^- + 1/2 \ HCO_3^- + 1/2 \ CO_2 + 1/2 \ H_2O$
$C(H_2O) = 1/2 \ CO_2 + 1/2 \ CH_4$

8.1.2 Trace Elements

Trace elements are associated with sedimentary phases, such as particulate organic substances and oxide minerals of iron and manganese (Salomons and Förstner 1984), which are decomposed during early diagenesis. Trace elements which are released from these solids into the porewater may be bound into authigenetic oxide, carbonate, and sulfide minerals. The trace elements Cd, Cu, Co and, As, which are considered in Chapter 8 may be separated into two groups.

Cd and Cu. The elements of the first group, Cd and Cu, are preferentially bound to organic substances in an oxic environment (Calmano 1989). In the presence of dissolved sulfide, i.e., in an anoxic environment, they are precipitated as unsoluble sulfide minerals. Due to their stability, these sulfide minerals may already be formed at very low dissolved sulfide concentrations (Wallmann 1992). The

highest concentrations of dissolved Cd and Cu are usually found in the oxic zone at the sediment surface, where the organic material is degraded at high rates and the associated trace elements are released into the porewater (Klinkhammer et al. 1982; Gobeil et al. 1987; Schwedhelm et al. 1988). In the deeper anaerobic sediment layers the degradation of organic material is limited by the supply of less effective oxidizing agents. Moreover, the concentrations of dissolved Cd and Cu are lowered by the precipitation of unsoluble sulfide minerals (Wallmann 1990). Cd and Cu are transported via diffusion from the oxic layer into the overlying bottom water and into the deeper anaerobic layers. The remobilization of these elements from the surface sediment into the water column is a threat to the aquatic ecosystem, because these toxic elements may be taken up by organisms and may thus return into the food chains.

Co and As. Forming the second group, Co and As have an high affinity to the oxide minerals of manganese and iron and are thus preferentially bound to these minerals in an oxic environment (Balistrieri and Murray 1986; Farmer and Lovell 1986; Belzile 1988). The sulfide minerals of these elements are less stable and more soluble than those of Cd and Cu (Smith and Martell 1976). Therefore, they are only formed at high dissolved sulfide concentrations. In anoxic environments with low sulfide concentrations, As and Co are often associated with sedimentary pyrite (Huerta-Diaz 1989). Dissolved Co and As usually have their maximum concentrations in suboxic sediment layers, where the oxide minerals of manganese and iron are reductively dissolved and the associated trace elements are released into the porewater (Heggie and Lewis 1984; Peterson and Carpenter 1986). In the oxic zone at the sediment surface, Co and As concentrations are limited by adsorption on oxide minerals, which may be formed by the oxidation of Mn(II) and Fe(II). In anoxic sediments, dissolved Co and As may be transferred into authigenetic pyrite and other sulfide minerals. Co and As are transported along the concentration gradients from the suboxic sediment to the oxic sediment surface and the deeper anoxic layers.

In this chapter, porewater data from sediments of the river Elbe estuary are reported. The microbial redox reactions will be characterized and their influence on the distribution and solubility of Cd, Cu, Co, and As will be discussed. The fluxes of dissolved Cd and Cu from the surface sediment into the bottom water will be determined from porewater profiles to quantify the remobilization of trace elements caused by early diagenesis.

8.2 Materials and Methods

Sediments were collected in March and November 1991 from the freshwater tidal mud flat "Mühlenberger Loch" which is located in the Elbe estuary 5 km down-

stream from Hamburg Harbor. The area is partially sheltered by a groyne and is characterized by low water velocities and sedimentation rates of 5–10 cm per year (Petersen et al. 1990). The March core was taken near the groyne, whereas the November core was taken from a location which was more exposed to the tidal currents. Water temperatures were 4 °C and 7.5 °C in March and November, respectively. Sediments were transported to the laboratory within a few hours. In the lab, the sediment cores were transferred into a glove box which was filled with nitrogen gas to prevent oxidation reactions within the anaerobic sediment. The sediment was cut in 0.1 cm to 5 cm segments and the porewater was squeezed out with a porewater press. Pore water samples were acidified and stored at -15 °C in polyethylene vials. The porewater press and vessels were prewashed in HCl to minimize trace element contaminations. The sediment core taken in March 1991 had a diameter of 10 cm and was cut down to a depth of 10 cm into the following segments: 0–0.5 cm, 0.5–1 cm, 1–2 cm, 2–3 cm, 3–4 cm, 4–5 cm, 5–7 cm, and 7–10 cm. In November 1991 a sediment core of 20 cm in diameter was taken and divided with a higher spatial resolution into 10 segments of 0.1-cm depth at the sediment surface. Further sediment samples were taken down to a depth of 15 cm.

8.2.1 Analysis of Pore Water Samples

Dissolved sulfate concentrations were determined with ion chromatography (Dionex, model 14). Dissolved silicate was quantified with the photometric molybdenum-blue procedure. Trace element concentrations (Cd, Cu, Co, As) were measured with graphite furnace AAS and Zeeman background correction (Perkin Elmer, model 3030). Mn and Fe concentrations were determined with AAS (Perkin Elmer, model 2380) using an air/acetylene-flame. In November 1991, dissolved oxygen in the oxic surface sediments was determined with a micro-electrode in a sediment cores taken at the same location and at the same time as the core used for porewater analysis. The microelectrode was introduced into the sediment from above by using a micro-manipulator with a computer controlled motor drive. The polarographic electrode had a tip diameter of 0.01 mm. Due to the low oxygen consumption of the electrode, the stirring effect was negligible. The current in the measuring circuit was measured by a picoamperemeter.

8.2.2 Determination of Total Dissolved Sulfide

A new polarographic procedure was developed for the determination of total dissolved sulfide ($H_2S + HS^- + S^{2-}$). Dissolved sulfide species in sediment porewaters are easily lost during sampling. Hydrogen sulfide may be transferred into the gas phase during the squeezing of porewater in a porewater press. Sulfide species are rapidly oxidized by Fe(III) hydr(oxides) which are instantaneously formed when

anaerobic Fe(II)-bearing porewaters are exposed to the atmosphere. To minimize these effects, the separation of porewater from the solids was avoided in the new procedure. The procedure was based on polarography which allows a rapid and very sensitive determination of total dissolved sulfide in particle water suspensions.

Wet sediment samples were placed into a polarographic cell that was filled with degassed carbonate buffer containing 0.05 mol/l Na_2CO_3 and 0.05 mol/l $NaHCO_3$. The sediment was suspended with a magnetic stirrer and the total dissolved sulfide concentration was measured on a hanging mercury drop electrode. The measurements were done in the differential pulse modus and were repeated every 30 seconds. They showed a linear increase in total dissolved sulfide with time during the first 2 to 4 minutes. In the carbonate buffer Fe(II) was precipitated as siderite ($FeCO_3$) and the resulting low dissolved Fe(II) concentration caused a continuous dissolution of sedimentary iron sulfide minerals. The total dissolved sulfide concentration in the sediment suspension was plotted as a function of time and the initial concentration was determined by extrapolation to time zero, i.e., to the time when the wet sediment sample was added to the carbonate buffer, to eliminate the influence of sulfide mineral dissolution. Taking into account the dilution of the porewater in the carbonate buffer, the in situ concentration of total dissolved sulfide in the porewater was calculated from the initial concentration. Total dissolved sulfide was only determined in the sediment core taken in November 1991.

8.3 Results and Discussion

The concentrations of silicate, sulfate, manganese and iron in the porewater of the sediment taken in March 1991 and the corresponding trace element concentrations (Cd, Cu, Co, As) are plotted in Fig. 8.1. Figure 8.2 shows the porewater data of the sediment core taken in November 1991. The data points at zero depth represent the bottom water concentrations determined in the overlying water of the sediment cores.

In March, the silicate and sulfate profiles showed a much steeper gradient within the surface sediment than in November 1991. In November, dissolved iron was not detectable in the upper 4 cm of sediment, whereas high Fe concentrations were measured in the surface sediment taken in March 1991. The lower sulfate and higher Fe concentrations indicate a more reducing environment for the surface sediment taken in March. This may be due either to higher mineralization rates of the organic matter or to a slower transport of oxidizing agents from the bottom water into the sediment.

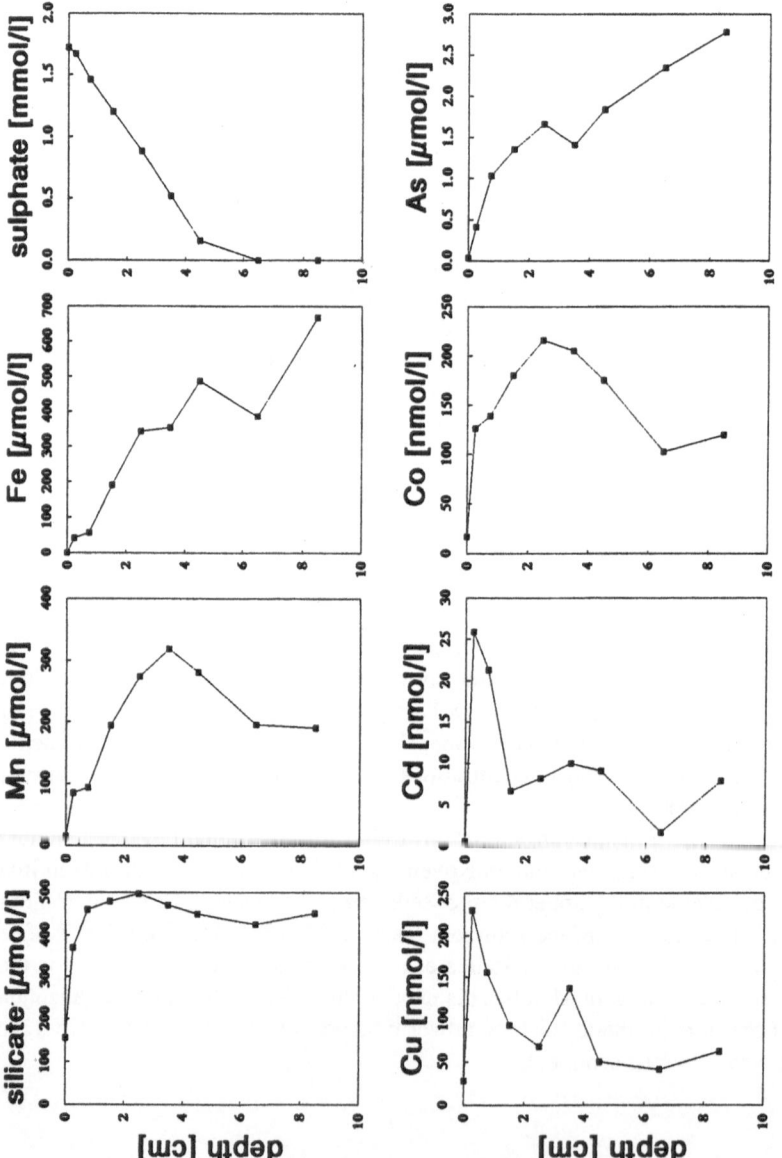

Fig. 8.1. Concentrations of dissolved silicate, manganese, iron, sulfate, copper, cadmium, cobalt, and arsenic in the porewater of the sediment core taken in March 1991

Fig. 8.2. Concentrations of dissolved silicate, manganese, iron, sulfate, sulfide, copper, cadmium, cobalt, and arsenic in the porewater of the sediment core taken in November 1991

8.3.1 Modeling of Silicate Profiles

Silicate profiles in sediment porewaters may be used to model the mass transport properties of surface sediments (Vanderborght et al. 1977). In contrast to the other dissolved compounds discussed in this chapter, the silicate concentrations are not controlled by redox reactions. They are determined by the dissolution of opal, by the precipitation of authigenetic silicate minerals and by the transport of dissolved silicate from the porewater into the overlying bottom water. The steep silicate gradient in the surface sediment in March may be taken as an indication for a slow transport of dissolved compounds across the sediment–water interface. In November, the silicate concentrations were much lower and a significant change in silicate and sulfate concentration gradients between the top 5 cm and the deeper sediment was observed (Fig. 8.2). Thus, the porewater data indicate an accelerated transport of dissolved species in the surface layer of the November core.

Vanderborght et al. (1977) described dissolved silicate profiles in sediments, which were composed of two layers with different transport properties, using the following equation:

$$D_i \cdot \frac{d^2[Si]}{dz^2} - w \cdot \frac{[Si]}{dz} + k_{Si} \cdot \left([Si]_{eq} - [Si]\right) = 0 \tag{8.1}$$

D_i represents the mass transfer coefficient in the layer considered, [Si] is the concentration of dissolved silicate, w is the sedimentation rate, z is the depth, k_{Si} is an apparent kinetic constant for the dissolution of opal, and $[Si]_{eq}$ is the equilibrium concentration of silicate. The differential equation can be analytically solved under the following boundary conditions: at $z = 0$, the concentration is equal to the concentration in the overlying water; at large z values, the concentration tends towards a finite value ($[Si]_{eq}$); at the interface between the upper and lower layer, the concentrations in each layer are equal and the flux across this interface is conservative. The analytical solutions for the upper layer and the lower layer are given in Vanderborght et al. (1977). In this chapter, they were used for the determination of the transfer coefficient D_1 in the upper sediment layer of the core taken in November 1991. The solutions of the differential equation (Eq. 8.1) were written on a LOTUS worksheet and the generated model curve was fitted to the experimental data by variation of the kinetic constant k_{Si} and the transport coefficient D_1. In this calculations we used a w value of 10 cm a^{-1}, a $[Si]_{eq}$ value of 450 μmol l^{-1}, a bottom water concentration of 125 μmol l^{-1} and a D_2 value of $3.4 \cdot 10^{-6}$ cm^2 s^{-1}. The transport coefficient D_2 for the lower layer was calculated from the diffusion coefficient of silicate in water D_0 and the mean porosity Φ of the sediment using the following equation:

$$D_2 = \frac{D_0}{-0.73 \cdot \Omega + 2.17} \tag{8.2}$$

Equation 8.2 is an empirical relationship between the diffusion coefficients of

dissolved species in freshwater sediments and the diffusion coefficients in water (Sweerts et al. 1991).

The model curve shown in Fig. 8.3 was generated with $D_1 = 2.3 \cdot 10^{-5}$ cm^2 s^{-1} and a k_{Si} value of $6 \cdot 10^{-7}$ s^{-1}. The k_{Si} value was varied within the range of $2-7 \cdot 10^{-7}$ s^{-1}, given by Vanderborght et al. (1977), until the data points measured in the lower layer were fitted by the model curve. Subsequently, the curve was fitted to the concentrations in the upper layer by variation of D_1. The diffusion coefficient of silicate in water (D_0) at the in situ temperature of 7.5 °C is $5.96 \cdot 10^{-6}$ cm^2 s^{-1} (Lerman 1979) and was thus smaller than the D_1 value. The model indicates that the transport in the top 5 cm of the sediment was faster than diffusion in pure water. The transport must have been accelerated by eddy diffusion or advective processes in the surface sediment. Movement of porewaters may either be caused by physical forces such as wind, waves and tidal currents or by the bio-irrigation of benthic organisms. The sampling position of the November core was more exposed to the tidal currents than the core taken in March 1991. Therefore, it is possible that physical factors caused the transport processes observed in the top 5 cm of the November core.

Fig. 8.3. Modeling of the dissolved silicate profile of the sediment core taken in November 1991. The *squares* indicate the measured concentrations and the *line* indicates the model curve

8.3.2 Modeling of Oxygen Profiles

The dissolved oxygen profiles shown in Fig. 8.4 indicate an oxygen penetration depth of 0.3 cm for the November core. Thus, the oxic zone was limited to the upper 3 mm of the sediment surface. The flux of oxygen into the sediment may be

Fig. 8.4. Dissolved oxygen concentrations in a sediment core taken in November 1991. The measurements were done without stirring of the bottom water (*squares*) and with intense stirring (*triangles*)

calculated from the profile measured in the laboratory without stirring the overlying water. This profile showed a concentration gradient above the sediment–water interface. Within this so-called diffusive boundary layer, the diffusion coefficient of oxygen in water may be used to quantify the transport of dissolved oxygen (Jorgensen and Revsbech 1985). Using Fick's first law of diffusion, the flux (J) was calculated from the concentration gradient (dC/dz) in the diffusive boundary layer and the diffusion coefficient of oxygen in water (D_0):

$$J = -D_0 \cdot \frac{dC}{dz} \tag{8.3}$$

A flux of 6.69 mmol m^{-2} d^{-1} was calculated using the measured gradient of 565 nmol/cm^4 and a D_0 value of $1.37 \cdot 10^{-5}$ cm^2 s^{-1} valid for the in situ temperature of 7.5 °C (Lerman 1979). The mean oxygen consumption rate in the oxic

layer was calculated from the ratio of flux to oxygen penetration depth as 2.23 mol m^{-3} d^{-1}. The rate of organic carbon mineralization has the same value as the oxygen consumption rate, if the oxygen is completely consumed by the mineralization of organic carbon in the oxidation state of carbohydrates.

8.3.3 Trace Element Profiles

The trace element profiles depicted in Figs. 8.1 and 8.2 show the typical patterns of early diagenesis. Cd and Cu had their highest concentrations at the sediment surface within the oxic zone. A second dissolved Cd maximum was found at 4–6 cm depth in the sediment core taken in November 1991. Dissolved Co and As reached high values in the manganese and iron reduction zones. The close similarity of Co and Mn profiles demonstrates that Co was preferentially associated with manganese oxides and was released into the dissolved phase during the reductive dissolution of these minerals. The As maximum concentrations were found below the Co maxima within the zone of iron reduction. This indicates, that the As mobilization was controlled by the reduction of iron oxides. In the November core, the Co and As concentrations decreased within the surface layer (Fig. 8.2). The zones of Co and As removal from the porewater were located from the change in concentration gradients at 0.6–0.7 cm and 2–4 cm depth, respectively. The Mn and Fe precipitation zones, as determined from the gradient changes, were located in the same layers, i.e., at 0.6–0.7 cm and at 2–4 cm depth, respectively. This indicates, that the Co and As concentrations in the surface sediment were controlled by adsorption onto or coprecipitation with authigenetic manganese and iron oxides.

The influence of sulfide mineral precipitation is visible in the total dissolved sulfide profile measured in the November core (Fig. 8.2). Concentrations were low within the surface layer and increased with depth in the deeper undisturbed sediment. The Cu, Cd and Co concentrations decreased with increasing total dissolved sulfide concentrations and reached approximately constant values at 5–6 cm, 6–7 cm, and 7–8 cm depth, respectively. The As concentrations decreased till the end of the sampled sediment column. The solubility of trace element sulfide minerals (Smith and Martell 1976) increases in the order:

$$Cu < Cd < Co < As$$

The total dissolved sulfide concentrations in the layers where trace elements were removed from the porewater increased in the same order. This indicates that the removal of trace elements within these layers was due to the precipitation of sulfide minerals.

Fig. 8.5. Concentrations of dissolved copper, cadmium, and sulfide in the top cm of the sediment core taken in November 1991

The concentrations of dissolved Cd, Cu and sulfide in the first cm of the sediment core taken in November 1991 are shown in Fig. 8.5. The concentration gradients indicate a flux of sulfide, Cd, and Cu into a sulfide mineral precipitation zone at 0.6 to 0.7 cm depth. The total dissolved sulfide maximum at 0.8–0.9 cm depth indicates active sulfate reduction within the top cm of the sediment column. Thus, sulfate reduction as well as Cd and Cu sulfide precipitation are not limited to anoxic sediments but may as well occur in oxic or suboxic surface sediments.

8.3.4 Benthic Flux of Cd and Cu

The flux of Cd and Cu from the surface sediment into the overlying water was calculated from Fick's first law of diffusion:

$$J = - \Phi \cdot D_s \cdot \frac{dC}{dz} \qquad (8.4)$$

In contrast to Eq. 8.3, which describes the diffusive transport in water, Eq. 8.4 is used for the calculation of the transport within sediments (Ullman and Aller 1982). The porosity Φ had a value of 0.9 at the sediment surface. D_s is the diffusion coefficient in the sediment. The concentration gradients between the bottom water and the porewater in the first mm of the sediment taken in November 1991 were used for the flux calculation. It was assumed that the concentrations measured in the 0–0.1 cm-depth interval indicate the concentration in the center of the interval, i.e., at a 0.05 cm depth. Furthermore, the concentration at 0 cm depth was equated with the concentration measured in the bottom water. This implies that the formation of a diffusive boundary layer was prevented by the fast tidal

currents in the field. With these assumptions, concentration gradients of 0.25 nmol cm^{-4} and 3.40 nmol cm^{-4} were calculated for Cd and Cu, respectively. The values of Cd and Cu diffusion coefficients in the surface sediment are difficult to access. These may be estimated from the diffusion coefficients in water of 7.5 °C which are given in Li and Gregory (1974) as $4.27 \cdot 10^{-6}$ cm^2 s^{-1} for Cd and $4.37 \cdot 10^{-6}$ cm^2 s^{-1} for Cu. Inserting these values and a porosity of 0.9 into Eq. 8.2 results in D_s values of $2.82 \cdot 10^{-6}$ cm^2 s^{-1} and $2.89 \cdot 10^{-6}$ cm^2 s^{-1} for Cd and Cu, respectively. The fluxes calculated with these diffusion coefficients and Fick's first law of diffusion are listed in the first row of Table 8.2. They describe the transport of Cd and Cu caused by molecular diffusion. The values are too low, because they do not consider the acceleration of transport which was demonstrated by the modeling of silicate profiles in the surface sediment of the November core. Disregarding the porewater movement, results in an underestimation of the transport because the effective transport coefficient is larger than the molecular diffusion coefficient and the gradients at the sediment surface are leveled off by the accelerated transport of dissolved species. More realistic fluxes are calculated when the D_1 value determined from the silicate profile is inserted into Eq. 8.4. The resulting values are one order of magnitude larger and are listed in the second row of Table 8.2.

The fluxes of dissolved trace elements from the sediment into the bottom water may be compared with the flux of particulate trace elements into the sediment. Flux (F) is determined by the concentration of trace elements in the suspended matter (C), the sedimentation rate (w), the density of dry solids (d), and the porosity (Φ) of the sediment surface. It can be calculated from the following equation:

$$F = -w \cdot d \cdot (1-\Phi) \cdot C \qquad (8.5)$$

The average concentrations of Cd and Cu in suspended matter of the Elbe river near the sampling site are 0.12 mmol Cd and 5 mmol Cu per kg of dry solids (Wilken et al. 1991). Fluxes of 8.2 μmol m^{-2} d^{-1} Cd and 340 μmol m^{-2} d^{-1} Cu were calculated using a sedimentation rate of 10 cm a^{-1} (Petersen et al. 1990), a d value of 2.5 g/cm^3, and a porosity of 0.9. The comparison of these values with the flux of dissolved Cd and Cu from the sediment into the bottom water, listed in the second row of Table 8.2, demonstrates that approximately 60% of the deposited Cd and 20% of Cu were remobilized as dissolved species into the bottom water from the surface of the sediment core taken in November 1991.

Table 8.2. Fluxes of dissolved Cd and Cu into the bottom water in μmol m^{-2} d^{-1} (November core)

	Cd	Cu
$D_s = D_0/1.51$	0.548	7.64
$D_s = D_1$	4.97	67.6

The fluxes of dissolved Cd and Cu from the sediment taken in March 1991 may be calculated from the concentration gradients at the sediment–water interface (Fig. 8.1). The concentrations in the first porewater interval and in the bottom water were similar to those in the November core (Fig. 8.2). The concentration gradients within the oxic zone were not determined, because of the low spatial resolution of 0.5 cm used in the segmentation of the March core. Therefore, the calculated gradients across the interface will have a much larger error than those determined for the November core with a spatial resolution of 0.1 cm. The silicate profile indicated that the transport in the March core was not accelerated by advection. Dissolved species were only transported via molecular diffusion. The fluxes of dissolved Cd and Cu from the surface of the March core into the bottom water were thus probably similar to those listed in the first row of Table 8.2.

8.3.5 Rates of Cd and Cu Mobilization

The rates of Cd and Cu mobilization in the top millimeter of the sediment core taken in November 1991 were calculated from the fluxes and the depth of the considered layer (1 mm). They are listed in Table 8.3 together with the rate of organic matter degradation, estimated from the mean oxygen consumption rate in the oxic zone. The concentrations of Cu and Cd in the organic matter mineralized at the sediment surface may be calculated from the rate ratios. The ratios indicate concentrations of 30 μmol Cu and 2 μmol Cd per mmol organic carbon. This implies concentrations of 1 mol Cu and 70 mmol Cd per kg organic matter with a carbon content of 40%. In the organic matter mineralized at the sediment surface, the Cu concentration is larger than the concentration in suspended matter by a factor of 200 (Wilken et al. 1991) and the Cd concentration exceeds the concentration in suspended matter by a factor of 600.

Table 8.3. Rates of organic matter degradation and dissolved Cu and Cd mobilization in the first millimeter of the surface sediment taken in November 1991 (mmol m^{-3} d^{-1})

C_{org}	Cu	Cd
2230	67.6	4.97

These extremely large enrichment factors seem to be rather unrealistic. It is possible that Cd and Cu are not only mobilized from organic matter but as well from oxidized sulfide minerals. The sulfide analysis demonstrated an active sulfate reduction in the surface sediment which was also found by Jorgensen and Bak (1991). Therefore, we postulate that the Cd and Cu fluxes into the overlying bottom water are predominantly controlled by the precipitation and reoxidation of

sulfide minerals. This delicate balance between precipitation and mobilization is influenced by the rates of sulfate reduction and sulfide reoxidation which will depend on the availability of degradable organic matter and dissolved oxidation agents. The effective transport coefficients of the electron acceptors oxygen and nitrate depend on the hydrodynamic conditions in the overlying bottom water, which are highly variable during the tidal cycle in the river Elbe estuary.

8.4 Conclusions

The concentration profiles of the dissolved trace elements Cd, Cu, Co, and As in the porewater of river Elbe sediments are controlled by the microbial mineralization of organic matter and associated redox reactions. Cd and Cu are released from decomposed organic substances and reoxidized sulfide minerals into the porewater within the oxic surface sediment. Co and As are mobilized from reductively dissolved manganese and iron oxides in the manganese and iron reduction zones. In the oxic zone, Co and As are bound to oxide minerals, which may be formed by the oxidation and precipitation of dissolved Mn(II) and ferrous iron. Dissolved sulfide is formed via the microbial sulfate reduction within anoxic sediments and in surface sediments. The trace elements are precipitated as sulfide minerals. The position of the precipitation zones in the sediment is controlled by the solubility of the trace element sulfide minerals.

Up to 60% of Cd and 20% of Cu deposited in the sediment via the settling of suspended matter are remobilized from the sediment surface into the bottom water. The flux of dissolved Cd and Cu is determined by the mineralization rate of organic matter at the sediment surface, by the concentration of Cd and Cu in degradable organic matter, by sulfide precipitation and reoxidation reactions and by the transport properties of the surface sediment.

The results presented in Chapter 8, strongly suggest that the early diagenesis of surface sediments has to be considered in the development of sediment quality criteria. Sediments with high trace metal concentrations, fast porewater movement and intense organic matter degradation are the ones most dangerous for the aquatic ecosystem.

References

Balistrieri LS, Murray JW (1986) The surface chemistry of sediments from the Panama Basin: the influence of Mn oxides on metal adsorption. Geochim Cosmochim Acta 50:2235–2243

Belzile N (1988) The fate of arsenic in sediments of the Laurentian Trough. Geochim Cosmochim Acta 52, 2293–2302

Berner RA (1980) Early diagenesis – a theoretical approach. Princeton University Press, Princeton, New Jersey, USA

Berner RA (1981) A new geochemical classification of sedimentary environment. J Sed Petrol 57:359–365

Calmano W (1989) Schwermetalle in kontaminierten Feststoffen. Verlag TÜV Rheinland GmbH, Köln

Farmer JG, Lovell MA (1986) Natural enrichment of arsenic in Loch Lomond sediments. Geochim Cosmochim Acta 50:2059–2067

Fossing H, Jorgensen BB (1990) Oxidation and reduction of radiolabeled inorganic sulfur compounds in an estuarine sediment, Kysing Fjord, Denmark. Geochim Cosmochim Acta 54:2731–2742

Froelich PN et al. (1979) Early oxidation of organic matter in pelagic sediments of the eastern equatorial Atlantic: suboxic diagenesis. Geochim Cosmochim Acta 43:1075–1090

Gobeil C, Silverberg N, Sundby B, Cossa D (1987) Cadmium diagenesis in Laurentian Trough sediments. Geochim Cosmochim Acta 51:589–596

Heggie D, Lewis T (1984) Cobalt in pore waters of marine sediments. Nature 311:453–455

Huerta-Diaz MA (1989) Geochemistry of trace elements associated with sedimentary pyrite from anoxic marine environments. PhD Thesis, Texas A and M University, Texas, USA

Irmer U, Knauth H-D, Weiler K (1985) Einfluß der Schwebstoffbildung auf Bindung und Verteilung ökotoxischer Schwermetalle in der Tideelbe. Vom Wasser 65:37–61

Jorgensen BB, Revsbech NP (1985) Diffusive boundary layers and the oxygen uptake of sediments and detritus. Limnol Oceanogr 30:111–122

Jorgensen BB, Bak F (1991) Pathways and microbiology of thiosulfate transformations and sulfate reduction in a marine sediment (Kattegat, Denmark). Appl Environ Microbiol 57:847–856

Klinkhammer GP, Heggie DT, Graham DW (1982) Metal diagenesis in oxic sediments. Earth Planet Sci Lett 61:211–219

Lerman A (1979) Geochemical Processes Water and Sediment Environments. John Wiley & Sons, New York

Li Y-H, Gregory S (1974) Diffusion of ions in sea water and in deep-sea sediments. Geochim Cosmochim Acta 38:703–714

Petersen W, Knauth H-D, Pepelnik R, Bendler I (1990) Vertical distribution of Chernobyl isotopes and their correlation with heavy metals and organic carbon in sediment cores of the Elbe estuary. Sci Total Environ 97/98:531–547

Peterson ML, Carpenter R (1986) Arsenic distribution in pore water and sediment of Pudget Sound, Lake Washington, the Washington coast and Saanich Inlet, B. C. Geochim Cosmochim Acta 50:353–369

Salomons W, Förstner U (1984) Metals in the hydrocycle. Springer, Berlin Heidelberg New York

Santschi PH (1988) Factors controlling the biogeochemical cycles of trace elements in fresh water and coastal marine waters as revealed by artificial radioisotopes. Limnol Oceanogr 33:848–866

Schwedhelm E, Vollmer M, Kersten M (1988) Bestimmung von Konzentrationsgradienten gelöster Schwermetalle an der Sediment/Wasser-Grenzfläche mit Hilfe der Dialysetechnik.
Fres Z Anal Chem 332:756–763

Smith RM, Martell AE (1976) Critical stability constants, Vol. 4. Plenum Press, New York

Sweerts J-PRA, Kelly CA, Rudd JWM, Hesslein R, Cappenberg TE (1991) Similarity of whole-sediment molecular diffusion coefficients in freshwater sediments of low and high porosity. Limnol Oceanogr 36:335–342

Ullman WJ, Aller RC (1982) Diffusion coefficients in nearshore marine sediments. Limnol Oceanogr 27:552–556

Vanderborght J-P, Wollast R, Billen G (1977) Kinetic models of diagenesis in disturbed sediments. Part I, Mass transport properties and silica diagenesis. Limnol Oceanogr 22:787–793

Wallmann K (1990) Die Frühdiagenese und ihr Einfluß auf die Mobilität der Spurenelemente As, Cd, Co, Cu, Ni, Pb und Zn in Sediment- und Schwebstoff-Suspensionen. PhD Thesis, TU Hamburg Harburg, Hamburg, FRG

Wallmann K (1992) Die Löslichkeit und die Bindungsformen von Spurenmetallen in anaeroben Sedimenten. Vom Wasser 78:1–20

Wilken R-D et al. (1991) Fakten und Hypothesen zum Schwebstoff- und Schadstofftransport in der Elbe. Vom Wasser 76:167–189

9 Mercury Distribution and Speciation in a Polluted Fluvial System

R. Ebinghaus and R.-D. Wilken

9.1 Introduction

Mercury is, besides lead, arsenic and cadmium, one of those elements which are well-known for their toxicity. In natural environments, mercury is widely dispersed and usually occurs at trace levels (Adriano 1986). However, this "natural background" can be exceeded in the presence of natural or anthropogenic emitters leading to accumulation of mercury in the biosphere.

Mercury reaches the biosphere both via natural processes and anthropogenic activities. A recent study has shown that the worldwide natural emissions can be estimated at 2500 ton per year. Anthropogenic emissions of mercury are higher and were estimated at 3500 ton per year (Pacyna and Münch 1991). Once accumulated in the biosphere, mercury can result in severe ecotoxicological consequences.

The mercury poisoning epidemics in Minamata, Japan, in the 1950s and in Iraq in the 1970s aroused the attention of the general public. Other, less well-known, incidents occurred in Pakistan and Guatemala (Gilmore and Henry 1991; Craig 1986; Merian 1984) and several other locations (Förstner and Wittmann 1979).

The poisoning in Iraq was the result of a direct uptake of an organomercury compound (ethylmercury p-toluenesulfonanilide) used in agriculture as a seed dressing (Förstner and Wittmann 1979). However, the so called Minamata disease was the result of an indirect intoxication. Inorganic mercury had been used as a catalyst by a factory producing acetaldehyde and the wastes were released into the Minamata Bay. The bay was used as a fishing ground by the local fishermen. The consumption of mercury-contaminated fish by the people of Minamata was soon determined as the reason for the disease, but it took many years of intensive research to find out that monomethylmercury was the specific cause of the "Minamata disease" (Tsubaki 1977). The compound CH_3HgSCH_3 was isolated

from fish tissues and identified as the causative agent. (Irukayama et al. 1962; Irukayama et al. 1961; Uchida et al. 1961a, b).

The methylmercuric compounds at Minamata could have been formed both by chemical and biological methylation. When chloride ions are present, acetaldehyde and mercuric ions can combine to form CH_3HgCl (Tsubaki 1977). To our knowledge it is not clear if compounds other than inorganic mercury and methylmercuric chloride had been discharged as industrial by-products into Minamata Bay.

Another important source for methylmercuric compounds is the methylation of inorganic mercury by bacteria via a process which was first reported by Wood et al. (1968).

In the past few years it has become clear that "hot spots" of anthropogenic mercury emissions are also located in the former German Democratic Republic (GDR). Effluents of chlor-alkali- and acetaldehyde plants have contaminated rivers, soils and ambient air. The Elbe river, for example, has been one of the most contaminated rivers with regard to mercury for many years. Compared to the natural background (0.05 mg Hg kg^{-1} dry weight; Turekian and Wedepohl 1961) in sediments, average concentrations in suspended material and sediments are in the range of 30 mg kg^{-1} (dr. wt.) and even high concentrations of up to 157 mg kg^{-1} (dr. wt.) have been found (Wilken and Weiler 1986).

In natural waters and sediments mercury usually occurs as divalent inorganic $Hg(II)$ (Banat et al. 1972). However, monomethylmercury, one of the most toxic mercury species and the one responsible for the Minamata disease, can also be determined in Elbe sediments. Concentrations of 1–1.5 mg CH_3Hg^+ kg^{-1} dr. wt. are very high both in absolute terms, and in relation to the total mercury content (8%; Wilken et al. 1990).

In the water column, mercury is primarily bound to suspended particulate matter (SPM; Wilken et al. 1990). Therefore, SPM is essential for the transport of mercury both in the horizontal and in the vertical direction. In the non-tidal section of the Elbe river the mercury transport in the horizontal direction was calculated to be 25 ton per year (Fanger et al. 1990; Michaelis et al. 1988; Wilken and Weiler 1986; Michaelis and Knauth 1985). The importance of SPM for the vertical transport of mercury in sedimentation processes is demonstrated by high concentrations of this element and its species in Elbe river sediments (Ebinghaus 1991).

Several characteristics of SPM can be important with regard to the binding of mercury. SPM isolated from the Elbe river is densely populated with bacteria which make up the most important part of the organic matrix of this material (Greiser 1988). Bacteria can produce and liberate polymeric substances which mediate their attachment to surfaces (DeFlaun and Mayer 1985). The concept of bacterial adhesion as an important bacterial survival strategy is widely accepted (Read and Costerton 1987), although the mechanism and physiological background is not completely understood (Koefoed 1988). The surface polymers consist mainly of polysaccharides (Sutherland 1981). Extracellular polymeric sub-

stances (EPS) commonly occur as discrete capsules firmly attached to the bacteria cells or as slime loosely bound or without contact to cell surfaces (Platt et al. 1985; Gehr and Henry 1983). In river systems like the Elbe, SPM is associated with bacteria and extracellular polymers forming microbial biofilms. The binding of trace metals by microbial biofilms is well-known (Brown and Lester 1979) and the high affinity for heavy metal ions such as copper has been described in the recent literature (Geesey et al. 1988; Mittelman and Geesey 1985).

The ability of bacteria to methylate inorganic mercury and to demethylate methylmercury has been described in detail (Robinson and Tuovinen 1984; Jeffries 1982 for example). Although the importance of SPM with regard to the transport properties of pollutants such as mercury in fluvial systems is commonly accepted, information about chemical and/or biological transformations during the processes of sedimentation and transportation through the water body is very limited (Ebinghaus 1991; Greiser 1988). Consequently, the consideration of "biologically active" SPM constituents such as micro-organisms and microbial biofilms is, therefore, of special interest.

9.2 Results

9.2.1 Microbial Influences on the Mercury Distribution in Sediment Samples

A sediment may be regarded as the final destination of SPM in a riverine system. Therefore, parameters which are important for the vertical transport of SPM and attached pollutants through the water body, should also be of importance for the mercury distribution in fresh sediment samples.

Assuming that the association of bacteria and their EPS to particulate matter is of special importance for the uptake of mercury and its species by SPM, depth profiles of natural sediment cores were investigated for their contents of mercury, carbohydrates, protein and LARH. The cores were collected with pre-cleaned polyacrylic cylinders downstream of the Hamburg harbour in the Elbe estuary. They were stored at 4°C and cut into 15 sections, each 2 cm in height. Several parameters were measured in each sediment section in parallel to assess the influence of bacteria and EPS on the mercury uptake by SPM flocs.

The protein content was measured according to Bradford (1967). This method allows the determination of living and dead biomass as well (Christensen and Packard 1977). Therefore, the protein content of sediment fractions can be used as a parameter for the microbial population in the matrix.

In addition to water which makes up approximately 95% of the wet weight of EPS, bacterial EPS consist mainly of polysaccharides (Hintelmann et al. 1993). The phenol-sulfuric acid method according to Liu et al. (1973) is a very fast and

Fig. 9.1. Depth profile of a natural sediment core. Mercury-, protein-, carbohydrate content and loss at red heat (LARH) were determined in parallel. The values for protein content and LARH were multiplied by a factor of 10 and the mercury content by a factor of 50 to obtain comparable bar heights in the diagram

reliable method for the determination of carbohydrates, especially hexoses. This method can be applied to quantify EPS or microbial biofilms (Liu et al. 1992). The LARH is commonly used to determine the quantity of organic matter of sediment samples or SPM material.

The mercury content was measured with Cold Vapour Atomic Absorption Spectroscopy (CVAAS) in all fractions. The results for the depth profile of a natural sediment core are shown in Fig. 9.1.

The mercury contents in the sediment depth fractions show distinct variations. Background levels could only be detected in the upper fractions which were in contact with the water column. In the deeper fractions the mercury contents were elevated: however, the highest values with 1.3 mg Hg kg^{-1} dr. wt. are much lower than results reported upstream of Hamburg (Wilken et al. 1990). Protein concentrations between 2 and 7 mg per g dry sediment were measured in the core fractions. Again, low protein contents have been found in the upper fractions. With the phenol-sulphuric acid method, clearly varying carbohydrate contents of between 15 and 55 mg g^{-1} dry sediment have been determined, where the lowest concentrations are again in the sediment–water interface fraction. The standard deviation of this method was approximately 8%.

The LARH is between 0.7 and 11% which is very low compared to measurements upstream of Hamburg (Wilken et al. 1991).

Looking for correlations between different analytical parameters, "optical evaluation" of the graphs can be helpful. To make sure that there is or is not a correlation, statistical treatment of the data sets is necessary (Doerffel 1990). F-tests were

Table 9.1. Correlation matrix for a natural sediment core involving mercury-, carbohydrate-, protein contents, depth of the fractions and loss at red heat (LARH)

	Carbohydrate	Protein	Mercury	LARH	Depth
Carbohydrate	1				
Protein	0.814	1			
Mercury	0.929	0.739	1		
LARH	0.675	0.479	0.824	1	
Depth	0.535	0.287	0.651	0.719	1

carried out to examine if differences between the standard deviations in the analysed cores were significant. When this was not the case, a parameter test (t-test) was used to verify if a comparison of the means was statistically valid (Doerffel 1990).

The correlation analysis was carried out with a Macintosh IIsi computer using SYSTAT®. The correlation matrix for the measured parameters is shown in Table 9.1.

According to Doerffel, a statistically significant correlation with a low error probability ($P = 0.99$) is given for data sets with 15 readings per parameter when the correlation coefficient is r>0.64 (Doerffel 1990). Table 9.1 shows that there are positive correlations between mercury-, carbohydrate-, protein contents and loss at red heat. There is no correlation between the depth of the fractions and the measured parameters. The most distinct correlation is given in the depth profiles for mercury and carbohydrates. Probably bacterial exudates are more important for the mercury accumulation by particulate matter than the microbes themselves. It is obvious that bacteria and EPS (and the whole organic matrix of these sediments) have an influence on the mercury distribution in the investigated sediment fractions. However, the correlation calculations cannot show if other sediment parameters (grain size for example) are more important. One possible explanation may be that the grain size is responsible for the microbial colonization and that both parameters have synergistic influence on the mercury distribution. Small particles with low densities have higher mercury contents because they are covered more effectively with microbial biofilms.

In order to check this, an artificial sediment core was prepared from another one which was also collected downstream of the Hamburg harbour at the same time. The natural core was resuspended with river water and filled into a precleaned polyacrylic cylinder. At 4°C a fairly slow sedimentation process took place. After completion of sedimentation the new core was divided into 12 sections, each 2 cm in height. The results for the determinations of mercury-, carbohydrate- and protein contents and LARH are shown in Fig. 9.2.

Again the values for protein content and LARH were multiplied by a factor of 10. The mercury content was multiplied by a factor of 50 to obtain comparable bar heights in the diagram.

Fig. 9.2. Depth profile of an artificial sediment core. Mercury-, protein-, carbohydrate content and loss at red heat (LARH) were determined in parallel. The values for protein content and LARH were multiplied by a factor of 10 and the mercury content by a factor of 100 to obtain comparable bar heights in the diagram

Figure 9.2 shows higher values for all measured parameters in the upper and lighter (slow sinking) fractions. The range of variation for all of these parameters is comparable with the original sediment cores. When the core was suspended in river water, protein could be found in the water phase, whereas no carbohydrates could be found in the river water. This means that bacteria were released from sediment particles when the core was resuspended and that carbohydrates or EPS are obviously more tightly bound to the surfaces.

Again correlation analysis was carried out to evaluate connections between the measured parameters. The correlation matrix for the artificial core is shown in Table 9.2.

Table 9.2. Correlation matrix for an artificial sediment core involving mercury-, carbohydrate-, protein contents, depth of the fractions and loss at red heat (LARH)

	Carbohydrate	Protein	Mercury	LARH	Depth
Carbohydrate	1				
Protein	0.71	1			
Mercury	0.826	0.65	1		
LARH	0.807	0.653	0.918	1	
Depth	-0.755	-0.585	-0.95	-0.94	1

For data sets with 12 readings per parameter a statistically significant correlation with a low error probability ($P = 0.99$) is given according to Doerffel when the correlation coefficient r> 0.71 (Doerffel 1990). Table 9.2 shows that mercury is correlated with carbohydrate and especially with the LARH in the artificial sediment core. The correlation coefficient (r) for mercury and protein is equal to 0.65. In the artificial core these parameters are not correlated. During the process of resuspension, bacteria seemed to be released from sediment particles as previously mentioned.

Mercury contents and losses at red heat in the core fractions are strongly correlated in a negative sense with the depth of the artificial profile. The lighter fractions have higher mercury contents and also higher amounts of the total organic matter. Protein, as one of the organic constituents, is not correlated with Hg or with the depth of the core. Carbohydrate, one of the other possible "constituents" of LARH shows a statistically significant positive correlation with Hg and a negative correlation with the depth.

The correlation analysis of the preceding data sets implies that particle density, total organic matter and EPS are important parameters for the adsorption properties of mercury to particles. With these data sets the significance of bacteria in these process could not be assessed, probably due to the release of microbes into the water phase during resuspension with river water.

The results indicate that surface properties and densities of inorganic suspended particles from the river Elbe can be decisive factors in microbial colonization. With regard to heavy metal accumulation the attachment of bacteria and their exudates to inorganic surfaces itself is probably more important than surface properties or grain sizes.

9.2.2 Mercury Uptake by Bacteria and Microbial Biofilms

Mercury Uptake by Bacteria. To assess the role of bacteria for the mercury accumulation in suspended material a suitable method for the isolation of SPM bacteria was needed. The preparation of bacterial cultures from suspended matter can be described as follows.

SPM was collected from a groyne field in the limnic part of the river Elbe, 50 km upstream of Hamburg, northern Germany. In these groyne fields (with low flow rates) high concentrations of total as well as methylmercury have been detected (Wilken et al. 1990).

SPM-bacteria were isolated according to a method described by Greiser (1988). Fractionated centrifugation was carried out to separate SPM from minerals and free bacteria cells. At 300 to 400 g SPM flocs could be harvested selectively, whereas microflocs up to 5 μm diameter, small algae and free bacteria cells occur at 1000 to 4000 g in the sediment of the centrifuge tubes. The mixed bacterial cultures were obtained according to a slightly modified method described by Liu and Thomson (1986). Centrifuged SPM was inoculated into 150 ml of growth

medium in 250-ml culture flasks with Kapsenberg caps on a rotatory shaker at room temperature.

When the stationary phase was reached, an innoculum of 1 ml was transferred twice into fresh growth medium to ensure young cultures with comparable metabolic activities. Bacterial growth was followed by optical density (O.D.$_{650nm}$) and simultaneously by determination of protein content.

Mercuric chloride and methylmercuric chloride were added to the enrichment cultures after bacterial growth had started, i.e. when O.D.$_{650nm}$ was greater than 0.1 [Extinction]. The concentration of each Hg-species was 350 ng ml^{-1}.

To measure the uptake of mercury by bacteria, the cells were isolated by centrifugation at 3500 g for 15 min. The supernatant was decanted and stored at 4°C. The bacterial pellet was treated with 1 ml of 20% tetramethylammoniumhydroxide solution (TMAH). TMAH is a very effective, non-oxidizing tissue-solubilizer and is suitable for the fast and complete digestion of bacteria (Ebinghaus et al. 1991). After digestion of the cells, 200 μl of a 0.05 M brominechloride were added in order to convert organic mercury compounds to Hg(II). Before CVAAS-measurements were carried out, àn aliquot of this solution was pre-reduced by the addition of 200 μl of a 20% hydroxylamino hydrochloride (NH$_2$OH·HCl). The results for the mercury accumulation by isolated SPM-bacteria is shown in Fig. 9.3.

During the 14-day incubation period both Hg(II)- and methylmercury ions were accumulated in comparable amounts by SPM-bacteria. Between 1 and 2 mg total mercury could be detected per g biomass (dr. wt.). A slight decrease of mercury in the biomass occurred for both species. This is probably due to the formation of volatile mercury species such as elemental mercury or methylmercury. River Elbe

Fig. 9.3. Uptake of mercuric and methylmercuric ions by SPM bacteria. Initial concentration of Hg: 350 ng ml^{-1}; incubation time: 14 days

bacteria have the ability to reduce Hg(II) to Hg(0) in very short time periods. This was shown in previous experiments in which $HgCl_2$ was added to bacterial cultures (Wilken et al. 1990).

Mercury Uptake by Microbial Biofilms. In the past few years microbial biofilms have increasingly attracted the attention of scientists and environmental engineers. Even newspapers and magazines reported problems as a result of the formation of microbial biofilms in water pipes and heat exchangers in power stations (DIE WELT 1988; DIE ZEIT 1988; GIT 1991). As an example, the "biocorrosion" of piping for drinking and waste water should be mentioned. The formation of acidic polymers is supposed to be an important factor for bacterial survival in natural environments. When attachment of bacteria to surfaces is mediated by these polymers, trace metal ions can be removed from the mineral surface by proton exchange and serve as nutrients for the microbes (Liu 1991).

In industrial water ducts the presence of acidic polymers leads to less effective water flow and damage in the pipe-lines. Thus, the prevention of microbial biofilms is of economic importance and consequently has been intensively discussed in recent literature (Flemming and Schaule 1988; Hejzlar and Chudoba 1986; Chudoba et al. 1986).

However, the usefulness of microbial biofilms has also been investigated (Ferris et al. 1989; Beveridge and Fyfe 1985; Beveridge et al. 1983; Sutherland 1972). Hsieh et al. (1985) presented a fixed-bed adsorption unit for the removal of heavy metal ions and other pollutants from waste waters .

The ability of extracellular polymers to bind heavy metal ions is well documented. However, due to fundamental problems with the basic unit for EPS and microbial biofilms a quantification of the accumulation has not been achieved (Hsieh et al. 1985; Rudd et al. 1984; Corpe 1975). Apart from water, carbohydrates are the main constituents of EPS (Hintelmann et al. 1993; Sutherland 1972). Following the use of the protein content as the basic parameter for bacterial biomass (Bradford 1967), Liu suggested to use the carbohydrate content as the basic parameter for extracellular polymers (Liu 1991). His phenol-sulphuric acid method has been successfully applied for the determination of hexoses and pentoses in sediment samples (Liu et al. 1973). Recent results indicate that this method is also applicable for the quantification of extracellular polymers or microbial biofilms (Liu et al. 1992).

Using the phenol-sulphuric acid method, the formation of a microbial biofilm on pre-cleaned glass slides can be followed photometrically at 485 nm. Five millilitres of a fresh bacterial culture were inoculated into 3.5 l of growth medium in 4- l crystallizing dishes for these experiments. The nutrient medium contained KH_2PO_4 (0.16 mg l^{-1}), $MgSO_4 \cdot 7H_2O$ (0.05 mg l^{-1}), $FeSO_4$ (0.01 mg l^{-1}), NH_4Cl (0.1 mg l^{-1}) and $NaOCOCH_3$ (1 mg l^{-1}) as described by Hintelmann et al. (1993).

The cultures were gently mixed on a multiple point stirrer with two PTFE coated stirring bars each. Double frosted glass slides (10 cm^2 surface area) were dipped into the culture. The slides were fixed with polyethylene clamping fixtures

Fig. 9.4. Growth curve of a microbial biofilm on a 10 cm² glass surface given by the carbohydrate content

and were totally covered by the growth medium. Five slides were removed from the culture at the time and were carefully drained off for each determination. A typical growth curve of a microbial biofilm on glass slides given by the carbohydrate content is shown in Fig. 9.4.

Besides the carbohydrate determination on the glass slides, the formation of biofilms was also followed using their wet and dry weight. This basic unit allows the comparison of the accumulation of pollutants in microbial biofilms with that in micro-organisms or higher organisms.

Approximately 40 mg microbial biofilms (wet weight) could be measured on one double-frosted slide with a surface area of 10 cm². The determination of dry weight was carried out for five parallel slides, because water is the main constituent of this matrix. The results are shown in Table 9.3.

Table 9.3. Dry weight determination of microbial biofilms on a surface area of 5 x 10 cm²

No. of Glass Slides	Wet weight (mg)	Dry weight (mg)	Water content %
5	236	7	97
5	202	10	95
5	256	5	98
5	198	<1	>99
5	242	17	93

About $97 \pm 3\%$ of a microbial biofilm consists of water. Hence, it follows that the determination of dry weight is accompanied by inaccuracies when the amount of material, e.g. the surface area of the slides, is too small. A "basic parameter", for example selected carbohydrates or uronic acids which are easily measured and directly convertible into the dry weight of microbial biofilms, should be found and examined. A good example for this strategy is the dry weight determination of bacteria using the protein content as the "basic parameter". The protein determination according to Bradford (1967) is a very simple and reliable method and the dry weight of bacteria can be calculated with sufficient precision (ca. 50% of the dry weight is protein).

To assess the accumulation of different mercury species in microbial biofilms, the following experiment was carried out. Mercuric chloride and methylmercuric chloride were added to the nutrient medium prior to inoculation with bacteria. Mercury contents in the biofilms were determined as total mercury with CVAAS after dissolving the organic matrix with 1 N hydrochloric acid. The uptake of mercuric chloride and methylmercuric chloride by microbial biofilms for an incubation period of 14 days is shown in Fig. 9.5.

Mercury(II) as well as methylmercuric ions were accumulated by microbial biofilms. When the concentration of mercury compounds in the bacterial culture was 350 ng ml^{-1} approximately 1 mg Hg was accumulated by 1 g of microbial biofilm (dry weight).

According to Doerffel (1990), differences in means are proven if

$$(x_1 + 1.96\ \sigma_1) < (x_2 + 1.96\ \sigma_2) \tag{1}$$

where x_1 and σ_1 represent the mean and standard deviation of Hg(II) uptake, whereas x_2 and σ_2 are the mean and standard deviation of methylmercury uptake. Figure 9.5 shows that there was no significant difference in the accumulation of inorganic mercury(II) and methylmercury by microbial biofilms. Although a slightly higher uptake of methylmercury can be assumed according to the diagram,

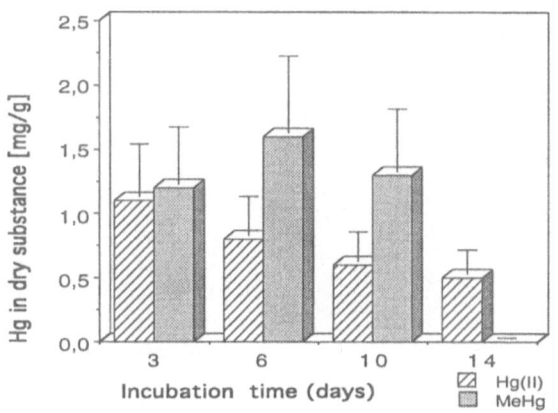

Fig. 9.5. Uptake of mercuric and methylmercuric ions by microbial biofilms. Initial concentration of Hg: 350 ng ml^{-1}; incubation time: 14 days

the standard deviation of 37% (mainly caused by the dry weight determination) was too high to make a definite valuation. The Hg(II) content of microbial biofilms on the glass slides seemed to decrease over 14 days, possibly due to the formation of volatile elemental mercury.

Accumulation of Mercury by Bacteria and Microbial Biofilms:

A Comparison. To compare the accumulation capacities of bacteria and microbial biofilms these results just previously mentioned were combined and are depicted in Figs. 9.6 and 9.7. Figure 9.6 shows the uptake of inorganic Hg(II) and Fig. 9.7 the uptake of CH_3Hg^+ by bacteria and microbial biofilms.

Figure 9.6 shows that the accumulation capacity of bacteria for inorganic mercury is higher than the one of microbial biofilms. Again the significance of differences has been examined by statistical means using Eq. 1 (Doerffel 1990). Inorganic mercury was enriched significantly higher by bacteria after 6, 10 and 14

Fig. 9.6. Uptake of mercury/II) ions ions by SPM bacteria and microbial biofilms. Initial concentration of Hg: 350 ng ml⁻¹; incubation time: 14 days

Fig. 9.7. Uptake of methyl-mercury ions by SPM bacteria and microbial biofilms. Initial concentration of Hg: 350 ng ml⁻¹; incubation time: 14 days

days with an error probability of $P = 0.05$. After 3 days there was a symptomatic difference with $P = 0.10$ and 1.64σ.

The result for the accumulation of methylmercuric ions, shown in Fig. 9.7, looks fairly similar. However, due to the fundamental problems with dry weight determination of microbial biofilms previously mentioned, a significantly higher uptake of CH_3Hg^+ was only statistically valid for an incubation period of 3 days. After 10 days of incubation the difference is symptomatic, whereas no statistical difference could be found after 6 days. No biofilms could be detected on the glass slides after 14 days of incubation.

A better dry-weight determination of microbial biofilms should be possible if glass slides or other materials with larger surface areas would be used. Under those conditions, significant differences in the accumulation capacities of bacteria and microbial biofilms should be detectable and statistically valid.

With the available results, a two- to three-times higher accumulation of both of the applied mercury species by bacteria may be estimated.

9.2.3 Mercury Speciation in Environmental Samples

In order to investigate the metabolism, transport and toxicity of pollutants in ecosystems it is important to determine their exact physico-chemical forms i.e. their speciation. The terms 'speciation' or 'species-analysis' are sometimes used in different ways and with different meanings. Therefore, IUPAC definition should be cited. Speciation is: "...the process yielding evidence of the atomic or molecular form of an analyte" (Broekaert et al. 1990).

Mercury forms covalent bonds with carbon and these compounds show a variety of different physical properties. For example CH_3HgCH_3 is volatile (b.p. $96°C$); CH_3Hg^+ as a salt is volatile with water vapour. Due to their toxicity these compounds are of great importance in environmental research. Over 50 other organic compounds of mercury have been synthesized by man as seed dressings and for other agricultural applications (Wilken 1992). However, just a fraction of the total mercury in environmental samples is usually of an organic form (Hintelmann and Wilken 1992). Therefore, an effective and non-destructive isolation of the organic species from the matrix is required.

Wilken, Hintelmann, Chau and Wong give a good overview on analytical techniques for the species analysis of mercury compounds (Wilken 1992; Hintelmann and Wilken 1993; Chau and Wong 1991). The most widely used technique for the determination of individual mercury species involves gas chromatography (GC) with electron capture detection (ECD; Westöö 1966; Cappon 1987), emission spectrometry (MIP; Greenway and Barnett 1989) or atomic absorption (AAS; Gui-Bin et al. 1989). However, without derivatization of the organomercurials the chromatographic behaviour of monoalkylmercurials can cause problems (Wilken and Hintelmann 1990). Therefore, the derivatization techniques have been developed to convert monoalkylmercury- to dialkylmercury compounds (Bloom 1989;

Bulska et al. 1991; Rapsomanikis and Craig 1991; Wilken 1992).

Liquid chromatography (LC) allows a more simple sample treatment and allows the separation of a great variety of organic mercury compounds. We developed an HPLC (high pressure liquid chromatography) procedure, which allows the detection of separated mercury compounds with the commonly used UV detector (Wilken et al. 1990). Although the extinction of organomercurials are below 210 nm (wavelengths maxima, at which the elutant also shows absorbance), the detection was also successful at 230 nm (Wilken and Hintelmann 1990). Results reported by Hempel et al. are showing that nine different organomercury species can be analysed with HPLC/UV if gradient elution is applied (Hempel et al. 1992).

Recently the development of new "hyphenated systems" with element specific detection has been reported. The coupling of LC to detectors such as:

- inductively coupled plasma atomic emission spectrometry (ICP-AES; Krull et al. 1986),
- atomic absorption spectrometry (AAS; Holak 1989),
- electrochemical devices (Evans and McKee 1988)
- and the very sensitive atomic fluorescence spectrometry (AFS; Hintelmann and Wilken 1993)

combines the advantages of LC for the separation of organomercurials with highly sensitive and element specific detection.

Biomethylation of Mercury. For the methylation of heavy metal cations, microorganisms require methyldonors. Methylcobalamine (MeCo), methyliodide (MeI) and S-adenosylmethionine (SAM) are the most important methylating co-enzymes (Vogt 1980) and were used as "biogenic methyldonors". MeCo was chosen as an example for a CH_3^--donor (Ridley et al. 1977). SAM is a CH_3^+-donor and methylates exclusively in the presence of enzymes (Ridley et al. 1977). A methyl group can be transferred from MeI to a metal cation non-enzymatically in the course of a redox reaction (Thayer 1989).

Methylation experiments were carried out with 1.75 μmol l^{-1} Hg(II) and 17.5 μmol l^{-1} of methyldonors: MeCo, MeI and SAM, respectively were added to 150 ml of the enrichment cultures after bacterial growth had started.

In order to follow the formation of methylmercury during the incubation period bacteria cells were digested with tetramethylammoniumhydroxide (TMAH) as described in the preceding. After the digestion was completed the extracts were acidified with hydrochloric acid to obtain CH_3HgCl which is relatively non-polar and has a high solubility in organic solvents. Methylmercury was extracted from the aqueous phase with toluene and subsequently back-extracted with sodiumthiosulphate solution (0.1 mmol l^{-1}). This final solution is miscible with the elutant in the HPLC and can be injected directly. The extraction procedure and the chromatographic conditions are described in detail by Wilken and Hintelmann (1990, 1991).

Fig. 9.8. Methylation of mercuric ions in sterile tap water with biogenic methyldonors S-adenosylmethionine (SAM), methyliodide (MeI) and methylcobalamine (MeCo)

All experiments were carried out in triplicates. Sterile tap water was treated in the same manner and used as a control. The results for the methylation experiments in sterile tap water are shown in Fig. 9.8.

In sterile tap water both MeCo and MeI can methylate Hg(II) ions. Within 1 week the CH_3^--donor MeCo methylated 10% of the mercury ions present (i.e. 0.18 μmol). In the presence of methyliodide 4% of the Hg(II) ions were transferred into CH_3Hg^+. No methylation was observed if SAM was used as the CH_3^--donor. A turnover of 0.5% Hg(II) ions into CH_3Hg^+ would have been detectable in these experiments.

In the presence of bacteria (enrichment cultures) no methylmercury could be detected in any experiment. Therefore, less than 0.5% of the present Hg(II) ions (i.e. 8.7 nmol l^{-1}) were methylated by biogenic methyldonors.

Transmethylation of Mercury. Transmethylation can play an important role as "abiotic methylation" for mercury (II) ions in aqueous environments (Chau 1986). Jewett et al. showed that both trimethylleadchloride and trimethyltinchloride are able to transfer methyl groups to Hg^{2+} (1975). Chau et al. (1987) found that methyl derivatives of lead(IV) compounds can transfer their methyl groups to tin(II) and tin(IV) salts. However, methylmercury does not donate its methyl group to either Sn(II) or Pb(II) salts. Methylarsenic acids can also methylate Sn(II) and Sn(IV) salts in aqueous solutions but the transfer of methyl groups to Pb(II) was not detected (Chau et al. 1987).

In addition to the high concentrations of inorganic and organic mercury species, the Elbe river is highly contaminated with organotin compounds and probably other organometallic species (Wilken et al. 1990; Wilken and Jantzen 1991; Jantzen 1992). Consequently, transmethylation reactions in sediments and in the water body may be responsible for the high methylmercury concentrations

Fig. 9.9. Transmethylation of mercuric ions in sterile water with organometallic compounds; trimethyllead chloride (Pb org.), trimethyltin chloride (Sn org.) and dimethylarsenic acid (As org.)

in Elbe sediments.

In the experiments described in the following, trimethylleadchloride, trimethyltinchloride and dimethylarsenic acid were examined for their methylation potential. Inorganic mercury (Hg(II): 1.75 μmol l^{-1} and the methyldonor (17.5 μmol l-l) were added to bacterial cultures and to sterile tap water as described in Section 9.2.3. All experiments were carried out in triplicates. The results for the formation of methylmercury in sterile water is depicted in Fig. 9.9.

Fig. 9.10. Transmethylation of mercuric ions in batch cultures with organometallic compounds: trimethyllead chloride (Pb org.), trimethyltin chloride (Sn org.) and dimethylarsenic acid (As org.)

In sterile tap water trimethylleadchloride is the most effective methyldonor for Hg(II) ions. The formation of CH_3Hg^+ was very fast: after 3 h a turnover of 40% could be detected. After 1 week, 100% of the Hg(II) ions present were methylated. In the presence of 45% trimethyltinchloride approximately 0.8 μmol l^{-1} and of 37% dimethylarsenic acid approximately 0.6 μmol l^{-1} of methylmercury could be detected in sterile water.

The results for the methylation experiments in the presence of SPM-bacteria are shown in Fig. 9.10.

In the presence of bacteria, the turnover of Hg^{2+} into CH_3Hg^+ is significantly lower than in sterile water, when trimethylleadchloride and trimethyltinchloride were used as methyldonors. After a 7-day incubation less than 10% of the Hg(II) ions present were methylated. The methylation activity of dimethylarsenic acid is much higher and is not affected by the presence of SPM-bacteria. The turnover of Hg^{2+} into CH_3Hg^+ in sterile water and in enrichment cultures was comparable. After 6 days approximately 33% of the Hg(II) ions were methylated by the organoarsenic compound.

With the exception of dimethylarsenic acid, the methylation activity of the examined methyldonors, biogenic as well as abiotic, was degraded in the presence of SPM-bacteria. At least three explanations for this fact can be put forward:

1. Hg(II) ions are attached to bacterial cell walls and are not available for methylation. Beveridge and Fyfe determined carboxyl groups to be very effective potential binding sites for metal cations using *B. subtililis* as test organisms (Beveridge and Fyfe 1985).

2. The methylating agents themselves are bound to cell walls and are not available for methylation.

3. SPM-bacteria are not only passive inhibitors of the methylation of mercury but, it is also possible that the methylmercury produced is demethylated by microbial activity (Ebinghaus and Wilken 1993).

The transmethylation of Hg^{2+} to CH_3Hg^+ in the presence of dimethylarsenic acid seems to reach a maximum at 72 h. This could possibly be explained by an inhibition of the microbial demethylation activity. SPM-bacteria needed an adaption period of 72 h probably due to the toxicity of dimethylarsenic acid. After this period the microbial decomposition of methylmercury began. However, in the presence of presumably more toxic trimethylleadchloride and trimethyltinchloride this phenomenon was not observed.

Demethylation of Methylmercury by Bacteria. In the presence of SPM-bacteria, the formation of methylmercury from inorganic Hg(II) and a suitable methyldonor was obviously hindered. Possible reasons for this are discussed in Section 9.2.3. To determine the demethylation activity of SPM-bacteria isolated from the river Elbe, methylmercuric chloride (1.75 μmol l^{-1}) was added to bacterial cultures (as described in Sect. 9.3.1). Again sterile tap water was used as a control to

Fig. 9.11. Demethylation of methylmercuric ions in batch cultures compared to sterile tap water. Total mercury and methylmercury were determined in bacterial cultures (Hg tot. and MeHg culture) whereas the MeHg control was determined in sterile water

identify external influences such as UV-light for example on the methylmercury decomposition. The results for the demethylation experiments are depicted in Fig. 9.11. In the presence of SPM-bacteria the methylmercury concentration decreased rapidly. After 24 h of incubation, 20% of the methylmercury added to the culture had disappeared. After 1 week, no methylmercury was detectable in the bacterial cultures. However, the total mercury content of the samples was constant over the same period. Therefore, it can be concluded that the decrease of CH_3Hg^+ concentration was not due to sedimentation processes where methylmercury was attached to sinking bacteria cells.

In the control samples the methylmercury was constant over the same 12-day period. No external factors such as UV light were responsible for the CH_3Hg^+ decomposition in the bacterial cultures. In order to determine the percentage of distribution and decomposition of methylmercury between the aqueous phase and the bacterial cells the following experiment was carried out: Methylmercury was added to the enrichment cultures as described in the preceding. Over a period of 12 days, samples were taken and centrifuged at 1500 g for 20 min at 4°C.

Methylmercury and total mercury were determined both in the aqueous phase and in the bacteria cells. The results are shown in Fig. 9.12.

After 24 h of incubation approximately 53% of the added methylmercury ions are attached to bacteria. Figure 9.12 shows that the total mercury concentration in the cells is constant, whereas a slight decrease is detectable in the aqueous phase. The loss of mercury may be due to the formation of volatile mercury species such as Hg(0) and dimethylmercury. Under the conditions examined, elemental mercury seemed to have been the more probable transformation product (Wilken et al. 1990).

Fig. 9.12. Demethylation of methylmercuric ions in bacterial cells and supernatant of batch cultures. Total mercury and methylmercury were determined in biomass (Hg tot. biomass and MeHg biomass) and in the supernatant (Hg tot. water and MeHg water)

From our results it is clear that SPM-bacteria were able to decompose methylmercury under oxic conditions in enrichment cultures. Methylmercury decreased in both compartments – the aqueous phase and the bacteria. However, the total mercury in the bacteria remained constant. That possibly means that methylmercury was decomposed by microbial activity and that it was deposited in the cells as inorganic mercury. The combination of the demethylation of organometallic mercury species followed by immobilization of inorganic Hg(II) is a good example for microbial detoxification.

9.2.4 Species Transformations of the Methylating Agent

In the previous experiments the transfer of a methyl group from biogenic or anthropogenic donor molecules to inorganic mercury ions had been followed by the determination of methylmercury formed in the course of the methylation reaction. With this procedure, the dynamic species transformation of one element, the *methylacceptor*, can be described.

In analogy, the loss of one or more methyl groups also means a transformation of the *methyldonor* species involved. Consequently, the reaction between mercury(II) ions and trimethyltinchloride leads to less toxic tin species but to mercury species with substantial higher toxicity (Thayer 1984).

In order to determine organotin species, a method developed by Maguire and Chau (which was recently improved by Jantzen) was used (Maguire and Huneault 1981; Chau et al. 1982; Jantzen 1992). A gas chromatographic separation of

volatile organotin species is followed by an element-specific atomic absorption detection system. The extraction procedure and chromatographic conditions are described in detail by Wilken and Jantzen (1991).

Sterile, cell-free water samples were treated with 17.5 nmol ml^{-1} mercuric-chloride and trimethyltinchloride in equimolar amounts. For the determination of

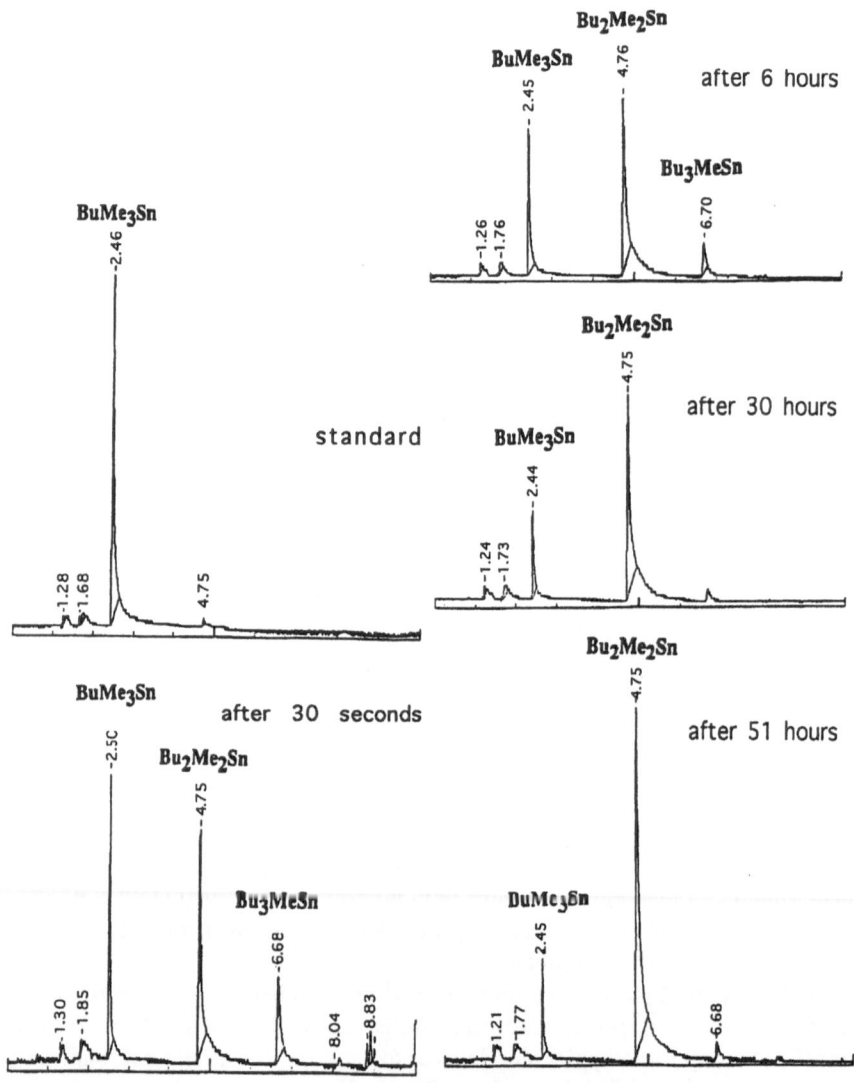

Fig. 9.13. GC-AAS chromatograms of organotin species in the course of the transmethylation reaction from trimethyltin chloride to mercuric ions. Tin compounds are perbutylated prior to GC-AAS analysis

organotin species during the reaction, subsamples were acidified with 10 μl 6N hydrochloric acid and extracted with 0.1% hexane/tropolone mixture. The extract was treated with butylmagnesium chloride in diethylether. As a result of this Grignard reaction, perbutylated tin species were obtained. Due to their volatility, these tetraalkyltin compounds can be separated by gas chromatography and detected by AAS.

The results of this experiment are shown in Fig. 9.13. The chronological sequence of the GC/AAS chromatograms was preferred to the representation of the transformation of tin species in a diagram.

Peaks in the GC/AAS chromatograms represent the following tin compounds:

- Trimethyltin*
- Dimethyltin*
- Monomethyltin*

*each in the perbutylated form.

The first chromatogram shows the extract of a trimethyltinchloride standard solution prior to the addition of mercury (II) ions. One tin compound, Me_3SnBu, was detected by atomic absorption at a retention time of 2.46 min. Thirty seconds after the addition of the methylacceptor Hg^{2+}, the trimethyltin peak was less intensive than in the standard solution. Dimethyltin, Me_2SnBu_2, could be identified at a retention time of 4.75 min. Another organotin species which was detected after 30 sec with a retention time of 6.68 min is monomethyltin, $MeSnBu_3$. The loss of two methylgroups of trimethyltinchloride is somewhat surprising. However, after 6 h of reaction monomethyltin could also be detected. After 30 h and 51 h, respectively, no monomethyltin species could be found in the solution, whereas the increase of the dimethyltin peak and the decrease of the trimethyltin peak continued.

To sum up: the final reaction products of the transmethylation reaction between Hg^{2+} and trimethyltinchloride are CH_3Hg^+ and dimethyltin. During short reaction times monomethyltin occurs as an intermediate which disappears in the further course of the reaction.

The quantitative evaluation of the chromatograms is given in Table 9.4. The standard deviation refers to triplicate determination in two parallel samples.

Table 9.4. Quantification of tin species during transmethylation to mercuric ions

Time of reaction	Trimethyltin [μg/ml]	Dimethyltin [μg/ml]	Monomethyltin [μg/ml]	Σ tin [μg/ml]
0 sec	4.1 ± 0.7	0.0 ± 0.0	0.0 ± 0.0	4.1 ± 0.7
30 sec	1.5 ± 0.3	2.1 ± 0.3	0.4 ± 0.03	4.0 ± 0.4
6 h	1.2 ± 0.2	2.2 ± 0.4	0.2 ± 0.03	3.6 ± 0.4
30 h	0.6 ± 0.1	2.8 ± 0.5	0.0 ± 0.0	3.4 ± 0.5
51 h	0.6 ± 0.1	4.1 ± 0.7	0.0 ± 0.0	4.7 ± 0.7

The decomposition of trimethyltin ions can be followed in detail by the chromatograms and the quantitative evaluation in Table 9.4. The demethylation to dimethyltin with monomethyltin is obvious. Total tin concentrations (Σ Sn) in the samples show that the applied method was fairly accurate and reliable. Differences in Σ Sn can probably be explained with varying losses in the course of the Grignard reaction and hydrolysis of the reaction mixture.

9.3 Conclusions

Sediment samples, especially depth profiles, can be used for the time-integrated documentation of environmental pollution. Contrary to that, water samples or samples collected from ambient air are only valid for the actual time of sampling. Mixing and diffusion processes in these matrices destroy their "memory".

However, not only external factors such as accidents or spillage are responsible for increased levels of different pollutants in sediment samples. Characteristics of the sediment facies itself can be important for the distribution of pollutants in depth profiles. Besides the physico-chemical factor "grain size", the microbial population and the production and excretion of microbial polymers are important factors in the concentration of mercury in sediment fractions. In natural sediment cores, a significant correlation of the vertical concentration profiles of mercury, protein, carbohydrate and loss at red heat could be shown. Obviously, the (living) biomass and/or their polymeric exudates have an influence on the mercury binding capacity of sediment fractions. However, it cannot be excluded that other characteristics of the sediment were also responsible for the mercury distribution. It is well known that both the natural background and the heavy metal content of polluted sediments are negatively correlated with the grain size, a characteristic of the sediment facies (Lichtfuss and Brümmer 1979; Förstner 1976). In general, the highest heavy metal concentrations can be detected in the fine fractions of sediments (Förstner and Wittmann 1979).

The results for artificial, density-dependent sedimented cores also show a significant negative correlation of mercury concentrations with the depth of the core. Mercury is more effectively enriched in the upper and lighter fractions with lower sedimentation velocities. The physico-chemical causes for different interactions between pollutants and sediment particles are determined by various surface properties. The results from natural sediment cores cannot prove if the mercury distribution was determined by the sediment facies, by microbial parameters or by a combination of both of them. However, looking at the results for the artificial cores it is thinkable that the sinking-velocity of sediment particles is linked to their microbial population and their mercury contents. The fractions which sank more slowly showed the highest protein-, carbohydrate- and mercury contents as well as the highest values for LARH. This implies that "inorganic particle

surfaces" are covered with bacteria and their exudates. This has been proven for suspended material by microscopic methods (Wilken et al. 1991). The contact between the (pore) water phase and particles is, therefore, mediated by "organic surfaces" which act as the potential adsorber material. Thus, it follows that the sediment composition was important for the microbial population which (itself) was responsible for the heavy metal content. It should be mentioned that these observations are the results for fresh samples of river Elbe sediments. It is very likely that the surface of inorganic particles itself is the effective adsorber for other river systems and probably for older Elbe sediment samples.

Suspended material and surface sediments are in permanent interaction. The sedimentation of SPM can be followed by resuspension of these particles from sediment material. Therefore, pollutants which had been "immobilized" by the sediment acting as a sink can enter the water phase again and interact with bacteria and other components of the water body. The adsorption to biological material as the initial step can be followed by bioaccumulation and biotransformation of the pollutant.

A comparison of the uptake of Hg^{2+} and CH_3Hg^+ by SPM bacteria and microbial biofilms shows that both species were more effectively accumulated by the microbes than by polymeric exudates. A 2–2.5-fold higher uptake could be determined when the dry weight was used as the basic parameter. The dry weight determination of bacteria is fairly simple and reliable, whereas this method can cause problems for microbial biofilms. The protein content can be determined for bacteria, and this is directly correlated with dry weight. The reliability and comparability of the quantification of the uptake of pollutants by microbial biofilms would be greater if a parameter were found which could be linked to the dry weight.

Bacteria and their EPS have a high capacity for different mercury species. In addition, bacteria are able to convert ionic mercury into volatile elemental mercury, into organic species as methylmercury or to decompose the mercury-carbon bond of organomercury compounds. With regard to transformation reactions mediated by bacteria, a high accumulation capacity may facilitate a high capability for microbial species transformations.

In the presence of biogenic methyldonors, only methylcobalamin, as a carbanion donor, and methyliodide transfer CH_3-groups to Hg^{2+} ions. However, the formation of monomethylmercury can only be observed in the absence of bacteria isolated from SPM. S-Adenosylmethionine does not methylate inorganic mercury.

Besides biogenic methyldonors, organometal compounds can transfer methyl-groups to mercury in the course of a transmethylation. Significant turnovers of Hg^{2+} to CH_3Hg^+ were observed in the presence of trimethyllead chloride, trimethyltin chloride and dimethylarsenic acid. The most effective methyldonor is $(CH_3)_3Pb^+$. This observation was also reported by Chau who found that both trimethyllead- and trimethyltin ions can methylate Hg^{2+}. However, the organolead compound can methylate $Sn(IV)$; the transmethylation from organotin to lead(IV) could not be detected (Chau et al. 1987).

SPM-bacteria are able to decompose methylmercury under oxic conditions in batch cultures. The cleavage of the mercury-carbon bond can be detected in the biomass and in the water phase. However, the total mercury in the bacterial cells remains constant. Therefore, the combination of demethylation of organometallic mercury species followed by immobilization of inorganic mercury can be regarded as a good example for microbial detoxification.

A fluvial system, especially a tidal one, is a highly complex system which cannot be treated experimentally as a whole. Therefore, simplifying laboratory experiments can be necessary and helpful in the investigation of single processes in selected areas or compartments of the ecosystem. The conception of the experiment should include environmental conditions as far as possible and should exclude similar disturbances to those which occur in field experiments.

Because of their importance for mercury transport and transformation in fluvial systems, bacteria isolated from the riverine system Elbe were selected as the compartment of interest. By cultivating isolated bacteria in nutrient media, the resulting population is certainly not identical to the original one. High concentrations of both nutrients and pollutants do not usually represent natural conditions. Although the applied mercury concentrations were significantly lower than necessary to inhibit bacterial growth in the enrichment cultures, it should be taken into consideration that mercury resistant strains are probably predominant in the laboratory experiments. Therefore, it is certainly not permissible to make quantitative assessments of mercury transformations in the ecosystem Elbe on the basis of batch cultures. Nevertheless, it seems clear that an active microbial methylation of inorganic mercury can be excluded under aerobic conditions in the water phase. In all experiments where the formation of CH_3Hg^+ was observed, the turnover decreased drastically in the presence of SPM bacteria. The remarkable reaction with dimethylarsenic acid as the methyldonor has been discussed above. None of the experiments gave any indication of the microbial formation of methylmercury, whereas the decomposition of that species is carried out rapidly in batch-cultures.

It therefore seems likely that bacteria, attached to suspended material of the river Elbe, are not responsible for elevated methylmercury concentrations in Elbe sediments. In spite of the high affinity of microbes for ionic mercury species the de-methylation is the preferred transformation reaction mediated by bacteria in the water phase. The microbial formation of methylmercury during the process of sedimentation of SPM is consequently unlikely.

The transmethylation of ionic mercury with organometals like trimethyltin- or trimethyllead compounds seemed to be very effective with respect to the formation of CH_3Hg^+. Besides the methylation of mercury by sulphate-reducing bacteria (SRB) or methanogenic bacteria, the transmethylation may play an important role for the genesis of monomethylmercury. Especially for areas where high concentrations of organotin- and organolead species can be detected, these completely "abiotic" reactions can be the decisive factor in the genesis of methylmercury. The knowledge about organotin- and especially organolead compounds in Elbe sediments is still very limited. However, there are distinct in-

Fig. 9.14. Mercury pathways in a polluted fluvial system

dications of "hot spots" of organotin pollution in the Mulde mouth and elsewhere (see Chapter 10, Kuballa et al.).

Figure 9.14 shows transport and transformation pathways of mercury and its species in surface waters.

More specifically, Fig. 9.14 shows the transition of mercury and its species between the compartments: atmosphere, water and sediment. The transition processes are very often accompanied by species transformations. The very effective binding of mercury to suspended particulate matter is of great importance for the transport properties of mercury in a riverine system. The sediment can act as a sink by immobilizing Hg(II) as HgS and $(CH_3Hg)_2S$ for example. The latter compound is supposed to be a possible intermediate in the formation of dimethylmercury. In the course of a re-arrangement involving mercury-carbon bonds, the following reaction can be observed (Thayer 1988):

In the upper anaerobic layers of the sediment, ionic mercury can be methylated. Both the biomethylation by sulphate-reducing and methanogenic bacteria and the abiotic methylation by organometallic compounds are possible methylation pathways. When surficial sediment layers are resuspended, monomethylmercury can be decomposed to ionic mercury.

Dissolved ionic mercury species can be reduced by microbial activity and released from the water phase. Once in the atmosphere elemental mercury can be oxidized again and adsorb to aerosols or especially to soot particles. This

"particulate mercury" is effectively wet and dry deposited, whereas elemental mercury can be transported over longer distances.

The direct introduction of mercury into surface waters via industrial waste water can be assessed with accuracy and this source for mercury contamination can be eliminated by improving or closing down factories. Knowledge about the atmospheric input of mercury into large river systems and their catchment areas is, however, very limited, even though this may be one of the most important sources of elevated mercury levels in the future.

References

Adriano DC (1986) Trace elements in the terrestrial environments. Springer, Berlin Heidelberg New York 298–328

Banat K, Förstner U, Müller G (1972) Schwermetalle in Sedimenten von Donau, Rhein, Ems, Weser und Elbe im Bereich der Bundesrepublik Deutschland. Naturwissenschaften 59:525–528

Beveridge TJ, Meloche JD, FyfeWS, Murray R G E (1983) Diagenesis of metals chemically complexed to bacteria: laboratory formation of metal phosphates, sulfides, and organic condensates in artificial sediments. Appl Env Microbiol 45:1094–1108

Beveridge TJ, Fyfe WS (1985) Metal fixation by cell walls. Can J Earth Sci 22:1893–1898

Bloom N (1989) Determination of picogram levels of methylmercury by aqueous phase ethylation followed by cryogenic gas chromatography with cold vapour atomic fluorescence detection. Can J Fish Aquat Sci 46:1131–1140

Bradford M (1967) A rapid and sensitive method for the quantification of microgram quantities of protein utilizing the principle of protein-dye binding. Anal Biochem 72:248–254

Broekaert JAC, Güçer S, Adams F (eds) (1990) Metal speciation in the environment. NATO ASI Series Vol G 23 Springer, Berlin Heidelberg New York

Brown M, Lester, JN (1979) Metal removal in activated sludge: the role of bacterial extracellular polymers. Water Res 13:817–837

Bulska E, Baxter DC, Frech W (1991) Capillary column gas chromatography for mercury speciation. Anal Chim Acta 249:545–554

Cappon CJ (1987) Uptake and speciation of mercury and selenium in vegetable crops grown on compost-treated soils. Water Air Soil Poll 34:353–361

Chau YK, Wong P TS, Bengert GA (1982) Determination of methyltin(IV) and tin(IV) species in water by gaschromatography/atomic spectrometry. Anal Chem 54:246–249

Chau YK (1986) Occurrence and speciation of organometallic compounds in freshwater systems. Sci Total Environ 49:305–323

Chau YK, Wong PTS, Mojeski CA, Carty AJ (1987) Transmethylation of metals in aquatic systems. Appl Organomet Chem 1:235–239

Chau YK, Wong PTS (1991) Recent developments in speciation and determination of organometallic compounds in environmental samples. Fresenius J Anal Chem 339:640–645

Christensen JP, Packard TT (1977) Sediment metabolism from the northwest upwelling system. Deep Sea Res 24:331–343

Chudoba J, Hejzlar J, Dolezal M (1986) Microbial polymers in the aquatic environment III: Isolation from river, potable and underground water and analysis. Water Res 20:1223–1227

Corpe WA (1975) Metal-binding properties of surface materials from marine bacteria. Dev Ind Microbiol 16:249–255

Craig PJ (1986) Organometallic compounds in the environment. Longman Harlow, pp 65–104

DeFlaun MF, Mayer LM (1985) Relationship between bacteria and grain surfaces in intertidal sediments. Limnol Oceanogr 28:873–881

DIE WELT (1988) Winzige Schädlinge trotzen selbst einer feindlichen Umwelt. 286: 7 Dezember

DIE ZEIT (1988) Vielseitige Biofilme. 51:16: Dezember

Doerffel K (1990) Statistik in der analytischen Chemie. Verlag für Grundstoffindustrie Leipzig 5. Auflage

Ebinghaus R (1991) Aufnahme und Umwandlung von Quecksilber(II)- und Methylquecksilberchlorid durch Schwebstoffbakterien und mikrobielle Biofilme – Untersuchungen zum Species-Verhalten des Quecksilbers in wäßrigen Systemen. PhD thesis University of Hamburg

Ebinghaus R, Wilken RD, Chau YK, Liu D (1991) Extraction of adenine nucleotides from bacteria. NWRI Contributions 91–53:1–18

Ebinghaus R, Wilken RD (1993) Transformations of mercury species in the presence of Elbe river bacteria. Appl Organomet Chem 7:127–135

Evans O, McKee G D (1988) Determination of mercury(II) and organomercury compounds by reversed-phase liquid chromatography with reductive electrothermal detection. Analyst, 113: 243–246

Fanger HU, Kappenberg J, Männing V (1990) A study on the transport of dissolved and particulate matter through Hamburg harbour. In: Michaelis W (ed) Coastal and estuarine studies. Estuarine water quality management. Monitoring, modelling and research. Springer, Berlin Heidelberg New York

Ferris FG, Schultze S, Witten TC, Fyfe WS, Beveridge TJ (1989) Metal interactions with microbial biofilms in acidic and neutral pH environments. Appl Env Microbiol 55:1249–1255

Flemming HC, Schaule G (1988) Biofouling on membranes – a microbiological approach. Desalination. 70:95–119

Förstner U (1976) Metal concentrations in freshwater sediments – natural background and cultural effects. Proc Int Symp Amsterdam, pp 94–103

Förstner U (1977) Metal concentrations in freshwater sediments – natural backgrounds and cultural effects. In: Golterman HL (ed) Interactions between sediments and fresh water. Pudoc/Junk BV Publ, Wageningen, The Hague, pp 94–103

Förstner U, Wittmann GTW (1979) Metal pollution in the aquatic environment. Springer, Berlin Heidelberg New York, pp 18–20

Geesey GG, Jang L, Jolley JG, Hankins MR, Iwaoka T, Griffiths P R (1988) Binding of metal ions by extracellular polymers of biofilm bacteria. In: The proceedings of the International Conference on Water and Wastewater Microbiology Newport Beach California February 8th–11th, pp 26.1–26.5

Gehr R, Henry JG (1983) Removal of extracellular material. Techniques and pitfalls. Water Res 17:1743–1748

Gilmore CC, Henry EA (1991) Mercury methylation in aquatic systems affected by acid deposition. Environ Poll 71:131–169

GIT (1991) Bakterienfilme als Auslöser der Metall-Korrosion. GIT Fachz Lab 4/91:372

Greenway GM, Barnett NW (1989) Optimisation of an atmospheric pressure microwave induced plasma coupled with capillary gas chromatography for the determination of alkyllead and alkylmercury compounds. J Anal At Spec 4:783–787

Greiser N (1988) Zur Dynamik von Schwebstoffen und ihren biologischen Komponenten in der Elbe bei Hamburg. Hamburger Küstenforschung 45:1–170

Gui-Bin J, Zhe-ming N, Shun-rong W, Heng-bin H (1989) Organic mercury speciation in fish by capillary gas chromatography interfaced with atomic absorption spectrometry. Fresenius Z Anal Chem 334:27–30

Hejzlar J, Chudoba J (1986) Microbial polymers in the aquatic environment II: isolation from biologically non-purified and purified municipal waste water and analysis. Water Res 20:1217–1221

Hempel M, Hintelmann H, Wilken RD (1992) Determination of organic mercury species in soils by high-performance liquid chromatography with ultraviolet detection. Analyst 117:669–672

Hintelmann H, Ebinghaus R, Wilken RD (1993) Accumulation of mercury(II) and methylmercury by microbial biofilms. Water Res 27:237–242

Hintelmann H, Wilken RD (1993) The analysis of organic mercury compounds using liquid chromatography with on-line atomic fluorescence spectrometric detection. Appl Organomet Chem 7:173–180

Holak WJ (1989) Liquid chromatography-atomic absorption spectrometry method for the determination of methyl mercury in sea food. Assoc Off Anal Chem 72:926–930

Hsieh KM, Lion LW, Shuler ML (1985) Bioreactor for the study of defined interactions of toxic metals and biofilms. Appl Env Microbiol 50:1155–1161

Irukayama K, Kondo T, Kai P, Fujiki M (1961) Origin of the causative agent of Minamata disease. I. Organic mercury compound in fish and shellfish from Minamata Bay. Kumamoto Med J 14:157–169

Irukayama K, Fujiki M, Kai P, Kondo T (1962) Origin of the causative agent of Minamata disease. II. Comparison of the mercury compound in shellfish from Minamata Bay with compounds experimentally accumulated in normal shellfish. Kumamoto Med J 15:1–12

Jantzen E (1992) Analytische Untersuchungen zur Beurteilung der Bindungsformen (Speziation) des Zinns in Elbesedimenten unter Zuhilfenahme von GC-AAS, GC-MS, TRFA und GFAAS. PhD thesis University of Hamburg

Jeffries TW (1982) The microbiology of mercury. Ind Microbiol 16: 22–75

Jewett KL, Brinckman FE, Bellama JM (1975) Chemical factors influencing metal alkylation in water. In: Church TM (ed) Marine chemistry in the coastal environment. ACS Symp Series 18:304–318

Koefoed BP (1988) Phytoplankton exudation of organic matter: why do healthy cells do it? Limnol Oceanogr 33:151

Krull IS, Bushee D S, Schleicher RG, Smith SB (1986) Determination of inorganic and organomercury compounds by high-performance liquid chromatography – inductively coupled plasma emission spectrometry with cold vapour generation. Analyst 111:345–349

Lichtfuss R, Brümmer G (1979) Geochemische Ausgangsgehalte von Schwermetallen in Sedimenten schleswig-holsteinischer Fließgewässer. Mitteilgn Dtsch Bodenkundl Gesellsch 29:549–554

Liu D, Wong PTS, Dutka BJ (1973) Determination of carbohydrates in lake sediments by a modified phenol-sulfuric acid method. Water Res 7:741–746

Liu D, Thomson K (1986) Biochemical responses of bacteria after short exposure to alkyltins. Bull Env Cont Toxicol 36:60–66

Liu D (1991) National Water Research Institute, Burlington, Canada, pers comm

Liu D, Lau YL, Chau YK, Pacepavicius GJ (1992) A simple technique for the determination of biofilm formation and growth. NWRI Contributions 92–102:1–24

Maguire J, Huneault H (1981) Determination of butyltin species in water by gas chromatography with flame photometric detection. J Chromatogr 209:458–462

Merian E (1984) Metalle in der Umwelt, Verteilung Analytik und biologische Relevanz. VCh Verlag, Weinheim

Michaelis W, Knauth H-D (eds) (1985) Das Bilanzierungsexperiment 1982 (BILEX 1982) auf der Unterelbe. Experimmentell und theoretische Hilfsmittel, Ergebnisse und ihre Bewertung. GKSS Bericht. GKSS 85/E/3

Michaelis W, Fanger H-U, Müller A (1988) Die Bilanzierungsexperimente 1984 und 1985 auf der oberen Tideelbe. GKSS Bericht. GKSS 88/E/22

Mittelman MW, Geesey GG (1985) Copper binding characteristics of exopolymers from a freshwater sediment bacterium. Appl Env Microbio 49:846–851

Pacyna JM, Münch J (1991) Anthropogenic mercury emissions in Europe. Water Air Soil Poll 56:51–61

Platt RM, Geesey GG, Davis JD, White DC (1985) Isolation and partial chemical analysis of firmly bound exopolysaccharides from adherent cells of a freshwater sediment bacterium. Can J Microbiol 31:675–680

Rapsomanikis S, Craig PJ (1991) Speciation of mercury and methylmercury compounds in aqueous samples by chromatography – atomic absorption spectrometry after ethylation with sodium-tetraethylborate. Anal Chim Acta 248:563–567

Read RR, Costerton JW (1987) Purification and characterization of adhesive exopolysaccharide from pseudomonas putida and pseudomonas fluorescens. Can J Microbiol 33:1080–1090

Ridley WP, Dizikes LJ, Wood JM (1977) Biomethylation of toxic elements in the environment. Science 197:329–332

Robinson JB, Tuovinen O (1984) Mechanisms of microbial resistance and detoxifcation of mercury and organomercury compounds: physiological, biochemical and genetic analysis. Microbiol Rev 48:95–124

Rudd T, Sterrit R M, Lester J N (1984) Complexation of heavy metals by extracellular polymers in the activated sludge process. J Water Poll Control Fed 56:1260–1268

Sutherland IW (1972) Advances in Microbial Physiology. 8: 143–213

Sutherland IW (1981) Polysaccharides in the adhesion of marine and freshwater bacteria. In: Berkeley RCW, Lynch JM, Melling J, Rutter PR, Vincent B (eds) Microbial adhesion to surfaces. Ellis Horwood Ltd, Chichester, pp 329–338

Thayer J (1984) Organometallic compounds in living organisms. Academic Press Inc, Orlando

Thayer J (1988) Organometallic chemistry – an overview. VCH Verlagsgesellschaft, Weinheim, Germany

Thayer J (1989) Methylation: its role in the environmental mobility of heavy elements. Appl Organomet Chem 3:123–128

Tsubaki T (1977) Epidemiology of methylmercury poisoning in Niigita. In: Tsubaki T, Irukayama K (eds) Minamata disease, Kodansha, Tokyo, pp 57–78

Turekian KK, Wedepohl K H (1961) Distribution of the elements in some major units of the earths crust. Bull Geol Soc Am 72:175–192

Uchida M, Hirakawa K, Inoue T (1961a) Biochemical studies on Minamata disease. III. Relations between the causal agent of the disease and the mercury compound in the toxic shellfish with reference to their chemical behaviours. Kumamoto Med J 14:171–179

Uchida M, Hirakawa K, Inoue T (1961b) Biochemical studies on Minamata disease. IV. Isolation and chemical identification of the mercury compound in the toxic shellfish with special reference to the causal agent of the disease. Kumamoto Med J, 14:181–187

Vogt HH (1980) Biomethylierung von Schwermetallen. Chemie für Labor und Betrieb 32:473–474

Westöö G (1966) Determination of methylmercury compounds in foodstuff, I. Methylmercury compounds in fish, identification and determination. Acta Chem Scand, 20:2131–2137

Wilken R, Weiler K (1986) Quecksilbermaxima im Schwebstoff der Elbe. Deutsche gewässerkundliche Mitteilungen 1:19–20

Wilken RD, Hintelmann H, Ebinghaus R (1990) Biologically induced mercury interconversions in the River Elbe. Vom Wasser 74:383–392

Wilken RD, Hintelmann H (1990) Analysis of mercury-species in sediments. In: Broekaert JAC, Güçer S, Adams F (eds) Metal speciation in the environment. NATO ASI Series, Vol G 23 Springer, Berlin Heidelberg New York

Wilken RD, Christiansen H, Fanger HU, Greiser N, Haar S, Puls W, Reincke H, Spott D, Vollmer M (1991) Fakten und Hypothesen zum Schwebstoff- und Schadstofftransport in der Elbe. Vom Wasser 76:167–189

Wilken RD, Hintelmann H (1991) Mercury and methylmercury in sediments and suspended particles from the river Elbe, North Germany. Water Air Soil Poll 56:427–437

Wilken RD, Jantzen E (1991) Organotin compounds in harbour sediments – analysis and critical examination. Vom Wasser 76:1–11

Wilken RD (1992) Mercury analysis: a special example of species analysis. Fresenius J Anal Chem 342:795–801

Wood J M, Kennedy F S, Rosen CG (1968) Synthesis of methylmercury compounds by extracts of a methanogenic bacterium. Nature (London) 220:173–174

10 Organotin Compounds in Sediments of the Rivers Elbe and Mulde

J. Kuballa, E. Jantzen and R.-D. Wilken

10.1 Introduction

10.1.1 Industrial Production and Applications of Organotin Compounds

The multifarious applications of organotin compounds (OTC) led to a rapid increase in production since their first use in 1936 (Zuckerman et al. 1978). The annual production of OTC was 63 000 ton in 1986 and is still on the increase (Blunden and Chapman 1986; Müller et al. 1989; Donard and Pinel 1989; and literature cited therein). The major application fields and the specific compounds are listed in Table 10.1.

About 70% of the produced organotins ends up in the form of less toxic products. Di- and monoalkyltin compounds, especially octyltins, are used in large amounts as heat and light stabilizers for rigid PVC. Dibutyltin compounds are also used as homogeneous catalysts in polymer syntheses.

Of great environmental concern are the biocidal applications. Well known is the use of tributyltin (TBT) in antifouling paints. As the toxic additive in marine paints, TBT has been found to be very effective in eliminating the fouling problem. Ship hulls and docks are the places where antifouling paints are commonly applied. The content of TBT in such paints can be as high as 30%.

In agriculture, highly toxic triorganotin compounds are used as fungicides, antifeedants and acaricides. Although restricted in many countries such as USA, Canada, Japan, Australia, New Zealand and the EC, they are still in use in countries like Singapore, China, etc. Triphenyl- and tricyclohexyltin compounds are applied as insecticides. Tributyltin compounds are not used as pesticides in agriculture due to their high phytotoxicity (Luijten 1972; Blunden and Chapman 1986; and literature cited therein). Tributyltin concentrations in cooling water

Table 10.1. Selected applications of organotin compounds (after Blunden and Chapman 1986)

Application	Active organotin compound [a]
Primary products in the manufacture of other organotin compounds	R_4Sn (Me, Bu, Oct, Ph_3,...)
Agriculture	
fungicides, antifeedants	Ph_3SnX
acaracides	c-Hex_3SnX
Antifouling paint biocides	Bu_3SnX, TBTO
	Ph_3SnX
Wood preservative fungicides	TBTO, Bu_3SnX
Disinfectants	Bu_3SnX, TBTO
Heat and light stabilizers	$R_2Sn(SCH_2COO$-i-$C_8H_{17})_2$
for rigid PVC	$RSn(SCH_2COO$-i-$C_8H_{17})_3$
	$(BuSn_{1.5})S_4$
	(R = Me, Bu, Oct)
Homogeneous catalysts for RTV silicones, polyurethane foams and transesterification reactions	Bu_2SnX_2., TBTO Bu_2SnX_2, $BuSnX_3$

[a] X=OH, OAc, F, Cl, PO_4Me_3N, neophyl, triazole, naphthenate (the anionic remainder, X, widely differs depending on the application), TBTO=$(Bu_3Sn)_2O$, Ph=phenyl, Bu=n-butyl, Me=methyl, Oct=n-octyl and c-Hex=cyclohexyl

circuits, used as slimicides, have been found in the range of 10 μg Sn/l (Fent and Müller 1991).

Tetraorganotin compounds are industrially synthesized in large amounts. They are important intermediates in the manufacture of tri-, di- and monoorganotin compounds. Tetraorganotins react with tin(IV) chloride in a redistribution reaction. The distribution of the resulting organotin compounds depends on the pH.

There are no further commercial applications of tetraorganotins. Therefore, an appearance of tetraorganotins in the environment is often caused by industrial discharges.

10.1.2 Toxicity

Organotin compounds, in particular triorganotin species, are highly toxic to the aquatic environment. The tributyltin species, the active ingredient in antifouling paint formulations, is perhaps the most acute toxic chemical to aquatic organisms

ever deliberately introduced into water (Maguire 1987). Studies have been carried out showing that sub-lethal effects to organisms are in the low ng/l range (Table 10.2). Tributyltin (TBT) is reported to induce shell malformations in oysters at concentrations of only 2 ng Sn/l in the water column, which led to a drastic decrease of the oyster population at the French Atlantic coast in the early eighties (Alzieu et al. 1989). Although in this case the concentration of TBT in water is close to the detection limit of 1 ng Sn/l, there are obvious biological effects to the aquatic life-forms. Batley et al. (1992) determined tributyltin concentrations in Sydney rock oysters. Bioaccumulation led to an enrichment of TBT in muscle tissues. Shell deformations on these oysters have been observed at TBT concentrations greater than 80 μg Sn/kg tissue.

The use of tributyltin compounds in antifouling paints since the early seventies has had negative impacts and disastrous consequences for the oyster industry (Sarradin et al. 1991). As a result of a drastic reduction of the oyster cultures in the Arcachon bay, France, the use of organotin compounds in antifouling paints has been prohibited in France since 1982 (Alizeu et al. 1986). To assess the contamination by organotin compounds, different sentinel organisms and bioindicator species have been suggested, e.g. neogastropodes (Stewart et al. 1992) and bivalves (Becker and Bringezu 1992).

It has been established that tributyltin species in dogwhelks induce imposex, the development of male characteristics in females (Bryan et al. 1986, 1987, 1988, 1989; Davis et al. 1987). Tributyltin-induced effects on early life stages of minnows were investigated by Fent (1991). Exposure to TBT concentrations from 0.81 up to 19.51 μg/l has led to histological and morphological changes and mortality. Thompson et al. (1985) reported that prolonged exposure of aquatic organisms to concentrations of 200 ng TBT/l can lead to serious effects on growth, reproduction and metabolic processes. Egg laying in freshwater snails is reduced at tributyltin concentrations of 1 ng /l (Ritchie et al. 1974).

A 48-h LD_{50}-test with golden orfes, *Indus indus*, suggests that tetrabutyltin (TTBT) is less toxic than TBT (Blunden and Chapman 1986). Laughlin et al. (1982) describe the delayed toxicity due to the loss of one butyl-group from TTBT and the subsequent conversion to toxic tributyltin. Therefore, TTBT has to be assessed as tributyltin in its toxic action on aquatic life.

A large variety of toxicological data on organotin compounds have been presented in literature. Lethal concentrations and doses (LC_{50}, LD_{50}), non-observable effect concentration (NOEC) and effect concentration (EC, EC_{50}) values of organotin compounds for various micro- and higher organisms have been reviewed by many authors (Blunden and Chapman 1986; Boyer 1989; Becker and Bringezu 1992; and literature cited therein). Some data on toxicity of bistributyltin oxide (TBTO) to different organisms are listed in Table 10.2.

The toxic action of organotin compounds to organisms and cell functions has not been examined in great detail (Arakawa and Wada 1993). Heywood et al.

Table 10.2. Toxicity of bistributyltin oxide (TBTO) to different organisms

Organism	Parameter	Exposure time	Conc [μg/l]	Reference
Oyster	Inhibit growth		0.004	Lawler and Aldrich (1987)
Marine algae	Growth stop		0.02	Beaumont and Newman (1986)
Mussel	Inhibit growth		0.08	Stromgren and Bongard (1987)
Oyster	Growth stop		0.3	Waldock and Thain (1983)
Shrimp, young	LC_{50}	96 h	0.3	Valkris et al. (1987)
Salmon	LC_{50}	96 h	0.3	Short and Thrower (1986)
Marine algae	Growth stop		0.4	Seinen et al. (1981)
Shrimp, adult	LC_{50}	96 h	0.7	Valkris et al. (1987)
Earth worm	LC_{100}	96 h	0.8	Short and Thrower (1986)
Marine algae	LC_{100}	2 d	1	Beaumont and Newman (1986)
Salmon	LC_{50}	12 h	4	Short and Thrower (1986)
Crayfish	LC_{50}	12 d	6.4	Laughlin et al. (1984)
Lobster larva	LC_{50}	24 h	8	Laughlin and French (1980)
Salmon	LC_{50}	6 h	11	Short and Thrower (1986)
Rainbow trout	LC_{50}	12 h	11.8	Seinen et al. (1981)
Guppy	LC_{50}	7 d	16	Polster and Halacka (1972)

(1989) have shown that inhibited enzyme activities are caused by organotin species. The trialkyltin cation binds to the active sites of enzymes such as sulphur, nitrogen and oxygen. Other toxic effects of trialkylated tin compounds are due to the:

- Prevention of enzyme activities (Heywood et al. 1989; Saxena 1987)
- Break-down of the mitochondrial functions (Yarlet and Lloyd 1981; Davis and Smith 1980)
- Disturbance of the ion transport (Martin et al. 1983)
- Prevention of the ATP-synthesis (Emanuel et al. 1984; Manabe et al. 1983)
- Damage of the MFO-system (Oehlmann et al. 1993)

There is a small number of studies that have investigated the genotoxicity of organotin compounds (Li et al. 1982; Dixon and Prosser 1986; Davis et al. 1987a; Hamasaki et al. 1992). Recently, Kuballa et al. (1995) investigated the genotoxicity of four butyltin compounds (tetrabutyltin, tributyltin chloride, dibutyltin dichloride, monobutyltin trichloride) with the SOS-Chromotest®. The SOS-Chromotest® is useful as a primary screening assay for genotoxicity

(Quillardet and and Hofnung 1985). All butyltin compounds show genotoxicity as an SOS-inducer. In particular dibutyltin dichloride and tributyltin chloride are genotoxic in low concentrations (5 and 20 μg Sn/l respectively). Important is the fact that Dibutyltin (DBT) has a higher genotoxic potential than TBT, which is more acut toxic than DBT.

10.1.3 Restrictions and Quality Objectives

Due to its high toxicity, the use of tributyltin as the active compound in antifouling paints for ships has been partially restricted in many countries. In 1982 the French government banned the use of antifouling paints containing tributyltin on boats of less than 25 m (Alzieu 1991). In June 1988 the US government introduced regulations on organotins in antifouling paints; paints containing tributyltin compounds may not be applied to boats of 25 m or less in length, but may be used on larger vessels and on aluminium-hulled boats (Wuertz et al. 1991). Aluminium boats were exempted because application of the alternative antifouling agent, copper, would result in corrosion.

In December 1988, the New Zealand government announced partial restrictions on organotin antifoulants, which prohibit the use of all high-release and non-copolymer organotin paints but permit the use of low-release co-polymer paints. The pesticide regulations came into force in July 1989 (Stewart et al. 1992). Also, in Germany and Japan, the use of antifouling paints based on tributyltin compounds is regulated. All of these restrictions are obviously aimed at the application of such paints on pleasure boats.

In the United Kingdom the environmental quality target value (EQT) was set at 20 ng Sn/l for freshwater in 1987 (Dowson et al. 1992). It has been recommended to reset the value at 2 ng Sn/l in order to achieve the complete protection of marine life (Quevauviller and Donard 1990). Since 1990 organotin compounds have been introduced into the assessment of quality criteria of natural waters in Germany. For the IKSE and IKSR (international commissions for protection of Elbe and Rhine rivers) tributyltin is a pollutant with special concern and is periodically measured. Based on the draft concept of the Federal Government/Federal States working group on quality targets (Becker and Bringezu 1992; and literature cited therein), values of 0.1 ng/l for tributyltin and 1 ng triphenyltin have been calculated as quality targets in water. The aim of a German DIN working group is to describe a method for the routine analysis of organotin compounds in water and wastewater.

Although there are regulations of TBT use in most western countries, important countries such as Singapore, Malaysia, Korea, Hongkong and China have not introduced restrictions for TBT to any vessels. Therefore, docking is now being shifted to those countries. To achieve a worldwide protection of marine life, it is necessary that even these countries also regulate the use of TBT-containing paints.

However, it is further necessary to set new quality criteria for surface waters

and to develop quality objectives for sediments in order to achieve the national and worldwide protection of aquatic organisms. Toxicological data are a conceivable basis for calculating quality targets. Because organotin compounds are highly toxic and have long lifetimes in sediments, it is imperative that organotin compounds are introduced into the assessment of sediment qualities.

10.2 Organotin Compounds in Elbe and Mulde Sediments

10.2.1 Fate and Pathways

Major pathways of organotin compounds in the aquatic environment are shown in Fig. 10.1. In general, alkyltin compounds can be released into water from special industrial discharges, effluents from municipal sewage, ships, boats and shipyards, PVC drain pipes. Biocides and fungicides from agricultural application can also be an anthropogenic source. A natural source for organotin species is the environmental methylation of inorganic tin (Ashby and Craig 1988). Several pseudo-natural butylmethyltins are also found in sediments (Maguire 1984; Maguire et al. 1986; Jantzen 1992). These compounds have no industrial application. They were probably formed by biomethylation of butyltin cations.

The release of volatile tetramethyltin and SnH_4 into the atmosphere has not been extensively investigated in the past. Inorganic tin can be converted to SnH_4 under anoxic conditions and can diffuse without oxidation through the overlaying water phase and may be released in this way into the environment (Donard and Weber 1988). Tetrabutyltin is less volatile and is strongly associated with particles and therefore it has probably no direct consequences to the atmosphere.

Tetrabutyltin, which is produced industrially in large amounts is only used as an intermediate in the organotin manufacture and it has no other application. Therefore, high tetrabutyltin burdens in sediments are a result of special industrial discharges. Normally, antifouling paints are mainly responsible for the contamination of sediments with tributyltin and in some cases triphenyltin compounds. In areas with high boating activities such as shipping routes, harbours and shipyards, the underlaying sediments are contaminated with tributyltin and its degradation products di- and monobutyltin. Surprisingly, highly toxic tetrabutyltin has also been found in the Elbe river (Jantzen and Wilken 1991).

In sediments butyltin compounds undergo slow degradation to inorganic tin by the loss of butyl groups. Cationic butyltin species (Bu_3Sn^+, Bu_2Sn^{2+} and $BuSn^{3+}$) are fairly persistent intermediates in the environmental decomposition of Bu_4Sn and Bu_3Sn^+. Di- and monobutyltin do not always originate from direct anthropogenic sources. A stepwise degradation of higher substituted butyltin compounds was assumed by Maguire et al. (1983).

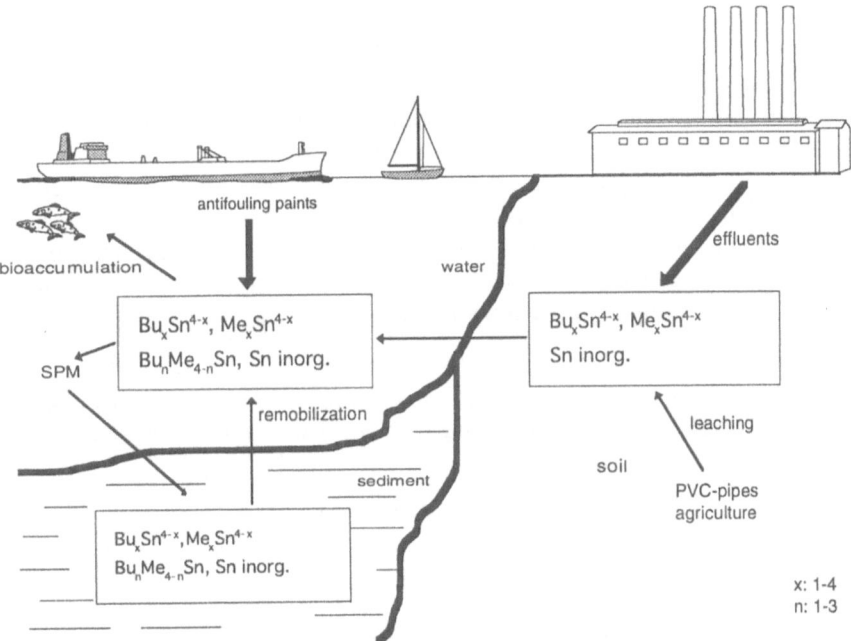

Fig. 10.1. Pathways of organotin compounds in the aquatic environments. Major anthropogenic sources for the Elbe river have been found to be plant effluents and the TBT release from antifouling paints

The stability of butyltin compounds in the water phase has been examined in several laboratory studies. Photolysis is a major factor in the degradation of organotin compounds in water. Besides chemical and physical degradation, a biodegradation pathway can also be assumed in sediments. Half-lives of OTC in water are in the order of weeks. In anaerobic sediments the half-lives can be up to several years long (Burns et al. 1987; Müller et al. 1989; Quevauviller and Donard 1991; and literature cited therein).

Sedimentation, e.g. via suspended particles leads to the enrichment of toxic organotins in sediments. Sediments acting as a sink for organotin compounds pose long-term problems for aquatic ecosystems. However, bioaccumulation of pollutants in the water phase by living organisms, e.g. fish, benthic organisms, etc., can occur from both directions:

direct intake \rightarrow bioaccumulation \leftarrow remobilization from sediment

Remobilization from the sediment into the water phase is also possible and can be caused by resuspension, tidal influence and high tide periods in spring time. Microbiologically induced remobilization is also presumed (Thayer 1989). Sediments contaminated with organotins may lead to a permanent pollution of pore water and the overlaying water phase.

10.2.2 Situation in Hamburg Harbour

Jantzen and Wilken (1991) have reported on first investigations of organotin compounds (OTC) in sediments of the Hamburg harbour. Sediment samples were taken from thirteen different sites of the harbour – five industrial and eight yacht harbours. Table 10.3 lists results from the sediment analysis. Tetrabutyltin (TTBT), tributyltin (TBT), dibutyltin (DBT) and monobutyltin (MBT) compounds were measured in all sediments. Of the organotin species, tributyltin is the highest in most of the samples. This indicates that tributyltin contents in Hamburg harbour sediments mainly arise from the use of antifouling paints (Table10.3).

Table 10.3. Butyltin concentrations in sediments from selected industry and yacht-harbours of the Hamburg harbour (sampling 1989, after Jantzen and Wilken 1991)

Location [a]	Bu_4Sn	Bu_3Sn^+	Bu_2Sn^{2+}	$BuSn^{3+}$	$Sn_{inorg.}$
		[μg Sn/kg dr.wt.]			
Neßkanal	10 ± 20	180 ± 27	30 ± 6	150 ± 30	710 ± 107
Köhlfleet	30 ± 6	120 ± 18	20 ± 4	40 ± 6	460 ± 69
Park hb.	50 ± 10	150 ± 23	40 ± 8	170 ± 34	610 ± 92
Kuhwerder hb.	280 ± 56	5200 ± 780	610 ± 120	430 ± 86	240 ± 36
Sea hb.	230 ± 46	280 ± 42	110 ± 22	200 ± 40	1050 ± 160
Wedel yhb.	100 ± 26	210 ± 32	40 ± 8	170 ± 34	480 ± 72
Mühlenberg	100 ± 26	250 ± 38	70 ± 14	210 ± 42	710 ± 107
Teufelsbrück	100 ± 20	420 ± 63	90 ± 18	190 ± 38	630 ± 95
State hb.	410 ± 82	380 ± 57	110 ± 22	250 ± 50	1440 ± 220
Diamantgraben	280 ± 56	220 ± 33	20 ± 4	70 ± 14	1420 ± 210
Arme Leute hb.	220 ± 44	190 ± 29	30 ± 6	< 10	660 ± 100
Oortkaten yhb.	250 ± 50	280 ± 57	70 ± 14	140 ± 28	1400 ± 210
Oortkaten	400 ± 80	360 ± 54	100 ± 20	170 ± 34	1600 ± 320

[a] hb. = harbour, yhb. = yacht harbour, $Sn_{inorg.}$ = inorganic tin that found with $NaBEt_4$

Extremely high tributyltin concentrations of up to 5200 μg Sn/kg dr.wt. were found in sediments near a shipyard in the Kuhwerder harbour. At this location tributyltin species were still found in sediments down to a depth of 2 m in amounts of up to 1400 μg Sn/kg dr.wt. The tributyltin concentrations of the Kuhwerder harbour sediments are about 20 times higher than the average contamination found in the Hamburg harbour. Organotin burdens of some international harbours are compared with Hamburg harbour in Table 10.4.

In addition to tributyltin all other butyltins are also present in sediments of Hamburg harbour. The amounts of di- and monobutyltin species are generally

lower than the tributyltin concentrations at the same site. They were probably formed by the degradation of tetra- and tributyltin species.

It is unusual to find tetrabutyltin in harbour sediments in concentrations of 50–410 μg Sn/kg dr.wt. There must be another source for tetrabutyltin than the leaching from the antifouling paints, which is discussed later in this chapter. Matthias et al. (1986) and Maguire et al. (1986) first reported the presence of tetrabutyltin in harbour waters. Maguire investigated more than 200 sediments. In only two samples, tetrabutyltin was found in concentrations of up to 20 μg/kg dr.wt. The occurrence of such low tetrabutyltin amounts in sediments is presumably due to their release from antifouling paints. The paints are technical products and can contain tetrabutyltin in low concentrations.

Table 10.4. Comparison of butyltin concentrations in sediments of international harbours

Location	Bu_4Sn	Bu_3Sn^+	Bu_2Sn^{2+}	$BuSn^{3+}$	Reference
		[μg Sn/kg dr.wt.]			
Hamburg, Germany	280	5200	610	430	Jantzen and Wilken (1991)
Osaka, Japan	n.a.	280	140	100	Harino et al. (1992)
Mainz, Germany	n.a.	182	44	27	Schebeck et al. (1991)
Toronto, Canada	n.a.	539	248	580	Maguire (1984)
Vancouver, Canada	n.a.	10780	8510	3360	Maguire et al. (1986)

10.2.3 Situation of the Elbe and Mulde River System

In 1991 and 1992 sediment samples were taken from different locations in the Elbe river in order to obtain an initial overview of the spatial distribution of organotin compounds in the river system. Beginning in Lutherstadt-Wittenberg samples were collected along the course of the Elbe river via Magdeburg, Hamburg harbour and as far as Cuxhaven/Elbe-estuary. Two more sediment samples were taken from Elbe river tributaries – Mulde and Saale rivers. The locations of the sampling sites are shown in Fig. 10.3. The sediment samples were analyzed for their organotin content (Table 10.5).

Organotin compounds were detected in all samples. Highly contaminated sites were found in the Hamburg harbour region, in particular in areas with restricted water exchange (i.e. Oortkaten harbour). A sediment sample collected in Dessau/Mulde contained 14 000 μg Sn/kg dr.wt. tetrabutyltin and 2700 μg Sn/kg dr.wt. tributyltin. In this location the total organotin concentration was greater than 20000 μg Sn/kg dr.wt. and thus represents one of the highest values which has ever been reported in literature.

Tetrabutyltin (TTBT) could be detected in Elbe sediments from the mouth of

Table 10.5. Butyltin concentrations in sediments of the Elbe river and its tributaries – Mulde and Saale (sampling 1991; see Fig.10.3)

Location	Bu_4Sn	Bu_3Sn^+ [μg Sn/kg dr.wt.]	Bu_2Sn^{2+}	$BuSn^{3+}$	$Sn_{inorg.}$
Cuxhaven	15 ± 3	75 ± 11	45 ± 11	< 2	n.a.
Wedel harbour	110 ± 22	30 ± 3	< 2	< 2	50 ± 10
Teufelsbrück hb.	540 ± 98	760 ± 114	250 ± 62	210 ± 42	120 ± 24
Kuhwerder hb.	280 ± 56	5200 ± 780	610 ± 120	430 ± 86	260 ± 39
Oortkaten hb.	3950 ± 790	2990 ± 790	1340 ± 335	590 ± 118	1600 ± 320
Wittenberge	1600 ± 320	1060 ± 159	450 ± 110	550 ± 110	330 ± 66
Glinde	260 ± 52	840 ± 130	530 ± 130	150 ± 30	120 ± 24
Barby	< 50	1230 ± 190	360 ± 90	290 ± 58	190 ± 38
Gr.Rosenbg/Saale	100 ± 20	1000 ± 150	140 ± 35	100 ± 20	220 ± 44
Breitenhagen	8350 ± 1670	2680 ± 410	2530 ± 633	2820 ± 560	330 ± 66
Dessau/Mulde	14270 ± 2850	2720 ± 410	2150 ± 540	1600 ± 160	14500 ± 2900
Roßlau	< 50	1350 ± 200	370 ± 92	50 ± 10	110 ± 22
Lutherst-Wittenbg	< 10	13 ± 4	14 ± 4	< 10	n.a.

[a] hb. = harbour, $Sn_{inorg.}$ = inorganic tin that found with $NaBEt_4$

the Mulde river all the way to Cuxhaven. As previously mentioned, tetrabutyltin has no commercial application, it is an intermediate in the production of the other butyltins. It is assumed that the anthropogenic source was a chemical plant for production of bisdibutyltin oxide in Bitterfeld, close to the Mulde river. In this plant dibutyltin compounds were formed by the redistribution reaction of tetrabutyltin and tin(IV) chloride. Tetrabutyltin, beside the other butyltins, was a major compound in the production effluents. The discharges were presumably released into the Mulde and thus entered the Elbe river system. The main source for the tetrabutyltin contamination of Hamburg harbour sediments is probably to be found in the effluents from this chemical plant.

Increased tributyltin concentrations, which were investigated in Roßlau, were presumably caused by a nearby shipyard. No tetrabutyltin was measured in Roßlau and further upstream the Elbe river in Lutherstadt-Wittenberg. This confirms the assumption that the tetrabutyltin concentrations found in the Elbe river are remobilized from Mulde river sediments. Therefore, tetrabutyltin can only be found in the Elbe river downstream of the Mulde mouth.

The lowest butyltin concentrations were found in tidally influenced areas of the Elbe river from Wedel to Cuxhaven and in Lutherstadt-Wittenberg (upstream of the Mulde mouth). The total organotin concentrations at these locations were 50–150 μg Sn/kg dr.wt.

10.2.4 "Natural" Organotin Species – Biomethylation

Methylation processes in sediments are mainly responsible for the occurrence of methylated organotin species in Elbe sediments. PVC-pipelines releasing leachates is an anthropogenic source for methyltins. Biomethylation reactions of tin and other metals have been well described and reviewed by several authors (Craig and Glocking 1988; Thompson et al. 1985, and literature cited therein). Several biotic and abiotic methylation agents are known. Methylcobalamin is a carbanion donor and is generally able to convert inorganic tin(IV) to methyltin compounds. Methyl iodide react only with ionic tin(II) to produce methyltin (Dizikes et al. 1978; Franchiang and Wood 1981; Chau et al. 1981). Certain Pseudomonas bacteria are able to methylate inorganic tin to the several methyltin species (Jackson et al. 1982; Hallas et al. 1982; Thayer 1989). The biotic and abiotic formation of tetramethyltin from trimethyltin in material taken from anaerobic sediment has been shown by Guard et al. (1981). Chau et al. (1981) determined tetramethyltin in sterile as well as in biologically active sediments. A disproportion mechanism for the formation of tetramethyltin is postulated by Rapsomanikis and Weber (1985) and Craig and Rapsomanikis (1985).

Methylated tin species are less toxic to algae and molluscs than the equivalent butyltin species. The IC_{50}^{4h}-values (IC = inhibition concentration) on freshwater algae are reported for trimethyltin chloride and tributyltin chloride as 2.6 mg/l and 0.16 mg/l respectively (Schebeck et al. 1991). Several authors describe transmethylation mechanisms of trimethyltin compounds to inorganic mercury (Huey et al. 1974; Chau et al. 1987; Ebinghaus 1991). Transmethylation reactions of methyltins with other heavy metals are of great ecological relevance, methylmercury in particular has a higher toxicity to aquatic organisms than the metal and inorganic mercury. There is a gap in the knowledge of transbutylation reactions with other metals. Possible transbutylation reactions of inorganic mercury and lead or other heavy metals i.e. with tetrabutyltin, which is found in high concentrations in the Elbe river, must be considered.

Since the beginning of the last decade a lot of data have been collected concerning the screening of naturally occurring methyltin compounds in water and sediments (Braman and Tompkins 1979; Hodge et al. 1979; Chau 1980; Brinckman et al. 1981; Chau et al. 1982; Maguire et al. 1985, 1986; Randall et al. 1986, Donard et al. 1986; Schebeck et al. 1991). Data on methyltin concentrations in the Elbe river have recently been reported by Jantzen (1992). Ten sediment samples from different locations along the Elbe river were analyzed for methyltin compounds. Methylated tin species were found in concentrations of up to 55 μg Sn/kg dr.wt. at seven sites. All methyltins were found in Teufelsbrück harbour. In Dessau (Mulde mouth) only tetramethyltin was determined (see Table 10.6). Among the various methyltin species, Me_3Sn^+ is present continuously from Barby to Wedel harbour in concentrations ranging from 10–55 μg Sn/kg dr.wt., whereas at Dessau/Mulde, this was at background levels (<2 μg Sn/kg dr.wt.). Me_4Sn is found in the areas of Glinde, Wittenberge and Dessau/Mulde.

Table 10.6. Methyltins in sediments of Elbe and Mulde rivers

Location	$MeSn^{3+}$	Me_2Sn^{2+}	Me_3Sn^+	Me_4Sn
		[μg Sn/kg dr.wt.]		
Wedel harbour	<2	<2	40 ± 20	<2
Teufelsbrück harbour	25 ± 12	5 ± 2	20 ± 10	15 ± 7
Oortkaten harbour	20 ± 10	10 ± 5	55 ± 27	<2
Wittenberge	10 ± 5	<2	25 ± 13	50 ± 25
Glinde	<2	<2	10 ± 5	20 ± 10
Barby	<2	<2	30 ±	<2
Dessau/Mulde	<2	<2	<2	20 ± 10

Me_2Sn^{2+} and $MeSn^{3+}$ are found mostly in the areas of Oortkaten and Teufelsbrück harbours. Due to the fact that methyltin compounds can methylate heavy metals, they must be considered in the assessment of sediment qualities.

The methylation products of butyltin compounds have also been observed in water, sediments and biological samples (Maguire 1984, 1991; Maguire et al. 1986; Quevauviller 1989). Jantzen (1992) investigated mixed butylmethyltin species in Elbe sediments. Such mixed butylmethyltin species are not introduced into the aquatic environment anthropogenically; they presumably arise from the microbiological methylation of butyltin cations (Matthias et al. 1986). A trans-alkylation reaction between butyl- and methyltins is also possible, but it could not be confirmed as yet. The toxicities of those mixed butylmethyltin compounds are still unknown.

The concentrations of mixed butylmethyltins found in Elbe sediments are presented in Table 10.7. The observed $BuMe_3Sn$ concentrations are generally higher than the Bu_2Me_2Sn and Bu_3MeSn concentrations. The fact that most $BuMe_3Sn$ values are significantly higher than the pure monobutyltin concentrations at the same locations is important (see Table 10.3). An enhanced biological activity in sediments during the summer period is probably responsible for this fact.

Table 10.7. Mixed butylmethyltin concentrations in some sediments of Hamburg harbour

Location	$BuMe_3Sn$	Bu_2Me_2Sn	Bu_3MeSn
		[μg Sn/kg dr.wt.]	
Teufelsbrück	490 ± 150	150 ± 50	130 ± 40
Oortkaten harbour	540 ± 160	300 ± 90	200 ± 60
Wedel yacht harbour	410 ± 120	<50	<50

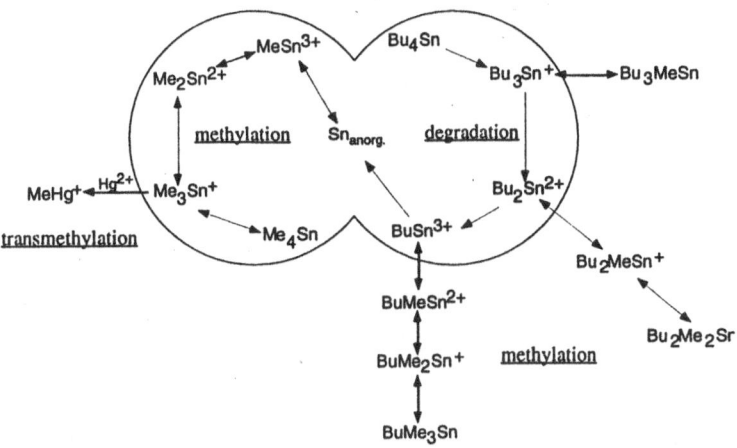

Fig. 10.2. Debutylation and methylation cycles of organotins. Methylation of cationic butyltins leads to mixed butylmethyltins. The ionic butyltins have not been found in the environment

Figure 10.2 shows the fate of butyl-, methyl- and mixed butylmethyltin compounds in sediments. Butyl- as well as methyltin compounds undergo slight decomposition to inorganic tin. Chemical as well as biological decomposition occurs in sediments. It is not known whether a microbial butylation of inorganic tin or butyltin cations does occur. Debutylation processes in sediments lead to the formation of inorganic tin, which can be methylated subsequently. All methyltin compounds can be formed by micro-organisms. There is probably an equilibrium in formation/degradation of methyltin compounds, which may be influenced by the biological activity of the sediment. Side reactions lead to butylmethyltin compounds (Fig. 10.2). These compounds are formed by further methylation of the several butyltin cations. The mechanism is supposed to be stepwise and induced by micro-organisms. However, the intermediate ionic methylation products $BuMeSn^{2+}$, $BuMe_2Sn^+$ and Bu_2MeSn^+ have not been investigated as yet.

10.2.5 Organotin Patterns: An Aid to the Interpretation of the Origin and Fate of Organotins

Figure 10.3 shows the sampling sites and the spatial distribution of butyltins along the Elbe river. Four different patterns of the origin and fate of butyltins in sediments of the Elbe and Mulde rivers are evident (Jantzen et al. 1993):

- Pattern 1: antifouling paints
- Pattern 2: shipyard
- Pattern 3: industry influenced
- Pattern 4: overlay-pattern of patterns 1, 2 and 3

Fig. 10.3. Distribution of butyltin coumpounds on selected sites of Elbe and Mulde rivers. The organotin patterns indicate various anthropogenic sources. A chemical plant in Bitterfeld was the main source of organotins

Pattern 1: Antifouling Paints – example: Roßlau and Lutherstadt-Wittenberg
The distribution of the butyltin species in Elbe sediments upstream the mouth of the Mulde river indicates that the main inputs of tributyltin, along with the degradation products di- and monobutyltin, were caused by antifouling paints. Tri-butyltin in concentrations of 1350 μg Sn/ kg dr.wt. were detected in Roßlau, which is probably connected with effluents from a shipyard. Boating activities are low in the middle section of the Elbe river. Consequently, the sediments in Lutherstadt-Wittenberg are less contaminated with tributyltin compounds.

Pattern 2: Shipyard – example: Kuhwerder harbour
Very high tributyltin contaminations were measured in sediments from Hamburg harbour, which is mainly due to the release from antifoulants. The butyltin-pattern of the Kuhwerder harbour points to fresh discharges of tributyltin and it is probably caused by a nearby shipyard and the high shipping activities in this area. At this location considerable levels of tributyltin were found (5200 μg Sn/kg dr.wt.). Di- and monobutyltin appear to be debutylation products from tributyltin.

Tetrabutyltin is also found in sediments from Kuhwerder harbour, but it does not originate from the antifoulants; it has been transported from the Mulde river into the Elbe river.

Pattern 3: Industry Influences – example: Dessau/Mulde
The Mulde river is not used as a shipping route. Therefore, the sediment organotin burden cannot come from the antifoulants. In comparison to the antifouling-related patterns 1 and 2, tetrabutyltin (TTBT) and mobile inorganic tin levels are dominant. As mentioned earlier, tetrabutyltin and tin(IV) chloride are primary products in the manufacture of dibutyltin. Besides tetrabutyltin and mobile inorganic tin, very high concentrations of all other butyltins, tributyl-, dibutyl and monobutyltin, were detected in sediments collected from the Mulde river near Dessau. However, the untreated discharges from a chemical plant in Bitterfeld have led to a high contamination of the Mulde sediments with organotins.

Pattern 4: Combination Pattern of 1, 2 and 3 – location dependence
The characteristic butyltin distribution, as shown under patterns 1, 2 and 3, changes along the Elbe river. The influences by the individual pattern types are pronouncedly different in each part of the Elbe river. The typical pattern of the Mulde river (pattern 3) has changed at the sampling site Breitenhagen. Mobile inorganic tin, resulting from tin(IV)-chloride in effluents of an organotin plant in Bitterfeld, decreased significantly in its concentration. This is due to the rapid transformation of tin(IV)-chloride to tin-hydroxide-oxides and finally to tin dioxide. These compounds are no longer mobile. They cannot be eth ylated by sodium tetraethylborate.

The groyne fields in Glinde and Wittenberge do not act as a good sink for organotin compounds. The fields are flooded during high tide periods and the organotin contents are therefore low. The Oortkaten harbour, located close to the Hamburg harbour within the tidal-influenced part of the Elbe river, represents a typical sink for suspended particulate matter (SPM). The concentrations of tributyltin species are continuously increased due to inputs from the antifouling paints. It is noteworthy that there is still an impact from the contaminated Mulde river sediments to the Elbe river. Tetrabutyltin, as an indicator for the Mulde river influence, is still found in SPM in the Elbe river downstream the Mulde mouth.

The organotin concentration in sediments decreases in the tidal-influenced part of the Elbe river from Hamburg harbour to the Elbe estuary (Teufelsbrück-Wedel-Cuxhaven). However, tetrabutyltin was found in Cuxhaven/Elbe estuary (Table 10.5). That can be explained by the discharges of a chemical plant in Bitterfeld.

10.3 Analytical and Chemical Methods – Species Analysis

10.3.1 Sediment Sample Preparation Techniques

Sample preparation techniques for organotin compounds are well documented in the literature (Zhang et al. 1991; Chau and Wong 1991; and literature cited therein). In order to achieve reproducible chromatographic separation, a complete alkylation/hydride generation of the organotin cations is required. The described methods differ mainly in the initial extraction and derivatization steps. The extraction yields vary with the method applied. With each extraction method it is only possible to isolate a part of the occurring organotin compounds with sufficient yields. Some extraction techniques of butyltin compounds in sediments have been evaluated by Zhang et al. (1991). With the hexane/tropolone-method either methyltin species or butyl-, octyl- and phenyltin compounds, depending on the pH used, can be extracted (Chau et al. 1982; Burns et al. 1987). Monoalkylated tin species (monobutyl- and monooctyltin) are generally less easy to extract (52–71%; Jantzen 1992).

In situ hydride generation followed by a cryogenic-trapping step considers only methyl- and cationic butyltins with appropriate recovery rates. Compounds with high boiling points such as octyl- and phenyl-tin compounds cannot be analyzed by this method.

Jantzen and Wilken (1991) described a new method considering all occurring organotin compounds in sediments and with generally good extraction yields. Jantzen used sodium tetraethylborate for the direct ethylation of the organotin cations in wet sediments. The advantages of this new method are:

- In situ ethylation of the organotin cations,
- Complete determination of all occurring OTC in sediments
 (methyl-, butyl-, octyl-, phenyl-tins, etc.)
- Good and proportionate extraction yields (77–83%)
- Formation of stable tetraorganotin compounds
- Extraction of the formed tetraorganotins into hexane without
 additional complexing agent

A comparison of the two extraction techniques are described by Jantzen and Wilken (1991):

Sodium Tetraethylborate-Method: Complete is the method by direct-ethylation with sodium tetraethylborate; methyl-, butyl- and phenyl-tin species as well as mobile inorganic tin(IV) can be extracted with high yields (77–83%). This is a method for the single-step investigation of all organotin compounds occurring in sediments. The wet acidified sediment is buffered at pH 4 with NaAc/AcH. Added are 20 ml of hexane and 1 ml of 10% aqueous sodium tetraethylborate solution. The mixture is shaken for 1 h and the organic layer is separated. After a clean-up procedure the volume is reduced to 1 ml. Subsequently, 1 μl is injected into the GC-AAS.

Extraction with Hexane/Tropolone: This method considers either methyltin species or butyl- and higher organotin compounds depending on the pH value used in the extraction. It is necessary to acidify the samples for analyzing butyltins (Dirkx et al. 1989). Under optimal conditions the extraction yields are within the range of 52–79%. The extraction yields depend on the used tropolone concentration and on the other hand, they are different for the several butyltin compounds. An increase of the tropolone concentration to 0.3% (w/w) increases the extraction yields of monoalkytin species but decreases the yields of tetrabutyltin. Good conditions were found to be at tropolone concentrations of 0.1% (w/w). The extracted organotin-tropolone-complexes are then derived by propylation. Pentylation is also possible but the chromatograms are longer and the peaks show more tailing.

10.3.2 Chromatographic Units and Detectors

The method of choice for the organotin analysis is a chromatographic separation technique in combination with an element or compound specific detector. The systems conventionally used differ in the choice of detector, detection limit and the sample preparation technique applied (Chau and Wong 1989). GC and HPLC systems are used for separating the individual organotin compounds. The basic detector units for gas chromatographs are electron capture (ECD) and the non-specific flame ionization detectors (FID). The ECD is usable only for halogenated organotin species (Tsuda et al. 1987); tetraalkylated tin species are not detectable. However, ECD and FID are less usable for analyzing organotins in complex matrices such as sediment and sewage sludge. Tin-specific flame photometric detectors (FPD) are also an option (Tolosa et al. 1992).

The most specific detector is a mass spectrometer. Organotins undergo extensive fragmentation and tin has a complex isotopic cluster. However, the GC-MS sig-nal is highly susceptible to matrix interferences, which leads to reduction of the detection limit. Atomic absorption and atomic emission detectors are element-specific and not sensitive to matrix interference. Therefore, analyzing organotin compounds in a complex matrix by GC-AAS has hardly any consequences on the detection limit.

Liquid-liquid chromatography is used less in the organotin analysis, but new HPLC-MS systems might be useful. Table 10.8 provides a survey of different coupled chromatographic systems and detector units applied in the determination of organotin species. Recent developments in speciation and determination of organometal compounds have been reported by Chau and Wong (1991).

Gas chromatographic techniques coupled to an atomic spectrometric detector are still the most widely used systems in organotin speciation. Several derivatization procedures were developed to obtain volatile organotins for GC analysis, which are hydride generation (Donard et al. 1986), in situ ethylation with sodium tetraethylborate (Jantzen and Wilken 1991) and Grignard reaction

Table 10.8. Coupling systems used in organotin analysis

Detection unit	Limit of detection	Reference
GC-ECD/FPD	200 pg Sn	Sonderquist and Crosby (1978)
GC-MS	500 pg Sn	Meinema et al. (1978)
	1–10 ng Sn/l	Müller (1984),
		Matthias et al. (1986),
		Jantzen and Wilken (1991),
		Tolosa et al. (1991)
GC-FPD	100 pg Sn	Müller (1987)
	3–15 pg Sn	Tolosa et al. (1991)
purge trap GC-FPD	1–100 pg Sn	Braman et al. (1979),
		Jackson et al. (1982)
GC-AED	1–10 pg Sn	Wylie and Quimby (1990),
		Scott et al. (1991)
	5 pg	Liu et al. 1993
	0.5 pg Sn	Chau et al. 1994
GC-QFAAS	10–100 pg Sn	Chau et al. (1982),
		Maguire et al. (1986),
		Ashby and Craig (1989),
		Jantzen and Wilken (1991)
HG-GC-QFAAS	1–2 ng Sn/g	Quevauviller a. Donard (1990)
	120 pg Sn	Martin et al. 1994
GC-IPC-MS	0.05 pg Sn	Prange and Jantzen (1995)
HPLC-GF-AAS	160–200 pg Sn	Brinckman el al. (1977),
		Parks et al. (1988),
		Nygren et al. (1988)
	5 pg Sn	Kadokami et al. (1988)
HPLC-HG-AAS/AES		Burns et al. (1981),
		Blais and Marshall (1989)
HPLC-AAS	2 pg Sn	Marshall et al. (1989)
HPLC-ICP-MS	0.4–1 pg Sn	Suyani et al. (1989)

(Maguire and Huneault 1981). The converted tetraalkylated tin or alkyltin hydride compounds are volatile. Subsequently, the organotin compounds can be chromatographically separated. The compounds will reach the detector unit via a transfer line.

The absolute detection limits are 10–20 pg tin (Jantzen and Wilken 1991). The

main components of the GC-AAS system are a gas chromatograph with an on-column injector, a transfer line, a modified quartz furnace and an AA-detector. The organometallic compounds have to be destroyed in the quartz furnace unit to obtain tin atoms. Hydrogen and air are the reactive gases which are mixed with the organotin compounds in the pre-furnace. Tin hydrides probably are formed and afterwards are destroyed with air-oxygen. Subsequently, tin is quantified using atomic absorption.

10.4 Conclusions

Pollution of harbours, lakes and inland waters with organotin compounds (OTC) is a worldwide problem. Due to the high toxicity and long-time stability of OTC in sediments, they are a permanent risk for the aquatic ecosystem. Improved sediment preparation techniques and low detection limits during the last 15 years have revealed the real organotin burden in the environment. This has led to restrictions in the application of antifouling paints in many countries.

Investigations of the Hamburg harbour sediments revealed unexpectedly high tetrabutyltin levels, which do not originate from antifouling paints. After the German reunification, it was possible to investigate the organotin burdens in sediments in the catchment area of the Elbe river.

Characteristic butyltin patterns, which have been found in sediments, have made it possible to investigate the presumed sources. Industrial discharges of a plant in Bitterfeld are probably the cause of extremely high organotin contaminations of the Mulde river, in particular with tetrabutyltin (TTBT).

Remobilization processes from Mulde sediments and transport through the water phase lead to a spatial distribution of organotin compounds in the Elbe sediments. Tetrabutyltin has been detected in Elbe sediments all the way to the Cuxhaven/Elbe estuary. In the years to come the extreme high organotin levels in the Mulde sediments will be a potential source of organotin compounds.

The biological activity of sediments is presumably the cause for the appearance of several methyltin species in the Elbe sediments. Besides pure methyltins, mixed butylmethyltin compounds have also been detected in the Elbe sediments. In some locations the concentrations of $BuMe_3Sn$ species were higher than those for the equivalent butyltin species. This is probably due to an enhanced biological activity in the respective sediments. A stepwise methylation of the butyltin cations is assumed, although the intermediate butylmethyl cations have not yet been found. Furthermore, redistribution reactions or butylation processes in the formation of butyltin species could not be confirmed.

The quality criteria for water and sediments, therefore, should consider the concentration, distribution and transformation of organotin species in a general assessment of aquatic ecosystems.

Important in species analyses are careful sediment sample preparation, pretreatment, chromatographic separation and sensitive detection. By in situ ethylation with sodium tetraethylborate in wet sediments, it is now possible to record all organotin compounds including mobile inorganic tin. The extraction yields are consistently good for all organotin compounds and are in the range of 77–83%.

As shown with the organotin pattern, it is now possible with species analysis to differentiate between point sources (i.e. industrial effluents) and diffuse sources (i.e. antifouling paint treated ships entering and leaving harbours). It could be shown that organotin compounds are released into the Elbe and Mulde rivers via two major pathways – effluents of a chemical plant in Bitterfeld and tributyltin-based antifouling paints.

The development of new coupling systems of chromatographic and detection units as well as new sample preparation strategies were the breakthrough in organometal speciation. It is now possible to distinguish between the species of an element and to estimate the amount in different matrices as air, water, soil, sediment or biological specimens. For the future, the aim of organometal species analyses is to:

- Determine matrix species (i.e. Bu_3Sn-protein, MeHg-S-humic acid)

- Understand more about transformation processes in the environment (i.e. methylation, transalkylation, degradation, diagenesis and aging processes)

- Consider ecotoxicological aspects (i.e. bioavailability and toxic action of organometal species on organisms)

Acknowledgements. The authors thank K. Wiltshire and M.V.M. Desai for the help in preparing the English version of the manuscript. Special thanks to Y.K. Chau and K.K. Kwan from the National Water Research Institute, Burlington, Canada, for giving us (J.K. and E.J.) the opportunity to work in their laboratories. The stay was financed by the German-Canadian Science and Technology Cooperation Program.

References

Alzieu C, Snajuan J, Detreil JP, Borel M (1986) Tin contamination in Arcachin Bay: effects on oyster shell anomalies. Mar Poll Bull 12: 494-498

Alzieu C, Sanjuan P, Michel P, Borel M, Dreno J (1989) Monitoring and assessment of butyltins in Atlantic coastal waters. Mar Poll Bull 20:22–26

Alzieu C (1991) Environmental problems caused by TBT in France: assessment, regulations, prospects. Mar Env Res 32:7–17

Arakawa Y, Wada O (1993) Biological properties of alkyltin compounds. In: Metal ions in biological systems. Sigel H, Sigel A (eds) Marcel Dekker Inc, New York, pp 101–136

Ashby JR, Clark S, Craig PJ (1988) Methods for the production of volatile organometallic derivates for application to the analysis of environmental samples. J Anal Atom Spectr 3:735–736

Ashby JR, Craig PJ (1988) Environmental Methylation of tin: an assessment. Sci Total Environ 73:127–133

Ashby JR, Craig PJ (1989) New method for the production of volatile organometallic species for analysis from the environment. Some butyltin levels in UK sediments. Sci Total Environ 78:219–232

Beaumont AK, Newman PB (1986) Low levels of tributyltin reduce growth of marine micro-algae. Mar Poll Bull 17:457–661

Batley GE, Scammell MS, Brockbank (1992) The impact of the banning of tributyltin-based antifouling paints on the Sydney rock oyster, Saccostrea commercialis. Sci Total Environ 122:301–314

Becker EC, Bringezu S (1992) Belastung von Binnengewässern durch biozide Organozinnverbindungen – Immissionen, Wirkungen, Qualitätsziele, Anwendungsverbote. Z Wasser- Abwasser-Forsch 25:40–46

Blais JS, Marshall WD (1989) Determination of ionic alkyllead compounds in water, soil, and sediment by high-performance liquid chromatography-quartz tube atomic absorption spectrometry. J Anal Chem 4:271–277

Blunden SJ, Chapman A (1986) Organotin compounds in the environment. In: Organometallic compounds in the environment. Craig PJ (ed) Longman, pp 111–159

Boyer IJ (1989) Toxicity of dibutyltin and tributyltin, and other organotin compounds to humans and to experimental animals. Toxicology 55, 253–298

Braman RS, Tompkins MA (1979) Separation and determination of nanogramme amounts of inorganic tin and methyltin compounds in the environment. Anal Chem 51:12–19

Brinckman FE, Blair WR, Jewett KL, Iverson WP (1977) Application of a liquid chromatograph coupled with a flameless atomic absorption detector for speciation of trace organometallic compounds. J Chromatogr Sci 15:493–503

Brinckman FE, Jackson JA, Balin WR, Olsen GJ, Iverson WP (1981) Ultratrace speciation and biogenesis of methyltin species transport in estuarine waters. NATO symposium on 'Trace Metals in Sea Water', Erice, Sicity, Italy 30 March– 3 April 1981

Bryan GW, Gibbs PE, Hummerstone GC, Burt GR (1986) The decline of the gastropod Nucella lappilus around southwest England: evidence for the effect of tributyltin from antifouling paint. J Mar Biol Ass UK 66:611–640

Bryan GW, Gibbs PE, Burt GR, Hummerstone GC (1987) The effects of tributyltin (TBT) on adult dogwhelks, Nucella lapillus: longterm field and laboratory experiments. J Mar Biol Ass UK 67:525–544

Bryan GW, Gibbs PE, Burt GR (1988) A comparison of the effectiveness of tri-n-butyltin chloride and five other organotin compounds in promoting the development of imposex in the dogwhelk, Nucella lapillus. J Mar Biol Ass UK 68:733–744

Bryan GW, Gibbs PE, Huggett RJ, Curtis LA, Bailey DS, Dauer DM (1989) Effects of tributyltin pollution on the mud snail Ilyanassa obsoleta, from the York river and Sarah creek, Chesapeake Bay. Mar Poll Bull 20:458–462

Burns DT, Glocking F, Harriot M (1981) Investigation of the determination of tin tetraalkyls and alkyltin chlorides by atomic-absorption spectrometry after separation by gas-liquid of high-performance liquid-liquid chromatography. Analyst 106:921–930

Burns DT, Harriott, Glocking F (1987) The extraction, determination and speciation of tributyltin in seawater. Fresenius Z Anal Chem 327:701–703

Chau YK (1980) Biological methylation of tin compounds in the aquatic environment. 3rd Int Conf Organomet Chem Germanium, Tin, Lead. University Dortmund, Germany, July 1980

Chau YK, Wong PTS, Kramar O, Bengert GA (1981) Methylation of tin in the aquatic environment. In: Ernst WHO (ed) Heavy metalls in the environment, Proc 3rd International Conference (Amsterdam). CEP Consultants Ltd, Edinburgh, pp 641–644

Chau YK, Wong PTS, Bengert GA (1982) Determination of methyltin(IV) and tin(IV) species in water by gas chromatography/atomic absorption spectrometry. Anal Chem 54:246–249

Chau YK, Wong PTS, Mojesky CA, Carty AJ (1987) Transmethylation of metals in aquatic systems. Appl Organom Chem 1: 235-239

Chau YK, Wong PTS (1989) Applications of gas chromatography to trace element speciation. In: Batley GE (ed) Trace element speciation: analytical methods and problems. CRC Press Inc, Boca Raton, Florida, pp 219–244

Chau YK, Wong PTS (1991) Recent developments in speciation and determination of organometallic compounds in environmental samples. Fresenius Z Anal Chem 339:640–645

Chau YK, Yang F, Maguire RJ (1994) Speciation of organotin compounds in water, sediment and sewage samples by gas chromatography/atomic emission detection. NWRI Contribution No. 94-70

Clark S, Craig PJ (1988) The analysis of inorganic and organometallic antimony, arsenic and tin compounds using an on-column hydride generation method. Appl Organom Chem 2:33–46

Craig PJ, Rapsomanikis S (1985) Reactions of methylcobalamin with tin and lead compounds. Inorg Chim Acta 107:39–43

Craig PJ, Glocking F (1988) The biological alkylation of heavy elements. The Royal Socieaty of Chemistry, London, ISBN 0-85186-716-2

Davis A, Barale R, Brun G, Forster R, Günther T, Hautefeuille H, van der Heiden CA, Knaap AGA, Krowke R, Kuroki T, Loprieno N, Malaveille C, Merker HJ, Monaco M, Mosesso P, Neubert D, Norpa H, Sorsa M, Vogel E, Vooged CE, Umeda M, Bartsch H (1987a) Evaluation of the genetic and embryotoxic effects of bis(tributyltin) oxide (TBTO) a broad-spectrum pesticide in multiple in vivo and in vitro short-term test. Mutation Res 188:65–95

Davis AG, Smith PJ (1980) Recent advances in organotin chemistry. IV. Biological and environmental aspects. Adv Inorg Radiochem 23:41–51

Davis IM, Bailey SK, Moore DC (1987) Tributyltin in Scottish sea lochs, as indicated by degree of imposex in the dogwhelk Nucella lapillus. Mar Poll Bull 18:400–404

Dirkx WR, Van Mol WE, Van Cleuvenbergen RJA, Adams FC (1989) Speciation of organotin compounds in water by gas chromatography/atomic absorption spectrometry. Fresenius Z Anal Chem 335:769–774

Dizikes LJ, Ridley P, Wood JMA (1978) Mechanism for the biomethylation of tin by reductive Co-C bond cleavage of alkycobalamins. J Am Chem Soc 100:1010–1012

Dixon DR, Prosser H (1986) An investigation of the genotoxic effects of an antifouling compound (bis(tributyltin) oxide) on the chromosomes of the edible mussle, Mytilus edulis. Aqua Toxicol 8:185–195

Donard OFX, Rapsomanikis S, Weber JH (1986) Speciation of inorganic tin and alkyltin compounds by atomic absorption spectrometry using electrothermal quartz furnace after hydride generation. Anal Chem 58:777–781

Donard OFX, Weber JH (1988) Volatilization of tin as stannane in anoxic environment. Nature 332:339–341

Donard OFX, Pinel (1989) Tin and germanium. In: Harrison RM, Rapsomanikis S (eds) Environmental analysis using chromatography interfaced with atomic spectrometry. John Wiley & Sons, Halsted Press, New York, pp 189–257

Dowson PH, Pershke D, Bubb JM, Lester JN (1992) Spatial distribution of organotins in sediments of lowland river catchments. Environ Poll 76:259–266

Ebinghaus R (1991) Aufnahme und Umwandlung von Quecksilber(II)- und Methylquecksilberchlorid durch Schwebstoffbakterien und mirobielle Biofilme-Untersuchungen zum Spezies-Verhalten des Quecksilbers in wäßrigen Systemen. Dissertation, Hamburg

Emanuel EL, Carver MA, Solani GC, Griffiths DE (1984) Different inhibition of F_0F_1-ATPase-catalysed reactions in bovine-heart submitochondrial particles by organotin compounds. Biochim Biophys Acta 766:209–214

Fent K, Müller MD (1991) Occurrence of organotins in municipal wastewater and sewage sludge and behaviour in a treatment plant. Environ Sci Technol 25:489–493

Fent K (1992) Embriotoxic effects of tributyltin on the minnow Phoxinus phoxinus. Environ Poll 76: 187-194

Fanchiang Y-T, Wood JM (1981) Alkylation of tin by alkylcobalamins: kinetics and mechanisms. J Am Chem Soc 103:5100–5103

Guard HE, Cobet AB, Coleman WM (1981) Methylation of trimethyltin compounds by estuarine sediments. Science 213:770–771

Hamasaki T, Sata T, Nagase H, Kito H (1992) The genotoxicity of organotin compounds in SOS chromotest and rec-assay. Mutation Res 280, 195–203

Harino H, Fukushima M, Tanaka M (1992) Simoultaneous determination of butyltin and phenyltin compounds in the environment by gas chromatography. Anal Chem Acta 264:91–96

Hallas LE, Means C, Cooney JJ (1982) Methylation of tin by estuarine microorganisms. Science 215:1505–1507

Heywood BR, Molloy KC, Waterfield PC (1989) Organotin biocides XV: modelling the interactions of triorganotins with cell membranes. Appl Organomet Chem 3:443–450

Hodge VF, Seidel SL, Goldberg ED (1979) Determination of tin(IV) and organotin compounds in natural waters, coastl sediments and macro algae by atomic absorption spectrometry. Anal Chem 51:1256–1259

Huey C, Brinckman FE, Grim S, Iverson P (1974) The role of tin in bacterial methylation of mercury. In: Microbial biochemistry, Proc Int Conf Transport of persistant chemicals in aquatic ecosystem II, August 1993, National Research Council, Ottawa, Canada 10-12

Jackson JAA, Blair WR, Brinkman FE, Iverson WP (1982) Gas-chromatographic speciation of methylstannanes in the Chesapeake Bay using purge and trap sampling with a tin-selective detector. Environm Sci Technol 16:110–119

Jantzen E, Wilken R-D (1991) Zinnorganische Verbindungen in Hafensedimenten – Analytik und Beurteilung. Vom Wasser 76:1–11

Jantzen E (1992) Analytische Untersuchungen zur Beurteilung der Bindungsformen (Speciation) des Zinns in Elbesedimenten unter Zuhilfenahme von GC-AAS, GC-MS, TRFA und GFAAS, Dissertation Hamburg

Jantzen E, Kuballa J, Wilken R-D (1993) Muster zinnorganischer Verbindungen in der Elbe. Vom Wasser 80:245–252

Kadokami K, Uehiro T, Morita M, Fuwa K (1988) Determination of organotin compounds in water by bonded-phase extraction and high-performance liquid chromatography with long-tube atomic absorption spectrometric detection. J Anal Atom Spec 3:187–191

Kuballa J, Wilken R-D, Jantzen, E, Kwan KK, Chau YK (1995) Speciation and genotoxicity of butyltin compounds. Analyst 120, 667–673

Laughlin RB, French WJ (1980) Comparative study of the acute toxicity of a homologoues series of trialkyltins to larval shore crabs, Hemigrapsus Nudus, and lobster, Homarus Americanus. Bull Environ Contam Toxicol 25:802–809

Laughlin RB, Johanesen RB, French WJ, Guard HG, Brinckman FE (1982) Structure-activity relationsships for organotin compounds. Environ Toxicol Chem 4: 343-351

Laughlin RB, French WJ, Johanesen RB, Guard HG, Brinckman FE (1984) Predicting toxicity using compound molecular topologies: the example of triorpanotincompounds. Chemosphere 13:575–584

Lawler IF, Aldrich JC (1987) Sublethal effects of bis(tri-n-butyltin)oxide on Crassostrea Gigas Spat. Mar Poll Bull 18:274–278

Li AP, Dahl AR, Hill JO (1982) In vitro cytotoxicity of dibutyltin dichloride and dibutylgermanium dichloride. Toxicol Appl Pharmacol 64:482–485

Liu Y, Lopez-Avila V, Alcaraz M (1993) Determination of organotin compounds in environmental samples by supercritical fluid extraction and gas chromatography with atomic emission detection. J High Res Chrom 16: 106-112

Luijten JGA (1972) Application and biological effects of organotin compounds. In: Sayer AK (ed) Organotin Compounds; Marcel Dekker Inc; New York; pp 931–974

Maguire RJ, Huneault F (1981) Determination of butyltin species in water by gas chromatograpgy with flame photometric detection. J Chromatogr 209:458–462

Maguire RJ, Carey JH, Hale EJ (1983) Degradation of tri-n-butyltin species in water. J Agric Food Chem 31:1060–1065

Maguire RJ (1984) Butyltin compounds and inorganic tin in sediments in Ontario. Environ Sci Technol 18:291–294

Maguire RJ, Tkacz RJ, Sartor DL (1985) Butyltin species and inorganic tin in water and sediment of the Detroit and St. Clair rivers. J Great Lakes Res 11:320–327

Maguire RJ, Tkacz RJ, Chau YK, Bengert GA, Wong PTS (1986) Occurrence of organotin compounds in water and sediment in Canada. Chemosphere 15:253–274

Maguire RJ (1987) Environmental aspects of tributyltin. Appl Organom Chem 1:475–498

Maguire RJ (1991) Aquatic environmental aspects of non-pesticidal organotin compounds. Water Poll Res J Canada 26:243–360

Manabe S, Wada O, Matsui H, Takada M, Kobayshi N, Maekawa T (1983) Triphenyltin fluoride "in vitro" inhibition of rabbit platelet collagen-induced aggregation and ATP secretion and blockade of archidonic acid mobilization from membrane phospholipids. Biochem Pharmacol 32:1627–1634

Marshall WD, Blais JS, Adams FC (1989) HPLC-AAS interfaces for the determination of ionic alkyllead, alkyltin, arsonium and selemium compounds. In: Broekaert JAC, Gücer S, Adams FC (eds) Metal speciation in the environment. Springer, Berlin Heidelberg New York, pp 253–306

Martin DC, Waniewski RA, Wolpaw EW (1983) Inhibitory effect of triethyltin on taurine transport by Glioma cells. Toxicol Appl Pharmacol 71:155–162

Martin FM, Tseng C-M, Belin C, Quevauvileer Ph, Donard OFX (1994) Interferences generated by organic and inorganic compounds during organotin speciation using hydrid generation coupled with cryogenic trapping, gas chromatographic separation and detection by atomic absorption spectrometry. Anal Chim Acta 286: 343-355

Matthias CL, Bellama JM, Olson GJ, Brinckman FE (1986) Comprehensive method for the determination of aquatic butyltin and butylmethyltin species at ultratrace levels using simultaneous hydratization/extraction with gas chromatography-flame photometric detection. Environ Sci Technol 20:609–615

Meinema HA, Burger-Wiersma T, Versluis-de Haan G, Chevers EC (1978) Determination of trace amounts of butyltin compounds in aqueous systems by gas chromatography/mass spectrometry. Environ Sci Technol 12:288–293

Müller MD (1984) Tributyltin detection at trace levels in water and sediments using GC with flame-photometric detection and GC-MS. Fresenius Z Anal Chem 317:32–36

Müller MD (1987) Comprehensive trace level determination of organotin compounds in environmental samples using high-resolution gas chromatography with flame photometric detection. Anal Chem 59:617–623

Müller MD, Renberg L, Rippen G (1989) Tributyltin in the environment – sources, fate and determination. An assessment of present status and research needs. Chemosphere 18:2015–2042

Nygren O, Nilsson CA, Frech W (1988) On-line interfacing of a liquid chromatograph to a continuously heated graphite furnace atomic absorption spectro-photometer for element-specific detection. Anal Chem 60:2204–2208

Parks EJ, Brinkman FE, Jewett KL, Blair WR, Weiss CS (1988) Trace speciation by HPLC-GFAA for tin- and lead-bearing organometallic compounds, with signal increases induced by transition-metal ions. Appl Organom Chem 2:441–450

Polster M, Halacka K (1972) Hygenic-toxicological problems of some anticrobially used organotin compounds. Ernährungsforschung 16:527–535

Prange A, Jantzen E (1995) Determination of organometallic species by gas chromatography inductively coupled plasma mass spectrometry. J Anal Atom Spetrom 10: 105-109

Oehlmann J, Stroben E, Bettin C, Fioroni P (1993) TBT-induczierter Imposex und seine physiologischen Ursachen bei marinen Vorderkiemerschnecken. SDN-Kolloquium Antifouling im Meer-Gefahren durch Schiffsanstriche? Schutzgemeinschaft Deutsche Nordseeküste eV (Herg) Heft Nr 2, Wilhelmshafen, ISSN 0943-9552, pp 58–72

Quevauviller Ph (1989) Organo-tins in sediments and Mussels from the Sado estuarine system (Portugal). Environ Poll 57:149–166

Quevauviller Ph, Donard OFX (1990) Variability of butyltin determination in water and sediment samples from European coastel environment. Appl Organom Chem 4:363–367

Quevauviller Ph, Donard OFX (1991) Organotin stability during storage of marine waters and sediments. Fresenius Z Anal Chem 339:6–14

Quillardet P, Hofnung M (1985) The SOS chromotest, a colorimetric bacterial assay for genotoxins: procedure. Mutation Res 147:65–78

Randall L, Han J S, Weber J H (1986) Determination of inorganic tin, methyltin and butyltin compounds in sediments. Environ Technol Lett 7:571–576

Rapsomanikis S, Weber JH (1985) Environmental implications of methylation of tin(II) and methyltin(II) ions in the presence of manganese dioxide. Environ Sci Technol 19:352–356

Ritchie LS, Lopez VA, Cora JM (1974) Prolonged applications of an organotin against Biomphalaria glabrata and Schistosoma mansoni. In: Cheng TH (ed) Molluscidides in Schistosomiasis control. Academic Press, New York, pp 77–78

Sarradin PM, Astruc A, Desauziers V, Pinel R, Astruc M (1991) Butyltin polution in surface sediments of Arcachan Bay after ten years of restricted use of TBT-based paints. Environm Technol 12: 537-543

Saxena AK (1987) Organotin compounds: toxicology and biomedicinal applications. Appl Organom Chem 1:39–56

Schebeck L, Andreae MO, Tobschall HJ (1991) Methyl- and butyltin compounds in water and sediments of the Rhine river. Environ Sci Technol 25:871–878

Scott BF, Chau YK, Rais-Fiouz A (1991) Determination of butyltin species by GC/atomic emission spectroscopy. Appl Organom Chem 5:151–157

Seinen W, Helder T, Vernij H, Penninks, Leeuwangh P (1981) Short term toxicity of tri-n-butyltin chloride in rainbow trout (Slmo Gairneri Richardson) York Sac Fry. Sci Total Environ 19:155–166

Short JW, Thrower FP (1986) Tri-n-butyltin caused morality of chinook salmon, Onorynchus Tshawytscha, on transfer to a TBT-treated marine net pen, Proc Organotin Symposium of the Oceans '86 Conf, Washington DC, Sept 23–25 1986, Vol 4, IEEE, New York, pp 1202–1205

Soderquist CJ, Crosby DG (1978) Determination of triphenyltin hydroxide and its degradation products in water. Anal Chem 50 1435–1439

Stewart C, de Mora SJ, Jones MRL, Miller MC (1992) Imposex in New Zealand neogastropods. Marine Poll Bull 24:204–209

Stromgren T, Bongard T (1987) The effect of tributyltin oxide on growth of Mytilus Edulis. Mar Poll Bull 18:30–31

Suyani H, Creed J, Davidson T, Caruso J (1989) Inductively coupled plasma mass spectrometry and atomic emission spectrometer coupled to high performance liquid chromatography for speciation and detection of organotin compounds. J Chromatogr Sci 27:139–140

Thayer JS (1989) Methylation: its role in the environment mobility of heavy elements. Appl Organom Chem 3:123–128

Thompson JRJ, Scheffer MG, Pierce RC, Chau YK, Cooney JJ, Cullen WR, Maguire RJ (1985) Organometallic compounds in the aquatic environment: scientific criteria for assessing their effects on environmental quality: NRCC No 22494; National Research Council Canada, Ottawa, Canada

Tolosa I, Baynona JM, Albaigés J, Alencastro LF, Tarradelas J (1991) Organotin speciation in aquatic matrices by CGC/FPD, ECD and MS, and LC/MS. Fresenius J Anal Chem 339: 646-653

Tolosa I, Dachs J, Bayona JM (1992) Tributyltin speciation in aquatic matrices by CGC-FPD and CGC-MS confirmation. Mirochim Acta 109:87–91

Tsuda T, Nakanishi H, Aoki S, Takebayashi J (1987) Determination of butyltin and phenyltin compounds in biological and sediment samples by electron-capture gas chromatography. J Chromatogr 387:361–370

Valkirs AO, Seligman PF, Olson GJ, Brinckman FE, Matthias CL, Bellama JM (1987) Di- and tributyltin species in marine and estuarine waters. Interlaboratory comparison of two ultratrace analytical methods employing hydride generation and atomic absorption or flame photometric detection. Analyst 112:17–21

Waldock MJ, Thain JE (1983) Shell thickening in Crassostrea Gigas: organotin antifouling of sediment induced. Mar Poll Bull 14:411–415

Wuertz S, Miller ME, Doolittle MM, Brennan JF, Cooney JJ (1991) Butyltins in estuarine sediments two years after tributyltin use was restricted. Chemosphere 22:1113–1120

Wylie PL, Quimby BD (1990) Application of gas chromatography with an atomic emission detector. J High Resolut Chromatogr 12:813–818

Yarlet N, Lloyd D (1981) Effects of inhibitors on mitochondrial adenosine triphosphatase of Crithidia Fasciculta: an unusual pattern of specificities. Mol Biochem Parasitol 3:13–17

Zhang S, Chau YK, Li WC, Chau ASY (1991) Evaluation of extraction techniques for butyltin compounds in sediment. Appl Organom Chem 5:431–434

Zuckerman JJ, Reisdorf RP, Ellis HV, Wilkinson RR (1978) Organotins in biology and the environment. In: Brinckman FE, Bellama JM (eds) Organometals and organometalloids occurrence and fate in the environment. American Chemical Society No 82, Washington DC, ACS Symposium, pp 388–422

11 Advective Transport of Interacting Solutes: The Chromatographic Model

Joachim Gruber

11.1 Introduction

In the past decade affordable computers have become capable of calculating transport of several chemically and biologically interacting components in porous media. Examples of such components are natural ions, ligands, nutrients and contaminants. Many transport codes have been written, as is evident from reviews (e.g. Kirkner and Reeves 1988; Reeves and Kirkner 1988; Kinzelbach et al. 1989). Numerical difficulties and long computation times have been addressed in various ways, depending on the choice of the geochemical problem. Subroutines, e.g. PHREEQE, MINEQL or other members of the MINEQL family, such as MINTEQ, GEOCHEM or HYDROQL, are called at each time step to establish chemical equilibrium in all cells of the spatial discretisation (nodes ; Walsh et al. 1984; Cederberg et al. 1985; Novak et al. 1988; Berninger et al. 1991). When applied to the nodes sequentially, these subroutines need more than 90% of the computation time. Soon excessive time is spent in these subroutines when the number of nodes is increased, unless the computer code has been vectorized, thus being able to establish chemical equilibrium in all nodes at once (Vogt 1990). Lichtner (1992) and Ortoleva and co-workers (1987) went in another direction, avoiding super- or mini-supercomputers by introducing very potent approximations in the transport equations for the case of mineral precipitation and dissolution.

The mathematics of non-linear sets of hyperbolic partial differential equations in one space co-ordinate and time (Lax 1973) and multicomponent chromatography (Helfferich and Klein 1970; Rhee et al. 1989) provides tools that help us understand and design numerical calculations as well as experiments when transport is purely advective and dominated by adsorption. The procedure to calculate the concentration profile is simpler and faster than with the previous

mentioned numerical models, largely because general properties of the solution of the problem are derived from mathematics rather than left to be found by the computer (Schweich et al. 1993). This way of proceeding has been used successfully in the field of chemical engineering (Helfferich 1967; Hwang et al. 1988). It has only recently entered into environmental sciences (Charbeneau 1988). The purpose of Chapter 11 is to demonstrate its potential in the latter field.

After having introduced the *transport equations* (Sect. 11.2) and the concept of the *centred waves* (Sect. 11.3), Chapter 11 will:

1. Present several solution techniques for hyperbolic differential equations, i.e.
 analytical integration of the transport equations (Sect. 11.4),
 the use of the coherence conditions (Sect. 11.5),
 the use of the Jacobi matrix of the isotherms (Sect. 11.6), and

2. Use this method to demonstrate the relationship that exists between the chemical interactions on the one hand – as represented by some adsorption isotherms and complex formation constants- and on the other hand, the concentration profiles (Sect. 11.7). In this context, isotherms will be presented for:
 - *ligand exchange* (Sect. 11.7.1),
 - *surface bridge formation* (Sect. 11.7.1),
 - *adsorption on variable charge surfaces at low ionic strength* (Sect. 11.7.1).

 The *properties of the waves* in systems with some of these isotherms will be used to solve the system of transport equations by matrix inversion (Sect. 11.7.2), introducing a generalized chemical component, the *Riemann invariant* (Sect. 11.7.2).

3. Exemplify the presented analytical solutions with *two-component systems* (see Sect. 11.8) because reliable adsorption data have been compiled mostly for such systems.

Chemists have been using speciation plots side by side with numerical speciation models to gain insight into the influence of complex formation constants in batch systems (see e.g. Chaps. 4.1–4.3 in the textbook by Stumm and Morgan 1981). Multicomponent chromatography provides us a similar graphical tool to help us understand the influence of thermodynamic constants on advective transport and to give us information we cannot get from a numerical transport model. This will be the topic of Sections 11.2 and 11.3.

11.2 Transport Equations

Migration of N_0 (the number of chemical components) A_j ($j = 1, 2,..., N_0$) in porous media is described by a set of N_0 mass conservation laws. In the absence of precipitation or dissolution of solids, the conservation laws have the form:

$$\frac{\partial}{\partial t} \{\phi c_j(x,t) + C_j(x,t)\} + \frac{\partial}{\partial x} (v_w - D \frac{\partial}{\partial x}) c_j(x,t) = 0 \quad (j = 1, ..., N_0) \, . \tag{1}$$

c_j is the soluble and C_j is the adsorbed concentration of component A_j. The constants $\phi \approx 0.3$, $v_w \leq 0.5$ m/a and $D = 0.16$ (m^2/a) $\tau + \delta \xi$ are the volumetric fraction of water, the flux of water and the hydrodynamic dispersion/diffusion coefficient of A_j in the porous medium, respectively. τ (≈ 0.3) is the tortuosity, δ (units: m) is the dispersivity, ($\approx L/100$, L units: m) being the length over which transport is calculated) and ξ is the speed of the concentration c_j (for definition of ξ see Eq. 9).

The adsorbed concentration C_j of component A_j relaxes to the value C_j^*, the so-called isotherm, with a relaxation time ε_j (first order kinetics)

$$\frac{\partial}{\partial t} C_j(x,t) = \frac{1}{\varepsilon_j} \{ C_j^*(c_1, c_2, ..., c_{N_0}) - C_j(x,t) \} \quad (j = 1, 2, ..., N_0) . \quad (2)$$

This chapter will discuss concentrations profiles when the flux of water is typically $v_w < 0.5$ m/a. Adsorption kinetics is then fast enough and has a negligible effect on the shape of the profile if $\varepsilon < 1$ week (see Appendices A and B). Therefore $C_j = C_j^*$ will be assumed henceforth.

When also the diffusive term in (Eq. 1) is neglected, the calculated concentration profiles will lack smoothness, but are expected to remain unchanged otherwise. Because this research aims at finding trends caused by chemical interactions, the assumption $D = 0$ seems appropriate.

The resulting set of partial differential equations is hyperbolic and due to the isotherm $C_j^*(c_1, c_2..., c_{N_0})$- non-linear in the concentrations c_j.

$$\frac{\partial}{\partial t} \{ \phi c_j(x,t) + C_j^*(c_1,c_2, ..., c_{N_0}) \} + v_w \frac{\partial}{\partial x} c_j(x,t) = 0 \quad (j = 1, 2, ..., N_0) . \quad (3)$$

Without giving the proofs, Sections 11.3 and 11.4 will present two basic phenomena in hyperbolic systems:

- The *centred wave*, its representation in concentration space and its transport equation, and
- The *speed* or *retardation* of the centred wave.

The first method presented in Section 11.3 for finding solutions of the transport equations is connected with the equation defining the speed of a concentration. These definitions are called "coherence conditions" in chromatography. Three further solution methods will be presented thereafter.

11.3 Riemann Problem and Centred Waves

Solutions of (Eq. 3) can be built from so-called centred waves (Helfferich and Klein 1970).

There are two types of centred waves (Lax 1973):

- Rarefaction waves (called "diffuse waves" in chromatography). They meet the differential equations (see Eq. 3). This type will mostly be dealt with here.
- Shocks (called "self-sharpening waves" in chromatography). They meet the Rankine-Hugoniot relations (see Eq. 24), which are called "integral coherence conditions" in chromatography (Helfferich and Klein 1970).

Temple (1983) has shown that shocks and rarefaction waves coincide in two types of multicomponent systems with constant adsorption parameters and negligible soluble complexes: (a) when adsorption follows a multicomponent Langmuir isotherm (see Eq. 46); or (b) when adsorption can be described as an exchange process (see Eq. 41).

Centred waves are the solution of the Riemann problem, i.e. the response of the

Fig. 11.1 a–b. Centred waves in a system with component M and H **a** without and **b** with competition between M and H for adsorption sites. The concentration jump $c_- = \{M_-, H_-\}$ -> $c_+ = \{M_+, H_+\}$ at time $t = 0$ develops into two waves

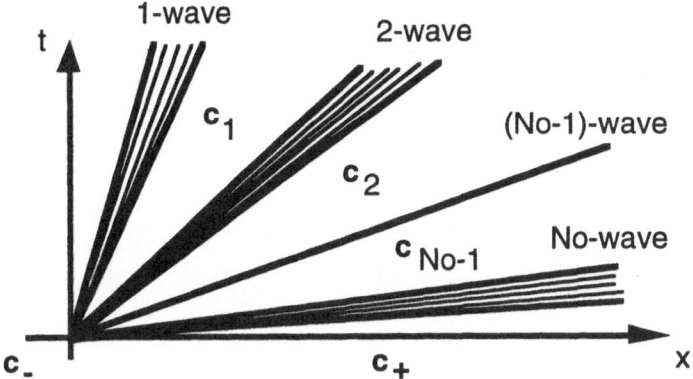

Fig. 11.2. In a system with N_0 components an initial discontinuity $c_- \rightarrow c_+$ decomposes with time into a fan of N_0 concentration changes (*centred waves*, bounded by *heavy lines*), separated by $N_0 - 1$ constant states $c_1, c_2,..., c_{No-1}$. Notation has been chosen according to conventions in mathematics

system (Eq. 3) to a single abrupt change of the chemical composition of the incoming ("feed") water (from c_+ to c_-). Let $x = 0$ be the initial location of this concentration jump, and assume that the pore water flows in positive x-direction:

$$c(x, t{=}0) = \left\{ \begin{array}{ll} c_+ \text{ for } x \geq 0 & \text{(pre-equilibrant)} \\ c_- \text{ for } x < 0 & \text{(feed)} \end{array} \right. , \qquad c = \left[\begin{array}{c} c_1 \\ ... \\ c_{N_0} \end{array} \right] . \qquad (4)$$

An example of centred waves in a two-component system (with components M and H) is shown in Fig. 11.1.

When there is no competition (Fig. 11.1a) between the components M and H e.g. for adsorption sites, H changes without M noticing it and vice versa. Thus, each wave corresponds to a component. In Fig. 11.1b, M and H compete for adsorption sites, and the concentrations of both M and H change in both waves.

The general structure of the solution of the Riemann problem has been described and proved by Lax (1957): the discontinuity (see Eq. 4) develops into a sequence of constant states separated by centred waves (Fig. 11.2). The former are regions of fixed concentration. The heavy lines in Fig. 11.1 are the boundaries of the constant states. The concentration changes from c_k to c_{k+1} across the k-wave, either continuously as shown in Fig. 11.2 for the 1-, 2- and N_0-waves, which are rarefaction waves, or abruptly as shown for the (N_0-1)-wave, a shock.

Because all waves originate from the initial discontinuity at $x = 0$, for times $t>0$ the vector of concentrations c does not depend on x and t separately, but instead on the speed, the ratio $\xi = x/t$.

$$c(x,t) = h(\xi). \qquad (5)$$

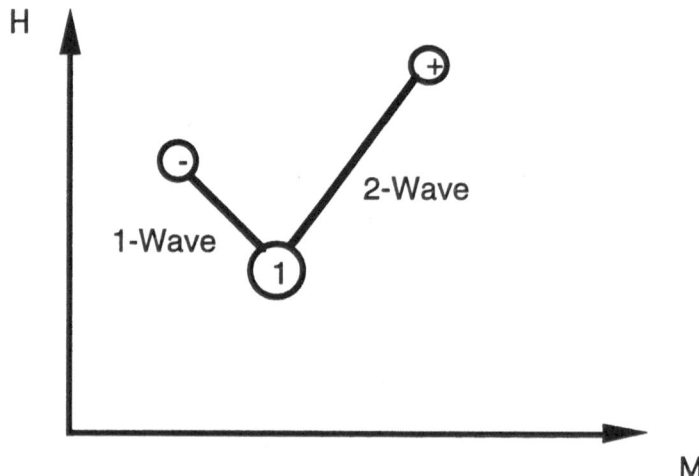

Fig. 11.3. Waves (shown arbitrarily as *lines*) and constant states: points -, *1* and + in 2-dimensional concentration space. *Point* - has the co-ordinates $\{M_-, H_-\}$, *point* + has the co-ordinates $\{M_+, H_+\}$, and *point 1* represents $\{M_1, H_1\}$

A concentration $\mathbf{c}(x,t) = \mathbf{h}(\xi)$ is constant on the line $x/t = \xi$, the "characteristic" (light or heavy lines in Fig. 11.2).

It is helpful to represent the concentration profile $\mathbf{c}(x,t) = \mathbf{h}(\xi)$ in concentration space (called "composition space" by Helfferich and Klein 1970). As an example, Fig. 11.3 shows the two waves of Fig. 11.1b. The concentration space has two axes, the M- and the H-axis. The constant state $\mathbf{c}_- = \{M_-, H_-\}$ is connected by the 1-wave to the middle state $\mathbf{c}_1 = \{M_1, H_1\}$, which in turn is connected by the 2-wave to the constant state $\mathbf{c}_+ = \{M_+, H_+\}$.

By definition of the 1-wave, point $\{M_1, H_1\}$ moves on the 1-wave when point $\{M_+, H_+\}$ is varied and point $\{M_-, H_-\}$ remains fixed. The definition of the 2-wave is similar: It is the curve on which point $\{M_1, H_1\}$ moves when point $\{M_-, H_-\}$ is shifted in $\{M, H\}$ space while point $\{M_+, H_+\}$ stays immobile.

The shapes of the waves reflect the chemical interactions in the system. In Fig. 11.2 they are arbitrarily represented as lines. The correspondence between their shape and the thermodynamic constants is the subject of the second part of this chapter (Sects. 11.7–11.9).

For centred waves, the transport equations (Eq. 3) become ordinary differential equations. With $\xi = \dfrac{x}{t}$ the partial derivatives become total derivatives

$$\frac{\partial}{\partial x} = \frac{\partial \xi}{\partial x} \frac{d}{d\xi} = \frac{1}{t} \frac{d}{d\xi} \quad ,$$

$$\frac{\partial}{\partial t} = \frac{\partial \xi}{\partial t} \frac{d}{d\xi} = -\frac{x}{t^2} \frac{d}{d\xi} = -\frac{\xi}{t} \frac{d}{d\xi} \quad , \tag{6}$$

and with these the transport equations can be written as

$$(\frac{v_w}{\phi} - \xi) \, h_j' - \frac{\xi}{\phi} \, H_j' = 0 \qquad (j = 1, 2, \ldots, N_0) \,, \tag{7}$$

where the prime designates the derivative with respect to ξ, and the vectors of soluble and adsorbed concentrations, respectively, are defined as

$$
\mathbf{h}(\xi) = \begin{bmatrix} h_1(\xi) \\ \ldots \\ h_{N_0}(\xi) \end{bmatrix} = \begin{bmatrix} c_1(x,t) \\ \ldots \\ c_{N_0}(x,t) \end{bmatrix}; \quad \mathbf{H}(\mathbf{h}(\xi)) = \begin{bmatrix} H_1(\mathbf{h}(\xi)) \\ \ldots \\ H_{N_0}(\mathbf{h}(\xi)) \end{bmatrix} = \begin{bmatrix} C_1^*(c(x,t)) \\ \ldots \\ C_{N_0}^*(c(x,t)) \end{bmatrix}. \tag{8}
$$

11.4 Analytical Integration of Transport Equations

If the sum of the adsorbed component concentrations H_j is a constant, e.g. a (constant) multiple of the concentration of adsorption sites,

$$\sum_{k=1}^{N_0} H_k = n \, X_T \tag{9}$$

or

$$\sum_{k=1}^{N_0} H_k' = 0 \,,$$

the sum of the set of equations (see Eq. 7) is simply

$$(\frac{v_w}{\phi} - \xi) \sum_{k=1}^{N} h_k' = 0 \,. \tag{10}$$

Possible solutions are

$$\xi = \frac{v_w}{\phi} \quad \text{together with} \quad H_j = 0 \quad \text{(non-retarded wave)}, \tag{11}$$

$$\sum_{k=1}^{N_0} h_k' = 0 \quad \text{(retarded waves)} \,. \tag{12}$$

When the number of components, N_0 is two, the waves are readily plotted as contour lines. The retarded wave is the line on which $h_1 + h_2$ is constant, the non-retarded wave is the line on which any adsorbed component concentration H_j is

constant. High level programming languages, e.g., Mathematica® (Wolfram Research Inc. 1992), make it easy to plot contour lines of these conserved quantities, $h_1 + h_2$ and H_j, in concentration space (see Fig. 11.8).

11.5 Coherence Conditions, Rankine-Hugoniot Relation and Retardations

Solving Eq. 7 for $\xi = \xi(\mathbf{h})$ we get

$$\xi = \frac{v_w}{\phi} \; \frac{1}{1 + \frac{1}{\phi}\frac{H_j'}{h_j'}} = \frac{v_w}{\phi} \; \frac{1}{1 + \frac{1}{\phi}\frac{dH_j/d\xi}{dh_j/d\xi}} = \frac{v_w}{\phi} \; \frac{1}{1 + \frac{1}{\phi}\frac{dH_j}{dh_j}} \qquad (j = 1, ..., N_0) .$$

The directional derivative dH_j/dh_j (i.e. the change of the adsorbed concentration across the wave, $dH_j/d\xi$, divided by the change of the soluble concentration, $dh_j/d\xi$) depends on the wave along which it is evaluated. Let k be the index for the k-wave, then, for any k $(1 \le k \le N_0)$

$$\xi_k = \frac{v_w}{\phi} \; \frac{1}{1 + \frac{1}{\phi}\frac{dH_j/d\xi_k}{dh_j/d\xi_k}} \qquad (j = 1, 2, ..., N_0) . \tag{13}$$

The k solutions of Eq.13, i.e. of

$$\frac{dH_1/d\xi}{dh_1/d\xi} = \frac{dH_2/d\xi}{dh_2/d\xi} = ... = \frac{dH_{N_0}/d\xi}{dh_{N_0}/d\xi} \tag{14}$$

or of

$$\frac{dH_1}{dh_1} = \frac{dH_2}{dh_2} = ... = \frac{dH_{N_0}}{dh_{N_0}} \tag{15}$$

are also solutions of the set of conservation laws (Eq. 3). Equation 15 has been given the name "differential coherence condition" in chromatography (Helfferich and Klein 1970).

If the isotherms are known functions $H_j(\mathbf{h})=C_j^*(\mathbf{c})$ of the vector of soluble concentrations, the directional derivative is

$$\frac{d}{d\xi} = \sum_{j=1}^{N_0} \frac{\partial}{\partial h_j} \frac{dh_j}{d\xi} \tag{16}$$

and Eq. 14 can be integrated. The k roots of Eq. 14 are the k solutions of the Riemann problem.

The procedure will be illustrated for the general two-component system. We start with the differential coherence condition in the form of Eq.14, and then apply the chain rule (see Eq. 16) to expand $d/d\xi$:

$$\frac{dH_1/d\xi}{dh_1/d\xi} = \frac{dH_2/d\xi}{dh_2/d\xi} \tag{17}$$

$$\frac{dH_1/d\xi}{dh_1/d\xi} = \frac{\partial H_1}{\partial h_1} + \frac{\partial H_1}{\partial h_2}\frac{dh_2/d\xi}{dh_1/d\xi} = \frac{\partial H_1}{\partial h_1} + \frac{\partial H_1}{\partial h_2}\frac{dh_2}{dh_1} \tag{18}$$

$$\frac{dH_2/d\xi}{dh_2/d\xi} = \frac{\partial H_2}{\partial h_1}\frac{dh_1/d\xi}{dh_2/d\xi} + \frac{\partial H_2}{\partial h_2} = \frac{\partial H_2}{\partial h_1}\frac{dh_1}{dh_2} + \frac{\partial H_2}{\partial h_2} \tag{19}$$

With Eqs. 18 and 19, Eq. 17 becomes a quadratic equation for dh_2/dh_1 with known coefficients p and q

$$(\frac{dh_2}{dh_1})^2 + p\ \frac{dh_2}{dh_1} + q = 0 \tag{20}$$

where

$$p = \left(\frac{\frac{\partial H_1}{\partial h_1} - \frac{\partial H_2}{\partial h_2}}{\frac{\partial H_1}{\partial h_2}}\right), q = - \left(\frac{\frac{\partial H_2}{\partial h_1}}{\frac{\partial H_1}{\partial h_2}}\right) \tag{21}$$

There are two roots of Eq. 20, one for the 1-wave and the other for the 2-wave

$$\frac{dh_2}{dh_1} = -\frac{p}{2} \pm \sqrt{\frac{p^2}{4} - q}\ . \tag{22}$$

High level programming languages such as Mathematica® (Wolfram Research Inc 1992) can be readily used to calculate the waves from Eq. 22. These languages give analytical functions $h_2(h_1)$ when they can find them and otherwise, give interpolating polynomials .

Similarly, the speed of the k-wave can be calculated, once the solutions $h_k(\xi) = \{h_1(\xi), h_2(\xi),..., h_{N_0}(\xi)\}$ are known. According to Eq. 13 the speed of the k-rarefaction wave is by a factor (the so-called retardation)

$$R_k = 1 + \frac{1}{\phi}\ \frac{dH_j/d\xi_k}{dh_j/d\xi_k} \qquad (j = 1, 2, ..., N_0) \tag{23}$$

smaller than the pore water velocity v_w/ϕ. Note that Eq. 23 states that all components A_j ($j = 1, 2,..., N_0$) experience the same retardation R_k in the k-wave.

Rarefaction waves are not the only type of solution of the mass conservation law (Eq. 3). The existence of the second type of centred wave, the shock, will not be proved. The proof of the existence of shocks (Lax 1957) will be replaced by a plausibility argument.

Suppose that the retardation $R_k(\mathbf{h})$ decreases with increasing concentration \mathbf{h}. When the rear end of a wave (e.g. point c_- of the 1-wave in Fig. 11.2) has a lower retardation than its front (point c_1 in Fig. 11.2), the rear will travel faster than the front. The corresponding concentration profile $c(x, t)$ will become steeper (sharper) as the wave proceeds. A concentration wave steepens until it has developed a discontinuity and is just about to break (in the sense a water wave breaks). From then on, all concentrations travel with the same speed. The resulting concentration discontinuity is the shock. Contrary to a concentration wave, the water wave develops beyond this point and actually breaks, i.e. the portions in the upper part of the wave overtake the ones in the lower part. The shock wave travels with a speed given by the conservation of mass across the shock.

Appendix C shows that the mass balance involves finite, rather than infinitesimal quantities, if the wave is a shock (Lax 1973). The speed of the shock is given by the Rankine-Hugoniot relation

$$\sigma_k = \frac{v_w}{\phi} \frac{\Delta c_j}{\Delta(c_j + \frac{1}{\phi} C_j^*)} = \frac{v_w}{\phi} \frac{1}{1 + \frac{1}{\phi}\frac{\Delta C_j^*}{\Delta c_j}} = \frac{v_w}{\phi} \frac{1}{R_k} \qquad (j = 1, 2, ..., N_0), \qquad (24)$$

where the symbol Δ denotes the change across the k-shock. In mathematics the symbol $\|\ \|$ is often used instead of Δ. Using h and H instead of c and C*

$$\sigma_k = \frac{v_w}{\phi} \frac{1}{1 + \frac{1}{\phi}\frac{\Delta H_j}{\Delta h_j}} \qquad (j = 1, 2, ..., N_0). \qquad (25)$$

The set of equations

$$\frac{\Delta H_1}{\Delta h_1} = \frac{\Delta H_2}{\Delta h_2} = ... = \frac{\Delta H_{N_0}}{\Delta h_{N_0}} \qquad (26)$$

is called "integral coherence condition" in chromatography and is used to calculate the concentration changes Δh_j across the shocks (Lax 1973).

The third solution method is based on the Jacobi-matrix of the adsorption isotherms $C_j^*(\mathbf{c})$. It is the most general of the solution methods.

11.6 The Jacobian and the Retardation Matrix

Because the adsorbed concentrations $H_j = C^*_j(c)$ depend only on the soluble concentrations, the derivatives with respect to ξ could be rewritten using the chain rule (see Eq. 16). The transport equation (see Eq. 7) becomes then

$$(\frac{v_w}{\phi} - \xi)\, h'_j - \frac{\xi}{\phi} \sum_{k=1}^{N_0} \frac{\partial C^*_j}{\partial c_k}\, h'_k = 0, \tag{27}$$

or – after division by ξ and rearrangement – in matrix form

$$\{\underline{I} + \frac{1}{\phi} \underline{Kd}\}\, h' = \frac{v_w}{\phi} \frac{1}{\xi} h' \ . \tag{28}$$

\underline{I} is the identity matrix and \underline{Kd} is composed of the partial derivatives of the adsorption isotherms C^*_j

$$\underline{I} = \begin{bmatrix} 1 & ... & 0 \\ ... & 1 & ... \\ 0 & ... & 1 \end{bmatrix}, \qquad \underline{Kd}(c) = \begin{bmatrix} \dfrac{\partial C^*_1}{\partial c_1} & ... & \dfrac{\partial C^*_1}{\partial c_{N_0}} \\ ... & ... & ... \\ \dfrac{\partial C^*_{N_0}}{\partial c_1} & ... & \dfrac{\partial C^*_{N_0}}{\partial c_{N_0}} \end{bmatrix} . \tag{29}$$

I will introduce the matrix $\underline{R}(c)$, the "retardation matrix":

$$\underline{I} + \frac{1}{\phi} \underline{Kd}(c) = \underline{R}(c) \ . \tag{30}$$

With the abbreviation (Eq. 30) the system of transport equations (Eq. 7) can be written as an eigenvalue problem

$$\underline{R}\, h'(\xi) = \frac{v_w}{\phi} \frac{1}{\xi} h'(\xi) . \tag{31}$$

This set of equations is satisfied when

$$\rho(h) = \frac{v_w}{\phi} \frac{1}{\xi(h)} \tag{32}$$

is an eigenvalue of the retardation matrix \underline{R} evaluated at the point $c = h$ in composition space, and when h', the change of h, is parallel to the right eigenvector corresponding to $\rho(h)$

$$h' = \gamma\ r(h) \ . \tag{33}$$

I will normalize \mathbf{r} such that $\gamma = 1$.

The system (Eq. 7) is genuinely non-linear, and the direction of \mathbf{r} is chosen such that it points in the direction of decreasing retardation

$\mathbf{r}\ \mathrm{grad}\,\rho(\mathbf{h}) < 0$

$\mathbf{h}'\ \mathrm{grad}\,\rho(\mathbf{h}) < 0$ (rarefaction wave), $\qquad\qquad$ (34)

where $\mathrm{grad} = \{\partial/\partial c_1, \partial/\partial c_2,..., \partial/\partial c_{N_0}\}$ is the change across the wave. For example, in Fig. 11.3 the change of the retardation at \mathbf{c}_- in direction to \mathbf{c}_1 is $\mathrm{grad}\,\rho_1(\mathbf{c}_-)$, where the index on ρ indicates the wave. Furthermore, if – as assumed in Fig. 11.3 – the waves are lines, $\mathbf{r}_1(\mathbf{c}_-)$ and $\mathbf{h}_1'(\mathbf{c}_-)$ are vectors that are co-linear with the 1-wave. If the retardation $\rho_1(\mathbf{h}_1)$ of the 1-wave decreases from \mathbf{c}_- to \mathbf{c}_1, the vectors $\mathbf{r}_1(\mathbf{c}_-)$ and $\mathbf{h}_1'(\mathbf{c}_-)$ point from \mathbf{c}_- to \mathbf{c}_1.

For a shock, the retardation increases in the direction of \mathbf{h}':

$\mathbf{h}'\ \mathrm{grad}\,\rho(\mathbf{h}) > 0$ (shock) .

The retardations $\rho_j(\mathbf{c})$ will be arranged in decreasing order. None of the retardations is smaller than 1:

$\rho_1 \geq \rho_2 \geq \rho_3 \geq ... \geq \rho_{N_0} \geq 1$. $\qquad\qquad$ (35)

According to Eq. 33 the change \mathbf{h}_k' of a concentration \mathbf{h}_k is parallel to the k-eigenvector \mathbf{r}_k at point \mathbf{h}_k. The k-rarefaction wave is the entity of all such changes. It follows from that, the k-rarefaction wave is the curve that is everywhere tangential to \mathbf{r}_k and can be constructed with a suitable Runge-Kutta method (Zielke 1993). Such methods are incorporated in the high level computer languages AVS (Application Visualization System; International AVS Center) or Mathematica® (Wolfram Research Inc. 1992). AVS constructs the k-wave on the computer screen after having received the vector field $\mathbf{r}_k(\mathbf{c})$ and the user has pressed the corresponding button.

11.7 Topology of Waves for Multicomponent Langmuir-Type Isotherms

11.7.1 Isotherms

Three Langmuir-type isotherms will be presented here, i.e. for multicomponent ligand exchange, for adsorption via surface bridge formation, and for surface complexation on variable charge interfaces at low ionic strengths. Let N_0 be the number of adsorbed species $\underline{A}_1, \underline{A}_2,..., \underline{A}_{N_0}$. The adsorptive bond is described by a set of parameters that are called "separation factors" α_{ij} or adsorption constants k_j, depending on the adsorption process.

Ligand Exchange. Unlike ion exchange processes, in which the surface bond is electrostatic, surface complexation is a predominantly chemical process. Complex formation – be it in solution or on surfaces – can bind electrically charged species together even if they carry charges of the same polarity. The surface complexation process changes the charge of surfaces. Thus, superimposed on the fixed chemical bond is an electrostatic bond that varies with the surface coverage, and thus, with the composition c of the solution (Dzombak and Morel 1990).

Fig. 11.4. Schematic of surface when adsorption process is ion or ligand exchange. Sites are occupied by components (ligands) $A_1, A_2,..., A_{N_0}$ represented by the numbers $1, 2,...,$ N_0. The adsorbed concentrations are described by isotherm (see Eq. 41). If c_N or the sum of the soluble concentrations of the adsorbing species is constant, the isotherm degenerates to a Langmuir-type isotherm (see Eq. 46)

Separation factors α_{iN} describe the exchange on the surface of ions or ligands A_N and A_i (see Fig. 11.4):

$$\underline{A}_N + A_j = \underline{A}_j + A_N, \quad (j = 1, 2, ..., N_0, j \neq N). \tag{36}$$

An example is the surface complexation of heavy metal cations M^{2+} on hydrous ferric oxides ($M^{2+} = Ag^{2+}, Ca^{2+}, CrOH^{2+}, Cu^{2+}, Ni^{2+}, Pb^{2+}, Sr^{2+}$ and of $SrOH^+$) (Dzombak and Morel 1990). Let X represent the crystal lattice of the oxide and let O represent the layer of oxygen atoms at the solid–liquid interface. The surface species involved in the adsorption process are then X-OH and X-OM$^+$. The adsorption reaction is

$$X\text{-OH} + M^{2+} = X\text{-OM}^+ + H^+. \tag{37}$$

Using the symbol C_j^* for the concentration of the surface species \underline{A}_j and c_j for the concentration of the adsorbing species, the mass action law for the adsorption reaction (Eq. 36) is

$$C_j^*(c) = \alpha_{jN} \frac{C_N^*(c)}{c_N} c_j, \quad (j = 1, 2, ..., N_0). \tag{38}$$

Note that $\alpha_{NN} = 1$. Because all surface sites are assumed to be filled, the sum of the surface species concentrations is equal to the concentration of surface sites:

$$\sum_{l=1}^{N_0} C_l^*(c) = X_T = \frac{C_N^*}{c_N} \sum_{l=1}^{N_0} \alpha_{lN} c_l \tag{39}$$

or

$$\frac{C_N^*}{c_N} = \frac{X_T}{\sum\limits_{l=1}^{N_0} \alpha_{lN}\, c_l} \; . \tag{40}$$

After replacing C_N^*/c_N in Eq. 38 with the expression in Eq. 40 we get the isotherm

$$C_j^*(c) = X_T \frac{\alpha_{jN}\, c_j}{\sum\limits_{l=1}^{N} \alpha_{lN}\, c_l}, \qquad (j = 1, 2, \ldots, N_0). \tag{41}$$

Note that the selectivity coefficients α_{jN} depend on the electric surface charge. If the latter is variable – as it is on metal oxides – it will be separated from the α's and the selectivity coefficient will be written as $\alpha_{jN}\, P_L^{\Delta z}$ with

$$P_L = \exp\left(-\frac{F\psi_L}{RT}\right). \tag{42}$$

Δz is the change of the charge of the adsorbing layer L (in units of the electric charge, $e = F/N_A$, N_A being the number of particles in a mole) in the surface complexation reaction. Δz is $+ 1$ for reaction 37 involving the exchange of a doubly charged metal ion with the proton. ψ_L (unit: Volt = Joule/Coulomb) is the electrostatic potential in the adsorbing layer L on the oxide, F (96485 Coulomb/mol) is the Faraday-constant, R, where 8.31 Joule/(mol K) is the general gas constant and T (unit: K) is the absolute temperature.

The electrostatic correction factor is a simple function of pH at low ionic strengths and small metal concentrations in solution. In Section 11.8 it will be shown that the isotherms assume a particularly simple form in that case.

Surface Bridge Formation. Adsorption can be the formation of bridges between a surface ligand A_N and components A_j (see Fig. 11.5). The surface species \underline{A}_N is called "empty surface site". It is thus not counted among the N_0 surface species $\underline{A}_N\underline{A}_j$.

$$\underline{A}_N + A_j = \underline{A}_N\underline{A}_j, \qquad (j = 1, 2, \ldots, N_0) \tag{43}$$

Fig. 11.5. Schematic of surface when adsorption process is bridge formation between a component (ligand) A_N and components A_1, A_2,..., A_{N_0}. The surface sites \underline{A}_N are sometimes called "empty". Adsorption isotherms for bridge formation are of Langmuir-type (Eq. 46)

On oxides A_N is sometimes written as X-OH. An example is the surface complexation of $M^{2+} = Ba^{2+}$, Ca^{2+}, Sr^{2+} (Dzombak and Morel 1990)

$$X\text{-}OH + M^{2+} = X\text{-}OHM^{2+} . \tag{44}$$

Adsorbed species concentrations are proportional to C^*_N, the concentration of empty sites, and c_j, the soluble concentration of the adsorbed species. The mass action law here is

$$C^*_j(c) = k_j C^*_N c_j, \qquad (j = 1, 2, ..., N_0, j \neq N) . \tag{45}$$

The concentration C^*_N of the empty sites can be expressed as a fraction of the total concentration of sites, similarly as it was done in Eq. 38.

$$\sum_{l=1}^{N_0} C^*_l = X_T = C^*_N \left(1 + \sum_{\substack{l=1 \\ l \neq N}}^{N_0} k_j c_j\right)$$

or

$$C^*_N = \frac{X_T}{1 + \sum_{\substack{l=1 \\ l \neq N}}^{N_0} k_j c_j} .$$

Thus

$$C^*_j(c) = X_T \frac{k_j \, c_j}{1 + \sum_{\substack{l=1 \\ l \neq N}}^{N_0} k_l \, c_l}, \qquad (j = 1, 2, ..., N_0, j \neq N) . \tag{46}$$

Again, the adsorption constants k_j depend on the electrostatic potential in the adsorbing layer, and when the potential varies due to the variation of the surface charge, its effect will be separated from the constant and the adsorption constant will be written as $k_j \, P_L^{\Delta z}$:

The concentration C^*_N of empty sites decreases with increasing concentrations c_1, c_2,..., c_{N_0} as more and more surface bridges are built. The same happens obviously when adsorption is an exchange process (Eq. 36) and:

1. One adsorbing species, e.g. A_N, has a constant concentration in solution. In that case, the isotherm (Eq. 41) degenerates to the Langmuir isotherm (Eq. 46).
2. When the sum of the soluble concentrations of the exchanging species is a constant, c

$$\sum_{l=1}^{N_0} c_l = c . \tag{47}$$

Then according to Eq. 41 the adsorbed concentrations are as given in the following Eq. 48.

$$C_j^*(c) = X_T \frac{\alpha_{jN} c_j}{c_N + \sum\limits_{\substack{l=1 \\ l \neq N}}^{N_0} \alpha_{lN} c_l} = X_T \frac{\frac{\alpha_{jN}}{c} c_j}{1 + \sum\limits_{\substack{l=1 \\ l \neq N}}^{N_0} \frac{\alpha_{lN} - 1}{c} c_l} . \tag{48}$$

Adsorption on Variable Charge Surfaces at Low Ionic Strengths. The concentrations of surface species on variable surface charge oxides can be calculated from the following adsorption mass action laws, similarly as in the preceding sections. Let H and M be the free proton and metal concentrations in solution, X_T the total concentration of adsorbing sites on the variable charge oxide surface and k_1, k_2 and α the adsorption constants for one, two protons and the doubly charged metal respectively – all assumed to adsorb in the same plane (Dzombak and Morel 1990), then the mass action laws and the surface site balance are

$$XOH_2 = k_1 \, P \, XOH \, H , \tag{49}$$

$$XO = \frac{k_2}{P} \, XOH \frac{1}{H} , \tag{50}$$

$$XOM = \alpha \, P \, XOH \frac{M}{H} , \tag{51}$$

$$X_T = XOH_2 + XOH + XO + XOM . \tag{52}$$

Combining these four equations, we find the isotherms

$$XOH_2(M, H) = X_T \frac{k_1 P \, H^2}{\alpha \, P \, M + H \, (1 + k_1 P \, H) + k_2/P} , \tag{53}$$

$$XOH(M, H) = X_T \frac{H}{\alpha \, P \, M + H \, (1 + k_1 P \, H) + k_2/P} , \tag{54}$$

$$XO(M, H) = X_T \frac{k_2/P}{\alpha \, P \, M + H \, (1 + k_1 P \, H) + k_2/P} , \tag{55}$$

$$XOM(M, H) = X_T \frac{\alpha \, P \, M}{\alpha \, P \, M + H \, (1 + k_1 P \, H) + k_2/P} , \tag{56}$$

At trace concentrations of the metal, P is a function only of pH and the ionic strength $I = 1/2 \sum z_i^2 c_i$, where i designates any of the soluble species A_i with charge z_i present in the system. Dzombak and Morel (1990) have tabulated P.

If, at small ionic strengths $I \leq 10^{-4}$ M, the correction factor is approximated by

$$P = \frac{\sqrt{\frac{k_2}{k_1}}}{H} = \frac{PZC}{H} , \tag{57}$$

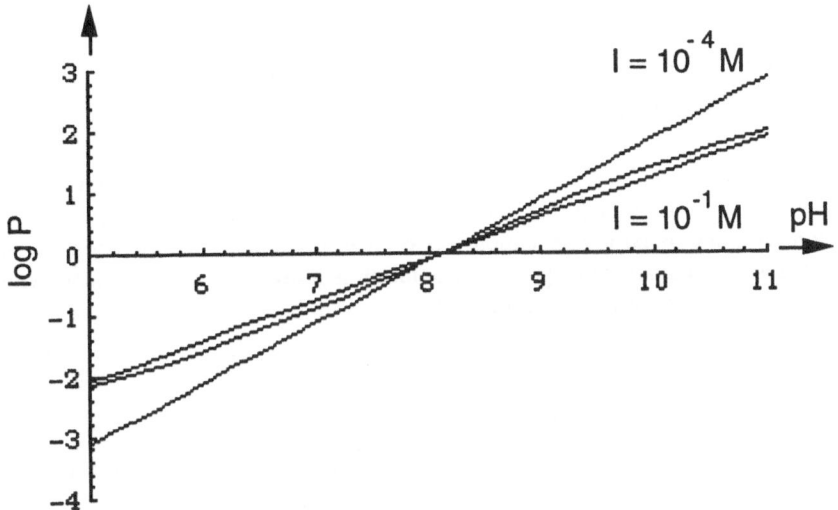

Fig. 11.6. Electrostatic correction factor P and its approximations P_a as a function of pH for two ionic strengths I (after Dzombak and Morel 1990). At $I = 10^{-4}$ M, P_a and P coincide. At $I = 10^{-1}$ M, P_a is the *straight line* close to it

the curves of the approximation and the true P coincide, as is demonstrated in Fig. 11.6. At the ionic strength $I = 10^{-1}$ M where many adsorption experiments have been made, the approximation

$$P = (\frac{PZC}{H})^{2/3} \qquad (58)$$

is good.

A property of approximation (Eq. 12) is the exact balance between positive and negative surface charges in pristine water,

$$XOH_2(M, H) = XO(M, H) . \qquad (59)$$

The isotherms resemble the ones for exchange of two protons with one metal ion (see Eqs. 36 and 41):

$$XOH_2(M, H) = X_T \frac{k_1 P H^2}{\alpha \sqrt{\frac{k_2}{k_1}} M + (1 + 2 \sqrt{k_1 k_2}) H^2} , \qquad (60)$$

$$XOH(M, H) = X_T \frac{H^2}{\alpha \sqrt{\frac{k_2}{k_1}} M + (1 + 2 \sqrt{k_1 k_2}) H^2} , \qquad (61)$$

$$XOM(M, H) = X_T \frac{\alpha \sqrt{\frac{k_2}{k_1}} M}{\alpha \sqrt{\frac{k_2}{k_1}} M + (1 + 2 \sqrt{k_1 k_2}) H^2} .$$

$$(62)$$

The exchange stoichiometry is reflected in the exponents of the free metal and proton concentrations, e.g. Eq. 41 being the isotherm for an exchange of one proton with one metal ion. Equations 60–62 lead to an analytical solution of the set of transport equations (Eq. 7). This solution is a useful guide to systems with H-M exchange, which will be shown in Section 11.8.

11.7.2 Properties of Waves

Non-Retarded Waves. A non-retarded wave appears under two conditions:

1. The sum of the adsorbed component concentrations is constant (see Eq. 9), or
2. One or several components A_j do not adsorb ($H_j' = 0$ in Eq. 7).

Condition 1 is characteristic of the isotherm, Eq. 41. Connected with this is a property of that isotherm: The surface speciation C_j^* ($j = 1, 2,..., N_0$) remains unchanged when all concentrations c_j vary by a common factor

$$C_j^*(c) = X_T \frac{\alpha_{jN} c_j}{\sum\limits_{l=1}^{N_0} \alpha_{lN} c_l} = X_T \frac{\alpha_{jN} a c_j}{\sum\limits_{l=1}^{N_0} \alpha_{lN} a c_l} = C_j^*(a c), \quad (j = 1, 2, ..., N_0) .$$

$$(63)$$

Because the equation of this non-retarded wave is $c = a\, c_+$, the wave lies on a line through the origin of composition space.

Condition 2 is necessary for a non-retarded wave to appear in systems in which adsorption of all components follows isotherms (Eqs. 46 or 48). Unlike for the exchange process (Eq. 36), the empty site A_N does not appear explicitly among the N_0 species in the isotherm (Eq. 46), although it is part of the system. Therefore, any variation of c causes C_j^* to change, or in other words: $H_j' \neq 0$ for any shift of c. Hence, the non-retarded wave, $v_w/\phi = \xi$, is a solution only if one of the H_j itself is zero.

Waves are Lines. Waves are particularly simple if all site types adsorb according to (Eq. 41) class and to (Eq. 46) isotherms (Temple 1983, Rhee et al. 1989)

$$C_j^*(c) = \sum_\alpha X_T^\alpha \frac{k_j^\alpha c_j}{\delta^\alpha(c)} \quad (j = 1, 2, ..., N \leq N_0) ,$$

$$\delta^\alpha(c) = b + \sum\limits_{j=1}^{N_0} k_j^\alpha c_j$$

$$(64)$$

where X_T^α is the concentration of adsorption site type α, the constant b has the

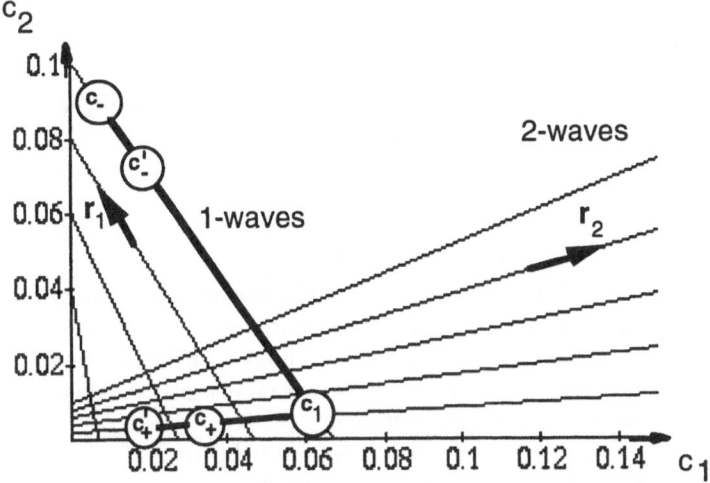

Fig. 11.7. Two-dimensional concentration space with some 1- and 2-waves. Adsorption follows Langmuir isotherm (Eq. 46). Wave velocity increases in the direction of right eigenvector r_k (see Eq. 33). The solution of the Riemann problem is the sequence of constant states $(c_-/c_1/c_+)$ connected by the 1- and 2- waves (*heavy lines*). According to property (c), $(c^*_-/c_1/c^*_+)$ is a solution of another Riemann problem

value 0 or 1, and the adsorption constants k_j^α are independent from c (see Appendix C for proof of the first two properties and Fig. 11.7 for an illustration):

1. The waves (shocks and rarefaction waves) coincide in concentration space.
2. The waves (shocks or rarefaction waves) are lines (in concentration space).
3. The waves (shocks or rarefaction waves) are invariant manifolds. This means that the k-wave is represented by the same line independently from its starting point c_{k-1} as long as c_{k-1} lies on the line.

Analytical Solution for the Multicomponent System: The Riemann Invariant.
The k-Riemann invariants are functions $w_{j \neq k}(c)$ of the concentrations and are defined as follows:

$$r_k \; \text{grad} \, w_j = 0 \quad (j = 1, 2, ..., k-1, k+1, ..., N_0), \tag{65}$$

where $\text{grad} = \{\partial/\partial c_1, \partial/\partial c_2, ..., \partial/\partial c_{N_0}\}$.

Definition (Eq. 65) means that across a k-wave only w_k changes, while the remaining $N_0 - 1$ invariants $w_{j \neq k}$ are constant. For example, w_1 changes only across the 1-wave and remains constant across all subsequent waves. More generally

$$w_j(c_k) = \begin{cases} w_j(c_+), & (j \leq k) \\ w_j(c_-), & (j > k) \end{cases}, \tag{66}$$

which means that if all N_0 Riemann invariants w_j (with $j = 1, 2,..., N_0$) are known for the initial and final states, c_+ and c_- they are known for any of the constant states c_k.

Because w_k changes only across the k-wave, w_k can be visualized as a generalized chemical component. If the components did not interact on the adsorbing surfaces, the k-Riemann invariant were identical with the kth chemical components. The initial discontinuity $c_+ \to c_-$ would develop with time into a series of waves, one for each component, each experiencing its own retardation. Across each wave only one component would change, all others remaining constant (see Fig. 11.1a). Similarly, in the case of interacting components (see Fig. 11.1b), each wave experiences its own retardation, and across each wave only one Riemann invariant changes. In their textbook Helfferich and Klein (1970) use the concept of the Riemann invariant, but do not give the invariants this name, calling them H-function roots instead.

In Appendix C it is shown that the waves are (straight) lines if adsorption follows a multicomponent Langmuir isotherm or is an ion exchange process with constant selectivity coefficients. Appendix D shows that then the scalar product of the left eigenvector of the retardation matrix and the concentration vector is a Riemann invariant

$$w_j(c_k) = l_j(c_k)\, c_k \, , \tag{67}$$

where the jth left eigenvector (corresponding to eigenvalue ρ_j) for constant state c_k is defined as

$$l_j(c_k)\, \underline{R}(c_k) = \rho_j(c_k)\, l_j(c_k) \, . \tag{68}$$

It is shown in Appendix D that all left eigenvectors at any constant state c_k are known. Introducing a matrix \underline{L}

$$\underline{L}(c_k) = \begin{bmatrix} l_1(c_k) \\ \\ l_{N_0}(c_k) \end{bmatrix} \tag{69}$$

the rows of which are the left eigenvectors of the retardation matrix $\underline{R}(c_k)$ and the vector of Riemann invariants

$$w(c_k) = \begin{bmatrix} w_1(c_k) \\ \\ w_{N_0}(c_k) \end{bmatrix} , \tag{70}$$

we can restate the Definition (Eq. 67) as

$$w(c_k) = \underline{L}(c_k)\, c_k \, . \tag{71}$$

Because \mathbf{w} and \underline{L} are known, the concentration vectors of the constant states, c_k, can be calculated by inverting Eq. 71

$$c_k = \underline{L}^{-1}(c_k)\, \mathbf{w}(c_k) \ . \tag{72}$$

11.8 Waves in Two-Component Systems

Mass action laws depend non-linearly on the concentrations. In certain restricted areas of concentration space, the behaviour of a multicomponent system can therefore be interpreted in terms of two-component systems, the contribution of the other components being negligible. Multicomponent transport in porous media might exhibit a similar feature, and therefore, thermodynamic adsorption data – although mostly determined in two-component systems – provide helpful, albeit non-complete, information for multicomponent systems. Dzombak and Morel's consistent set of thermodynamic data (1990) describing adsorption of protons and a metal ion on hydrous ferric oxides will be used here as an example.

11.8.1 Analytical Solution of the Transport Equations

The proton concentration \underline{H} adsorbed on an oxide, e.g. a hydrous ferric oxide, is

$$\underline{H} = 2\ XOH_2(M,\ H) + XOH(M,\ H) \tag{73}$$

and the adsorbed metal concentration is

$$\underline{M} = XOM(M,\ H) \ , \tag{74}$$

thus

$$\underline{M} + \underline{H} = XOM + 2\ XOH_2 + XOH \ , \tag{75}$$

and with $XOH_2 = XO$ (see Eq. 59)

$$\underline{M} + \underline{H} = XOM + XOH_2 + XOH + XO = X_T \ . \tag{76}$$

To solve the transport equations (Eq. 7) for the (M-H)-system

$$(\frac{v_w}{\phi} - \xi)\, M' - \frac{\xi}{\phi}\, \underline{M}' = 0 \tag{77}$$

$$(\frac{v_w}{\phi} - \xi)\, H_t' - \frac{\xi}{\phi}\, \underline{H}' = 0 \ , \tag{78}$$

where H_t is the total proton concentration (see following), we add Eqs. 77 and 78

$$(\frac{v_w}{\phi} - \xi)\, (M + H_t)' - \frac{\xi}{\phi}\, (\underline{M} + \underline{H})' = 0 \tag{79}$$

and use the fact that the sum (Eq. 76) of the adsorbed metal and proton concentrations is constant

$$(\frac{v_w}{\phi} - \xi)\, (M + H_t)' = 0 \ . \tag{80}$$

There are two solutions of Eq. 80:

$$(M + H_t)' = 0 \tag{81}$$

and

$$\xi = \frac{v_w}{\phi} \ . \tag{82}$$

After using Eq. 82 in Eq. 62 we know that the adsorbed metal concentration does not change across this non-retarded wave: $XOM' = 0$ or

$$XOM(M, H) = XOM(M_+, H_+) \ . \tag{83}$$

The adsorption isotherm (Eq. 56) then fixes the relationship between M and H. With Eq. 57, the approximation of P for pristine water ($I \leq 10^{-4}$ M), the isotherm is Eq. 62, thus we get

$$XOM(M, H) = X_T \frac{\alpha \sqrt{\dfrac{k_2}{k_1}\dfrac{M}{H^2}}}{\alpha \sqrt{\dfrac{k_2}{k_1}\dfrac{M}{H^2}} + (1 + 2\sqrt{k_1 k_2})} \ . \tag{84}$$

This isotherm meets Eq. 83 when

$$H = H_+ \sqrt{\frac{M}{M_+}} \qquad \text{(2-wave, non-retarded)} \ . \tag{85}$$

Thus, the non-retarded wave is determined by the exchange stoichiometry, the adsorption constants do not enter.

The equation for the other wave – given by Eq. 81 – involves only soluble species. Equation 81 states that the sum of the soluble concentrations is a conserved quantity, i.e. it does not change across the wave

$$M + H_t = M_- + H_{t-} \ . \tag{86}$$

It is customary in speciation calculations to approximate H_t in pristine water by the difference between the proton and the hydroxyl concentrations, assuming that the concentration of the H_2O complex remains constant:

$$H_t = H - OH = H - \frac{K_w}{H} ,$$
(87)

where $K_w = 10^{-14}$ (mol / l)². With Eq. 87, Eq. 86 becomes

$$M = M. - (H - H.) - K_w \left(\frac{1}{H} - \frac{1}{H.}\right) \quad \text{(1-wave, retarded)} .$$
(88)

Figure 11.8 shows the grid of 1- and 2-waves, with $M_+ = M_- = 10^{-9}$ mol/l and $H_+ = H_- = 10^{-4}$ mol/l, 10^{-5} mol/l,..., 10^{-12} mol/l. Helfferich (1992) has called it the "street map" of the system, (see example in Fig. 11.8) because the only way to get from a state (-) to a state (+) is to first follow the 1-wave until point (1) where it meets a 2-wave along which one can get to state (+).

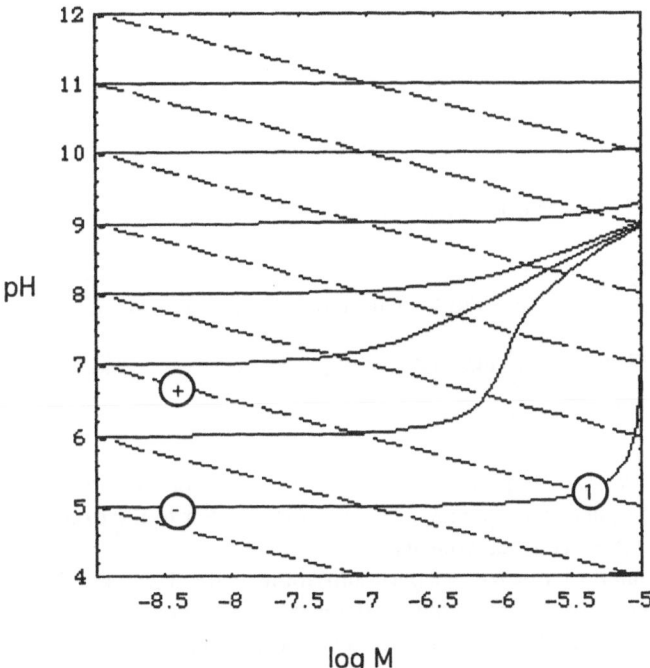

pH

log M

Fig. 11.8. 1- and 2-waves (Eqs. 85 and 88) in concentration space. 1-waves are *solid lines* and 2-waves are *dashed lines*. Units of M: mol/l. The *points* (-), (*1*) and (+) mark the solution of a particular Riemann problem. Note that the logarithmic plot exaggerates the angle between the waves at high pH

The retardation of the 1-wave is $\rho_1(M, H)$, the larger of the two eigenvalues of $\underline{R}(M, H)$ (see Eq. 31). It has the simple form

$$\rho_1(c) = 1 + \frac{1}{\phi} \frac{\partial XOM(c)}{\partial M} \left(1 + \frac{2HM}{H^2 + K_w}\right) \tag{89}$$

with

$$\frac{\partial XOM(c)}{\partial M} = X_T \frac{\alpha\ PZC\ (1 + 2\ k_1 PZC)\ H^2}{(\alpha\ PZC\ M + H^2\ (1 + 2\ k_1 PZC))^2}. \tag{90}$$

At low ionic strength the surface charge on the oxide is small, and therefore, approximation (Eq. 57) could be used to calculate a solution of the transport equations (Eqs. 77 and 78).

It is interesting to notice the similarity between the oxide system at low ionic strength and an ion exchange system with constant surface charge. For a two-component system ($N_0 = 2$) with an exchange isotherm (Eq. 41), the analytical solution method gives for the 1- and 2-waves $(c_1 + c_2)' = 0$ and $\xi = v_w/\phi$, respectively, or

$$c_1 + c_2 = (c_1)_- + (c_2)_-$$

i.e.

$$c_2 = -c_1 + (c_1)_- + (c_2)_-.\quad\text{(1-wave, retarded)}, \tag{91}$$

as compared to Eq. 88.

From $C_j{}^*(c_1, c_2) = C_j{}^*((c_1)_+, (c_2)_+)$ follows for the isotherm (Eq. 41)

$$c_2 = (c_2)_+ \frac{c_1}{(c_1)_+}\qquad\text{(2-wave, non-retarded)} \tag{92}$$

as compared to Eq. 85.

The independence of the shape of those waves (Eqs. 91 and 92) from the concentrations c_- and c_+ and the separation factor α_{12} is peculiar to the exchange (Eq. 35) in a system with only 2-components.

Contrary to the shape of the waves, the retardation of the 1-wave, i.e. the eigenvalue $\rho_1(c_1, c_2)$ of $\underline{R}(c_1, c_2)$, depends on the identity of the exchanging ions. As compared to Eqs. 89 and 90 it is a simpler function of the separation factors and concentrations

$$\rho_1 = 1 + \frac{X_T}{\phi} \frac{\alpha_{12}(c_1 + c_2)}{(\alpha_{12}c_1 + c_2)^2}. \tag{93}$$

11.8.2 Exchange Systems with Buffers

Both waves will be retarded whenever there is a buffer in the system that allows adsorption to break away from the rigid one-to-one exchange depicted in Fig. 11.4 or expressed in Eq. 81 as a consequence of Eq. 57, or of Eq. 91. We have such a buffered situation when:

1. The two components adsorb according to a Langmuir isotherm (Eq. 46) on constant charge surfaces ($P = 1$), the empty sites playing the role of the buffer.

2. There are three species and
 a) one of the species concentrations in solution is constant (isotherm, Eq. 46) or
 b) the sum of the species concentrations in solution is constant (isotherm, Eq. 48).

The initial discontinuity $c_+ \rightarrow c_-$ of the Riemann problem develops into a sequence of two retarded waves, a slow one, the 1-wave, and a fast one, the 2-wave. In composition space, these waves are lines tangential to a parabola (Helfferich and Klein 1970; Rhee et al. 1989), as depicted in Fig. 11.9. The proof can be found in Appendix D. The equation of the parabola is

$$\bar{c}_2 = 1 - \bar{c}_1 \pm 2\sqrt{-\bar{c}_1}, \text{ with } \bar{c}_j = \frac{c_j}{k} . \tag{94}$$

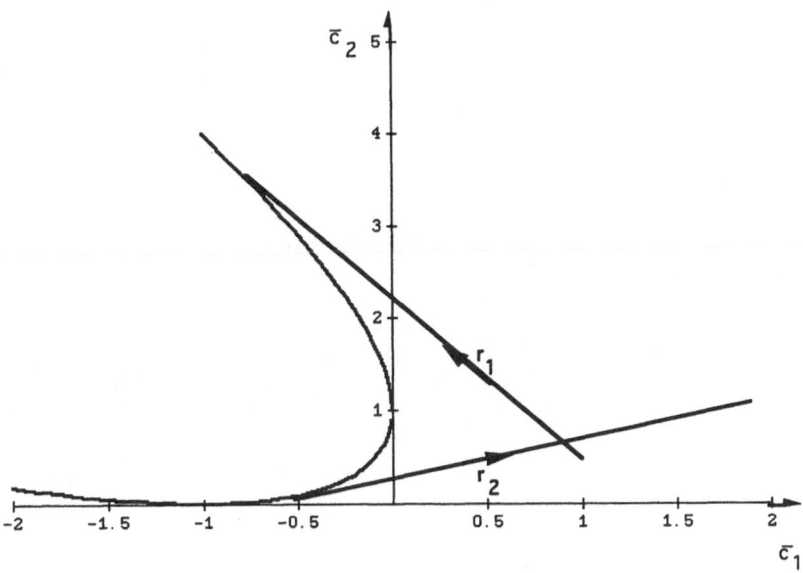

Fig. 11.9. Parabola (Eq. 93) with waves in concentration space, here $\{\bar{c}_1, \bar{c}_2\}$ space. r_1 and r_2 are the right eigenvectors parallel to the 1- and 2-wave (the *lines*), respectively

The value of k depends on the form of the isotherm. In the exchange system, as described in preceding situation 2b, in which all three concentrations may vary under the constraint $c_1 + c_2 + c_3 = c$ (c = constant, including $c = 0$), i.e. for

$$C_j^*(c) = X_T \frac{\alpha_{j3} c_j}{\alpha_{13} c_1 + \alpha_{23} c_2 + c_3} \qquad (j = 1, 2, 3 \text{ and } \sum_{i=1}^{3} c_i = c) \tag{95}$$

we get

$$k = \frac{1 - \alpha_{12}}{\alpha_{13} - 1} c \ . \tag{96}$$

For a system with surface bridge formation or an exchange system with constant c_3 (in the latter we introduce $\kappa_j = \alpha_{j3}/c_3$), i.e. for

$$C_j^*(c) = X_T \frac{\kappa_j c_j}{1 + \kappa_1 c_1 + \kappa_2 c_2} \qquad (j = 1, 2) \tag{97}$$

we get

$$k = \frac{1}{\kappa_1} - \frac{1}{\kappa_2} \ . \tag{98}$$

The parabola lies in the upper left quadrant when k is positive, and in the lower right quadrant when k is negative. The buffer disappears when $k \rightarrow 0$, which is equivalent to an isotherm (Eq. 41) with $N_0 = 2$,

$$C_j^*(c) = X_T \frac{\alpha_{j2} c_j}{\alpha_{12} c_1 + c_2} \qquad (j = 1, 2) \ . \tag{99}$$

For the 2-component system with bridge formation, Eq. 97, we get for the retardations of the 1- and 2-wave

$$P_{1,2} = 1 + \frac{X_T}{2} \left\{ \frac{\kappa_1 \left(1 + \kappa_2 (c_1 + c_2)\right) + \kappa_2}{\delta^2} \pm \sqrt{\frac{\kappa_1 \left(1 + \kappa_2 (c_1 + c_2)\right) + \kappa_2^2}{\delta^2} - \frac{4 \kappa_1 \kappa_2}{\delta^3}} \right\} \tag{100}$$

with

$$\delta = 1 + \kappa_1 c_1 + \kappa_2 c_2 \tag{101}$$

When $c_2 \gg k$, the 1-waves in the 2-component system with bridge formation approach asymptotically the 1-waves (Eq. 91) of the 2-component exchange system (Eq. 41). At these elevated concentrations almost all surface sites (in the

system with surface bridges) are occupied by either species A_1 or A_2, and the adsorption process degenerates into an exchange between these two species. This explains the asymptotic slope -1 of the 1-waves in Fig. 11.9.

11.9 Barrier Against Remobilization of a Contaminant

From Eq. 93 one can see that in a Langmuir or ion exchange system with two independent components A_1 and A_2:

1. The slope of the 1-wave approaches -1 as $c_2 \gg k$,
2. The slope of the 2-wave approaches 0 as $c_{2+} \ll k$,

if k is positive, i.e. component A_2 adsorbs more strongly ($\alpha_{12} < 1$ or $\kappa_1 < \kappa_2$). Assume that the other component, A_1, is a contaminant.

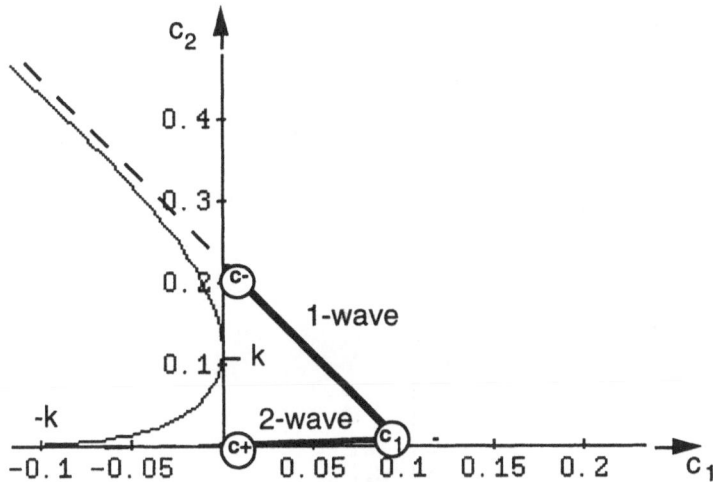

Fig. 11.10. Parabola (Eq. 93) and waves in $\{c_1, c_2\}$ space with Langmuir-type Isotherm (Eq. 97), $\kappa_1 = 5.0$ M^{-1}, $\kappa_2 = 12.0$ M^{-1}. Component A_1: contaminant; component A_2: remobilizing agent. $c_- = \{c_{1-}, c_{2-}\}$, $c_1 = \{c_{11}, c_{21}\}$ and $c_+ = \{c_{1+}, c_{2+}\}$ are the concentration vectors in the feed, the constant state intermediate between feed and equilibrant and in the equilibrant, respectively (see e.g. Fig. 11.1a or b)

Figure 11.10 displays the failure of the adsorption barrier: initially, the contaminant inventory is adsorbed on the soil particles ($c_{1+} \approx 0$). When the concentration of the other component in the feed is raised above the threshold k,

the contaminant is removed from the adsorbing sites and its inventory is carried between the waves. The contaminant concentration c_{11} between the waves is of the same order of magnitude as the concentration of the remobilizing solute in the feed ($c_{11} \approx c_{2-}$), because the slopes of the 1- and 2-waves are approximately -1 and 0, respectively.

Figure 11.11 shows the spatial variation of the corresponding concentration profiles, $c_i(x,t)$. The described adsorption barrier (Eq. 97) will not fail when the contaminant is bound more strongly to the adsorbing surfaces than to the competing species, A_2, i.e. when $\alpha_{12} > 1$ or $k_1 > k_2$ (see Fig. 11.12).

Whereas the 1-waves are not changed as compared to the case shown in Figs. 11.10 and 11.11, the slope of the 2-waves does not approach 0 as c_+ approaches 0. Therefore, the concentration of the contaminant between the waves cannot rise much above the concentration prior to remobilization.

Fig. 11.11. Failure of adsorption barrier. A_1 is contaminant, A_2 remobilizing agent. $c_- = \{c_{1-}, c_{2-}\}$, $c_1 = \{c_{11}, c_{21}\}$ and $c_+ = \{c_{1+}, c_{2+}\}$ are the concentration vectors in the feed, the constant state intermediate between feed and equilibrant and in the equilibrant, respectively (see e.g. Fig. 11.1 a or b)

The described effect has been applied in various areas to purify materials from unwanted components: Pfann (1958) has introduced the zone melting process to remove impurities from semiconductor materials. A heat wave (instead of a wave of a competing species) is remobilizing the impurity atoms. A similar process (using steam) has been proposed by Hunt et al. (1988) to remove gasoline trapped in soils. Harwell (1992) discussed remobilizing such non-aqueous phase liquids with a chlorinated hydrocarbon solvent or surfactant wave.

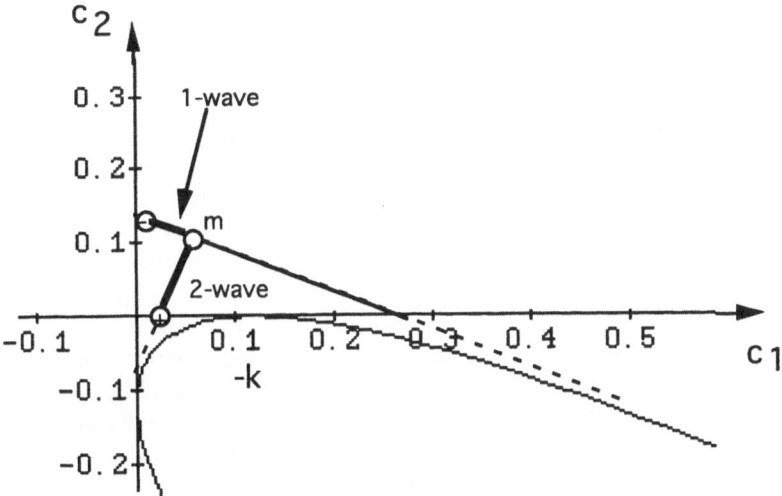

Fig. 11.12. Parabola (Eq. 93) and waves in $\{c_1, c_2\}$ space with Langmuir-type isotherm (Eq. 97), $k_1 = 12.0 \, M^{-1}$, $k_2 = 5.0 \, M^{-1}$. Component A_1: contaminant; component A_2: remobilizing agent. This adsorption barrier cannot fail because $c_{1m} \approx c_{1+}$ regardless of c_{2-}, where c_{1m} and c_{1+} are the contaminant concentrations in the state (m) between the waves and equilibrant (+), repectively, and c_{2-} is the concenztration of the remobilizing agent in the feed(-)

11.10 Conclusions

A single abrupt change of the water composition at the inlet of a column of a porous medium with a homogeneous concentration distribution develops with time into a series of centred waves, if hydrodynamic dispersion is neglected. Centred waves are thus a class of solutions of the chromatographic transport equations that can be considered the basic building blocks of general solutions. There are as many centred waves as there are chemical components in the system.

It is customary and helpful in chromatography to plot the variation of the concentration vector across a wave in concentration space. When this is done for a set of waves, a net of curves results which Helfferich has called the "Street Map of the System". It is characteristic for the chemical interactions in solution and on the adsorbing surfaces. Its computation does not exceed the limitations of an experimental laboratory. It requires similar capacities as equilibrium speciation calculations which are now standard tools in such labs.

Because of the mentioned chemical non-linearities, thresholds run through concentration space at which the net of waves changes in shape. Experience we gained on one side of a threshold is irrelevant beyond it. Once the net of centred waves is known for a set of thermodynamic parameters, the influence of parameter variations can be interpreted in terms of its deformations and the change of the threshold locations.

This helps to understand the impact of uncertainties in the chemical interactions more comprehensively than results of standard numerical models, e.g. of coupled finite-difference chemical-speciation models. The latter compute waves for initial and boundary conditions that are discrete, isolated points in concentration space. Since they do not reveal the general structure of the waves such as the locations of the thresholds, neither inter- nor extraplolation of numerical results is justified without prior knowledge of the thresholds.

The results of chromatographic calculations are thus, useful guides for design and interpretation of numerical computations as well as chemical experiments assessing the adequacy of the assumed chemical interactions. When the character of e.g. the chemical barriers against remobilization has been clarified in feedback between experiment and chromatographic model computations, follow-up numerical calculations can concentrate on site-specific critical problems.

Acknowledgement. Part of this research was done while the author was a visiting member of the Courant Institute for Mathematical Sciences of New York University, New York. Without the long discussions with F. Helfferich from Pennsylvania State University this work would not have been possible.

Appendix A
Relative Influence of Diffusion/Dispersion and First Order Kinetics in a Single-Component System

For a system (Eqs. 1 and 2) with only one component adsorbing

$$\frac{\partial}{\partial t} (\phi \, c(x,t) + C(x,t)) + \frac{\partial}{\partial x} (v_w - D \frac{\partial}{\partial x}) c(x,t) = 0 \tag{A1}$$

$$\frac{\partial C(x,t)}{\partial t} = \frac{C^*(c) - C(x,t)}{\varepsilon} \tag{A2}$$

it can be shown (Liu 1987, see also Appendix B) that the solution of the Riemann problem with $c_- > c_+$ is for large t $(t \to \infty)$
(*)a traveling wave $c_\infty (x, t)$ for $D=0$

$$c_\infty(x,t) = c(s) \tag{A3}$$

with

$$s = x - \sigma t, \tag{A4}$$

$$\sigma = \frac{v_w}{\phi R}, \tag{A5}$$

$$R = \frac{C^*(c_+) - C^*(c_-)}{c_+ - c_-} \tag{A6}$$

– the latter given by the Rankine-Hugoniot relation (Eq. 24).
Thus, the concentration jump in Eq. 4 develops into a steady profile c(s) when we wait long enough.
With Eq. A3, the time asymptotic transport equation is with d/ds abbreviated as '

$$\frac{d}{ds} [-\sigma(\phi \, c + C^*) + v_w \, c - 2\sigma^2 \varepsilon C^{*''} c'] = 0 . \tag{A7}$$

(*) the same traveling wave (Eq. A3) for $\varepsilon = 0$ or $C = C^*$.
With Eq. A3, the time asymptotic transport equation is with d/ds abbreviated as '

$$\frac{d}{ds} [-\sigma(\phi \, c + C^*) + v_w \, c - D \, c'] = 0 . \tag{A8}$$

The interpretation of the development of a traveling wave is this: Dispersion or adsorption kinetics tend to produce a smooth transition between c_+ and c_-, whereas the non-linear dependence of the adsorbed concentration C^* on c counteracts dispersion or kinetics and tries to produce a steep profile. The traveling wave is the stationary compromise.

It is assumed that also for $D \neq 0$ the transport equations (Eqs. A1 and A2) have a traveling wave as time asymptotic solution, and that its width is given by the dispersion process if

$$2\sigma^2 \varepsilon C^{*'} < D. \tag{A9}$$

Using the Definitions (Eqs. A5 and A6) and the separation of the dispersion into diffusion and hydrodynamic dispersion

$$D = D^* \tau + \delta \frac{v_w}{\phi} \tag{A10}$$

where D^* is the diffusion coefficient in water, τ the tortuosity (0.3), δ the dispersivity, and the approximation

$$C^{*'} = \phi(R - 1) \tag{A11}$$

the upper limit of ε determined after Eq. A9 can be written as

$$\varepsilon = \frac{R^2 \phi}{2v_w^2(R - 1)} D = \frac{R^2 \phi}{2v_w^2(R - 1)} (D^* \tau + \frac{\delta v_w}{\phi}) \approx 11 \text{ R days for R} > 5 \tag{A12}$$

The right hand side of Eq. A12 has been evaluated for $\phi = \tau = 0.3$, $\delta = 0$, $D^* = 0.16$ m²/a and $v_w = 0.5$ m/a.

The interpretation of Eq. A12 is this: if the adsorbed concentration C reaches 37% of its equilibrium value, i.e. C^*/e (e = 2.71), within a period ε or less, the kinetics of the adsorption process has a negligible influence – as compared to diffusion/dispersion – on the shape of the traveling wave.

Figure A1 shows more detail of Eq. A12 by demonstrating how the upper limit of ε depends on the speed v_w of the water, two retardations (R = 5 and 1.1) and various dispersivities δ, the latter being approximated by L/100.

To get an analytical expression for the influence of relaxation relative to dispersion, let us consider a moderately retarded species (e.g. R > 5), so that we can neglect the 1 in the denominator of Eq. A12. After multiplying both sides of Eq. A12 with $\sigma^2 \varepsilon = (v_w/(\phi R))^2 \varepsilon$ and taking the square root of both sides

$$\sigma \varepsilon < \frac{1}{\sqrt{2}} \sqrt{\frac{D}{\phi R} \varepsilon} = \frac{1}{\sqrt{2}} \sqrt{\tilde{D}\varepsilon}. \tag{A13}$$

$\tilde{D} = D/(\phi R)$ is the retarded dispersion coefficient.

Equation A13 compares two lengths calculated in two different one-component systems:

- Left side: Assume a purely advective system at equilibrium (Eq. 1 with $C = C^*$ and $D = 0$). $\sigma\varepsilon$ is the advective distance a wave travels during the relaxation time ε (*advection during relaxation*).
- Right side: Assume a purely dispersive system at equilibrium (Eq. 1 with $C = C^*$ and $v_w = 0$). $\sqrt{(D\varepsilon)}$ is the *width of the profile* $c(x, t = \varepsilon)$ for an initial concentration jump $(c(x, 0) = c_-$ for $x < 0$ and c_+ for $x \geq 0)$.

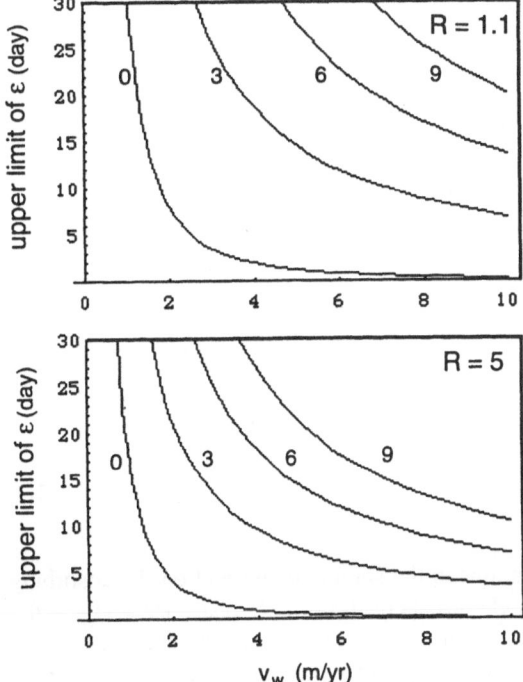

Fig. A1. Upper limit of relaxation time ε in Eq. A12, as a function of flux v_w of water for a migration distance, $L = 0, 3, 6$ and 9 m. $\delta = 0.01$ L has been used. R is the retardation of the wave

In this sense relaxation of adsorption can be neglected against dispersion if advection during relaxation ($\sigma\varepsilon$) is smaller than the width of the profile $\sqrt{(D\varepsilon)}$ divided by $\sqrt{2}$.

Evaluated for natural water flow ($v_w < 0.5$ m/a) through river or lake sediments in moderate latitudes this means: Relaxation times $\varepsilon < 1$ week have a negligible influence on the shape of naturally produced profiles (see also the very comprehensive analysis of van der Zee (1990).

Appendix B
Non-Linear Single-Component Transport

Appendix B.1
Non-Equilibrium Transport without Diffusion

The transport equation for a single component without diffusion is the conservation law (Eq. B1)

$$\frac{\partial c}{\partial x} + \frac{1}{v_w} \frac{\partial}{\partial t} (\phi c + C) = 0 , \tag{B1}$$

where

$$\frac{\partial C}{\partial t} = \frac{C^* - C}{\varepsilon} \tag{B2}$$

c and C are the species concentrations in the soluble phase and on the adsorbing soil surfaces, respectively. The soil consists of the soluble phase (water) and the solid phase (soil particles), the volume fractions of which are ϕ and $1 - \phi$, respectively. v_w is the flux of the water. ϕ and v_w are constants. The adsorbed concentration $C(x, t)$ relaxes to its equilibrium value $C^*(c)$. ε is the corresponding relaxation time due to adsorption kinetics.

Liu (1987) presented a solution for Eqs. B1 and B2 for the Riemann problem: This system (Eqs. B1 and B2) has the characteristics $l_1 = 0$ and $l_2 = v_w/\phi$. The equilibrium characteristic l^*, i.e. the one for Eq. B1 with $C = C^*$, is

$$\lambda^* = \frac{v_w}{\phi} \frac{1}{1 + \frac{1}{\phi} \frac{dC^*(c)}{dc}} = \frac{v_w}{\phi} \frac{1}{1 + \frac{X_T}{\phi} \frac{k}{(1+kc)^2}} . \tag{B3}$$

Thus, l^* is a subcharacteristic ($l_1 < l^* < l_2$).

The system (Eqs. B1 and B2) can be asymptotically approximated by

$$\frac{\partial c}{\partial x} + \frac{1}{v_w}\frac{\partial}{\partial t}(\phi c + C^*) - (\frac{\partial}{\partial t}\beta(c)\frac{\partial c}{\partial t}) = 0 \ , \tag{B4}$$

where

$$\beta(c) = \frac{2\varepsilon}{v_w}\frac{dC^*(c)}{dc} \ . \tag{B5}$$

Liu shows that a traveling wave

$$c(x,t) = c(x - \sigma t) = c(s) \tag{B6}$$

is a solution of Eqs. B1 and B2 for the Riemann problem (with c_+ and c_- being the concentrations right and left of the discontinuity, respectively). σ is given by the Rankine-Hugoniot condition (Eq. 24) for the equilibrium system $(C = C^*)$

$$\sigma = \frac{v_w}{\phi}\frac{\Delta c}{\Delta(c + \frac{1}{\phi}C^*)} = \frac{v_w}{\phi}\frac{1}{1 + \frac{1}{\phi}\frac{\Delta C^*}{\Delta c}} = \frac{v_w}{\phi}\frac{1}{R} \ , \tag{B7}$$

with $\Delta c = c_+ - c_-$, $\Delta(C^*) = (C^*(c_+)) - (C^*(c_-))$.

Using Eq. B6 in Eq. B4, we can integrate the transport equation to get the profile $s(c)$ of the traveling wave. With the abbreviations

$$\hat{c} = \frac{c - c_+}{\Delta c}, \ \hat{C}^* = \frac{C^* - C_+^*}{\Delta C^*}$$

the result is (for proof see Appendix B4)

$$\int_{c_+}^{c}\frac{ds}{dc}\,dc = s(c) - s(c_+) = \frac{2\sigma^2\varepsilon}{v_w(1 - \frac{1}{R})}\int_{c_+}^{c}\frac{\hat{C}^*d\hat{c}}{\hat{c} - \hat{C}^*} \ . \tag{B8}$$

Appendix B.2
Equilibrium Transport with Diffusion

When diffusion is included, the transport equation for a single component in local equilibrium with the soil surfaces is

$$\frac{\partial}{\partial t}(\phi c + C^*) + \frac{\partial}{\partial x}(v_w - D\frac{\partial}{\partial x})c = 0 \ . \tag{B9}$$

Limiting ourselves again to the asymptotic solution, the traveling wave (Eqs. B6, and B9) can be integrated resulting in the following equation for the profile

$$
s(c) - s(c_+) = \frac{D}{v_w(1 - \frac{1}{R})} \int_{c_+}^{c} \frac{d\hat{c}}{\hat{c} - \widehat{C^*}} \, ,
\tag{B10}
$$

For the Langmuir isotherm, Lake and Helfferich (1978) derived an analytical expression for the integral in Eq. B10. Van der Zee derived an analytical expression for the case of combined non-equilibrium and diffusion/dispersion (1990).

Appendix B.3
Linear Transport of a Single Component

If the adsorption isotherm is linear

$$
C^*(c) = \phi \, (R - 1) \, c
$$

the solution of the transport Eq. B1 for the Riemann problem is (Bear 1979)

$$
\frac{c[x,t]}{c_0} = \frac{1}{2}(\text{erfc}\frac{x - \frac{v_0 t}{R}}{2\sqrt{\frac{Dt}{R}}} + e^{2v_0 x D} \, \text{erfc}\frac{x + \frac{v_0 t}{R}}{2\sqrt{\frac{Dt}{R}}}) \, ,
\tag{B11}
$$

where

$$
v_0 = \frac{v_w}{\phi} \, .
$$

Appendix B.4
Proofs of Equations B8 and B10: Integration of Transport Equation Assuming a Traveling Wave

Integration of the transport Eq. B9 will be shown first. For the traveling wave solution (Eq. B 6), Eq. B 9 becomes an ordinary differential equation

$$\frac{d}{ds}[-\sigma(\phi c + C^*) + v_w c - Dc'] = 0 , \tag{B12}$$

from which $c' = dc/ds$ can be calculated.

$$c' = \frac{1}{v_w}\frac{d}{ds}[\sigma(\phi c + C^*) + Dc'] = 0 . \tag{B13}$$

Since $c'+ = 0$, the integral of Eq. B13 from c_+ to c

$$c - c_+ = \frac{1}{v_w}\{\sigma[\phi(c - c_+) + C^* - C_+^*] + D(c' - c'_+)\} \tag{B14}$$

can be simplified and solved for c'

$$c' = \frac{1}{D}\{(c - c_+)(v_w - \sigma\phi) - \sigma(C^* - C_+^*)\} =$$

$$= \frac{v_w - \sigma\phi}{D}\{(c - c_+) - \frac{\sigma}{v_w - \sigma\phi}(C^* - C_+^*)\} . \tag{B15}$$

Replacing (see Eq. B7)

$$v_w - \sigma\phi = v_w(1 - \frac{1}{R}) \text{ and } \frac{\sigma}{v_w - \sigma\phi} = \frac{\Delta c}{\Delta C^*} ,$$

Equation B15 appears as

$$c' = \frac{v_w(1 - \frac{1}{R})}{D}\{(c - c_+) - \frac{\Delta c}{\Delta C^*}(C^* - C_+^*)\} =$$

$$= \Delta c\frac{v_w(1 - \frac{1}{R})}{D}\{\frac{c - c_+}{\Delta c} - \frac{C^* - C_+^*}{\Delta C^*}\} . \tag{B16}$$

After introducing

$$\hat{c} = \frac{c - c_+}{\Delta c}, \quad \hat{C^*} = \frac{C^* - C_+^*}{\Delta C^*}$$

Eq. B16 becomes

$$\frac{dc}{ds} = \Delta c \frac{v_w(1 - \frac{1}{R})}{D} \{\hat{c} - \hat{C^*}\} .$$

(B17)

The reciprocal of dc/ds is integrated from c_+ to c

$$\int_{c_+}^{c} \frac{ds}{dc} dc = s(c) - s(c_+) = \frac{D}{v_w(1 - \frac{1}{R})} \int_{c_+}^{c} \frac{d\hat{c}}{\hat{c} - \hat{C^*}} .$$

(B18)

For the case of non-equilibrium transport with relaxation the asymptotic profile can be calculated similarly, since – after introducing the traveling wave solution into Eq. B4 – the resulting ordinary differential equation is similar to Eq. B12:

$$\frac{d}{ds} [-\sigma(\phi c + C^*) + v_w c - 2\sigma^2 \varepsilon C^{*'} c'] = 0 .$$

(B19)

Replacing D in Eq. B12 with $2\sigma^2 \varepsilon C^{*'}$ gives the profile for relaxation. (Note that $C^{*'}$ depends on c and has to placed under the integral.)

$$\int_{c_+}^{c} \frac{ds}{dc} dc = s(c) - s(c_+) = \frac{2\sigma^2 \varepsilon}{v_w(1 - \frac{1}{R})} \int_{c_+}^{c} \frac{C^{*'} d\hat{c}}{\hat{c} - \hat{C^*}}$$

(B20)

Appendix B.5
Example: Langmuir Adsorption

Silver adsorbs following a Langmuir-type isotherm

$$C^*(c) = X_T \frac{k c}{1 + k c} .$$

(B21)

The Langmuir constant k is composed of the exchange constant a for Ag/H exchange and the surface hydrolysis constant k_2, both of which are defined as

adsorption mass action constants. Referring to the adsorbing surface site on an oxide as XO, the site occupied by silver is XOAg and the ones occupied by one or two protons are XOH and XOH_2, respectively (Dzombak and Morel 1990). The adsorption reactions can be described by the following mass action laws, where the symbols $XOAg$, XOH_2, XOH and XO are used for the corresponding adsorbed concentrations, e.g. $XOAg = C^*$

$$XOAg = \alpha \, XOH \, \frac{M}{H} \, , \tag{B22}$$

$$XOH_2 = k_1 \, P \, XOH \, H \, , \tag{B23}$$

$$XO = \frac{k_2}{P} \frac{XOH}{H^+} \, . \tag{B24}$$

P is an electrostatic correction factor (Dzombak and Morel 1990)

$$P = \exp\left(-\frac{F\psi}{RT}\right) \, ,$$

where ψ (unit: Volt = Joule/Coulomb) is the electrostatic potential at the adsorbing surface, F (96485 Coulomb/mol) is the Faraday constant, R (8.31 Joule/(mol Kelvin)) is the gas constant and T (Kelvin) is the absolute temperature.

The electrostatic correction factor is generally a function of the composition of the aqueous solution. In dilute systems, i.e. at ionic strengths $I = \frac{1}{2}\Sigma z_i^2 c_i \leq 10^{-4} M$, where z_i and c_i are the charge and concentration of component i in solution, P can be approximated by

$$P = \frac{\sqrt{\dfrac{k_2}{k_1}}}{H} \, . \tag{B25}$$

It follows from Eq. B25 that

$$XOH_2 = XO = \sqrt{k_1 \, k_2} \, XOH \, . \tag{B26}$$

Using X_T, the concentration of adsorption sites,

$$X_T = XOH_2 + XOH + XO + XOM \, , \tag{B27}$$

where $X_T = XOH \left(1 + 2\sqrt{k_1 k_2} + \alpha \dfrac{M}{H}\right)$ because of Eq. 26 we can construct the isotherm for silver from Eqs. B22–B24.

$$XOM(M, H) = X_T \frac{\alpha M}{\alpha M + H (1 + 2 \sqrt{k_1 k_2})} \quad , \tag{B28}$$

which can be brought into the form of Eq. B21 when the proton concentration is held constant by buffers:

$$XOM(M, H) = X_T \frac{\dfrac{\alpha}{H (1 + 2 \sqrt{k_1 k_2})} M}{1 + \dfrac{\alpha}{H (1 + 2 \sqrt{k_1 k_2})} M} \quad . \tag{B29}$$

Thus,

$$k = \frac{\alpha}{H (1 + 2 \sqrt{k_1 k_2})} \quad . \tag{B30}$$

We get a similar expression for k for regions in {M, H} space in which P is constant, e.g.

$$P = 1$$

and either XOH_2 or XOH_2 + XOH being negligible. In the first case, i.e. when

$$XOH_2 \ll XOH .$$

the isotherm is

$$XOM(M, H) = X_T \frac{\alpha M}{\alpha M + H + k_2} = \frac{\dfrac{\alpha}{H + k_2} M}{\dfrac{\alpha}{H + k_2} M + 1} \quad , \tag{B31}$$

and, thus,

$$k = \frac{\alpha}{H + k_2} \quad . \tag{B32}$$

In the second case, i.e. at $H \ll k_2$, the isotherm is

$$XOM(M, H) = X_T \frac{\alpha M}{\alpha M + k_2} = \frac{\dfrac{\alpha}{k_2} M}{\dfrac{\alpha}{k_2} M + 1} \quad , \tag{B33}$$

and the Langmuir factor is

$$k = \frac{\alpha}{k_2} \quad . \tag{B34}$$

At $H = k_2$ all expressions for k, Eqs. B30, B32 and B34, are similar. For this example, expression Eq. B34 is chosen.

With the data from Dzombak and Morel (1990)

$$\alpha = 10^{-1.72} \tag{B35}$$

$$k_2 = 10^{-8.9} \text{ mole/litre} \tag{B36}$$

we get

$$k = \alpha/k_2 = 1.8 \ 10^7 \text{ litre/mole.} \tag{B37}$$

The profiles in Fig. 11.1 were calculated based on the following data (B38):

X_T = 5 10^{-3} HFO, concentration of sites adsorbing Ag (Dzombak and Morel 1990).

HFO = 10^{-3} mol hydrous ferric oxides (Fe_2O_3) per litre of soil = 0.09 g hydrous ferric oxides per litre of soil.

c_- = 5.0 10^{-4} mg Ag per litre of feed-water .

c_+ = 0, Ag concentration in undisturbed soil column.

ϕ = 0.3, fraction of soil filled with liquid phase (water).

v_0 = v_w/ϕ = 0.5 m/a, speed of liquid phase (water).

The traveling wave was calculated by integrating Eq. B10 and plotting c vs. s. The profiles $c(x,t)$ for the linear isotherm were calculated at $t = 0.1, 0.5, 1$ and 2 years, using Eq. B11.

Fig. B1. Profiles for linear and non-linear adsorption, calculated for the same retardation (R = 2.7, according to Eq. B7). Traveling wave – the curve entering the abscissa at a steep angle – was calculated using Eq. B10. Profiles for linear adsorption isotherm calculated for times $t = 0.1, 0.5, 1, 2$ year(s) using Eq. B11. Element and environment: silver in hydrous ferric oxide coated inert sediment matrix. Adsorption data from Dzombak and Morel (1990)

Appendix C

Langmuir- and Exchange-Type Isotherms: Shocks and Rarefaction Waves are Lines and Coincide

The first part of this proof concerns *rarefaction waves*, i.e. solutions of the set of differential equations (Eq. C7). Using c_j and C^*_j to denote soluble and adsorbed concentrations as functions of ξ, this equation reads

$$(\frac{v_w}{\phi} - \xi) \frac{dc_j}{d\xi} - \frac{\xi}{\phi} \frac{dC^*_j}{d\xi} = 0 \quad (j = 1, 2, ..., N_0) . \tag{C1}$$

The corresponding differential coherence condition (Eq. 15) is

$$\frac{dC^*_1}{dc_1} = \frac{dC^*_2}{dc_2} = ..., \frac{dC^*_{N_0}}{dc_{N_0}} . \tag{C2}$$

Because the change $c_k - c_{k-1}$ of the N_0 components of the concentration vector across a k-wave is restricted by $N_0 - 1$ conditions (Eq. C2), the concentration vector c_k can be expressed as functions of one parameter if c_{k-1} is given. Here we choose δ which is a measure of the sum of all adsorbed concentrations, the surface coverage

$$\frac{1}{X_T} \sum_{j=1}^{N_0} C^*_j = 1 - \frac{1}{\delta} = \text{surface coverage} . \tag{C3}$$

Correspondingly, the Langmuir isotherm reads now

$$C^*_k(\delta) = X_T \frac{k_k \, c_k(\delta)}{\delta} , \tag{C4}$$

It is this structure of the isotherm that causes the waves to be linear, as will be shown now (see Eqs. C5–C13).

The N_0 derivatives dC^*/dc_j in the direction across the k-wave in composition space are

$$\frac{dC_j^*}{dc_j} = \frac{C_j^{*\prime}}{c_j'} = \frac{X_T k_j}{\delta^2} \left(\delta - \frac{c_j}{c_j'} \right) \ . \tag{C5}$$

where the prime denotes the derivative with respect to δ. Using Eq. C5 in Eq. C2 gives us

$$k_1 \left(\delta - \frac{c_1}{c_1'} \right) = k_2 \left(\delta - \frac{c_2}{c_2'} \right) = \ldots = k_{N_0} \left(\delta - \frac{c_{N_0}}{c_{N_0}'} \right) \ . \tag{C6}$$

Differentiating each term in the equality again with respect to δ gives for the j-term

$$\frac{d}{d\delta} k_j \left(\delta - \frac{c_j}{c_j'} \right) = k_j \left(1 - 1 + c_j \frac{c_j''}{c_j'^2} \right) = \frac{k_j c_j''}{\frac{1}{c_j} c_j'^2} \ .$$

With that in mind, Eq. C6 can be rearranged

$$\frac{k_1 c_1'}{\frac{1}{c_1} c_1'^2} = \frac{k_2 c_2'}{\frac{1}{c_2} c_2'^2} = \ldots = \frac{k_{N_0} c_{N_0}''}{\frac{1}{c_{N_0}} c_{N_0}'^2} \ . \tag{C7}$$

In equalities like $a_1/b_1 = a_2/b_2 = a_3/b_3$ the following equation holds: $a_i/b_i = \Sigma \, a_k/\Sigma \, b_k$, where the sum extends over $k = 1$ to 3 and i is 1, 2 or 3. This is applied to Eq. C7:

$$\frac{k_1 c_1'}{\frac{1}{c_1} c_1'^2} = \frac{k_2 c_2'}{\frac{1}{c_2} c_2'^2} = \ldots = \frac{k_{N_0} c_{N_0}''}{\frac{1}{c_{N_0}} c_{N_0}'^2} = \frac{\displaystyle\sum_{j=1}^{N_0} k_j c_j''}{\displaystyle\sum_{j=1}^{N_0} \frac{1}{c_j} c_j'^2} \ . \tag{C8}$$

From the definition of δ we know that

$$\sum_{j=1}^{N_0} k_j c_j'' = 0 \ . \tag{C9}$$

Therefore, any solution of Eq. C8 must meet the following condition

$$c_j'' = 0 \tag{C10}$$

or

$$\sum_{j=1}^{N_0} \frac{c_j'^2}{c_j} = 0 .$$ (C11)

Integration of Eq. C10 tells us that in each k-wave all components c_j of the concentration vector **c** are linear functions of the common parameter δ:

$$c_j'' = 0$$

i.e.

$$c_j' = a_j, \text{ or } \mathbf{c}' = \mathbf{a} .$$ (C12)

with $\mathbf{a} = \{a_1, a_2,..., a_{N_0}\}$. Because – for any wave, e.g. the k-wave – the a_j are integration constants, **a** is independent of **c**, i.e. independent from the location on the k-wave. Consequently the k-wave is a (straight) line in composition space.

Comparing Eq. C12 with Eq. 33 one can see that **a** is the right eigenvector \mathbf{r}_k corresponding to the k-wave. Thus, the vector describing the change of **c** across the k-wave

$$\mathbf{c}' = \mathbf{r}_k ,$$

is independent of **c**.

Integration of Eq. C12 will give us the k-wave. With $\delta_{k-1} - \delta_k$ denoting the change of the parameter δ across the k-wave and with the nomenclature introduced in Fig. 11.1, we get the following expression for the solution of the Riemann problem

$$\mathbf{c}_k = \mathbf{c}_{k-1} + \mathbf{r}_k (\delta_{k-1} - \delta_k), \quad \text{with } \mathbf{r}_k \text{ being independent of } \mathbf{c} .$$ (C13)

Equation C11 gives another condition for the slopes c_j' when $c_j'' = 0$. In a 2-component-system, Eq. C11 gives the equation of a curve $c_2(c_1)$ having as tangents the lines which meet (Eq. C13; see also Appendix D).

It will now be shown that *k-shocks* coincide with k-rarefaction waves (Eq. C13). The concentration changes across a k-shock by a finite amount rather than by differential amounts as across a rarefaction wave. Accordingly, the mass conservation law involves a differences rather than differentials. Otherwise the conservation laws are the same as for rarefaction waves (i.e. Eq. 3)

$$\frac{\Delta}{\Delta t} \{\phi c_i(x,t) + C_i^*(c_1 c_2, ..., c_{N_0})\} + v_w \frac{\Delta}{\Delta x} c_i(x,t) = 0 \quad (i = 1, 2, ..., N_0) \quad (C14)$$

or after dividing by ϕ and multiplying with $\Delta\xi$

$$\Delta\left(c_i + \frac{1}{\phi}C_i^*\right)\frac{\Delta\xi}{\Delta t} - \frac{v_w}{\phi}\Delta c_i = 0 .$$

Solving for $\Delta\xi/\Delta t=\sigma_k$, the speed of the k-shock,

$$\sigma_k = \frac{V_w}{\phi}\,\frac{\Delta c_i}{\Delta(c_i + \frac{1}{\phi}C_i^*)} = \frac{\dfrac{V_w}{\phi}}{1 + \dfrac{1}{\phi}\dfrac{\Delta C_i^*}{\Delta c_i}}, \qquad (i = 1, 2, ..., N_0)\,. \tag{C15}$$

Equation C15 is the so-called *Rankine-Hugoniot relation*. It has N_0 physical solutions (Lax 1973).

The speed of the k-shock is by a factor R_k smaller than the speed of the water. R_k is called *retardation of the k-shock*.

$$R_k = 1 + \frac{1}{\phi}\frac{\Delta_k C_i^*}{\Delta_k c_i} \tag{C16}$$

The Rankine-Hugoniot relation (integral coherence condition) can be written in a form analogous to Eq. C2:

$$\frac{\Delta C_1^*}{\Delta c_1} = \frac{\Delta C_2^*}{\Delta c_2} = ..., = \frac{\Delta C_{N_0}^*}{\Delta c_{N_0}} \quad \text{for each k-shock (k = 1, 2, ..., } N_0)\,. \tag{C17}$$

As was shown in (Eqs. C5–C13), the relationship (Eq. C4) is satisfied when the concentrations c_j are linear functions of a parameter δ. Writing $c_j(\delta)$ as

$$c_j = a_j\,\delta + b_j\,, \tag{C18}$$

we get for the differential coherence condition

$$\frac{dC_j^*}{dc_j} = \frac{\dfrac{dC_j^*}{d\delta}}{\dfrac{dc_j}{d\delta}} = -X_{Tkj}\,\frac{b_j}{a_j\delta^2}\,, \tag{C19}$$

and for the integral coherence condition

$$\frac{\Delta C_j^*}{\Delta c_j} = -X_{Tkj}\,\frac{b_j}{a_j\,\delta(\delta + \Delta)}\,, \tag{C20}$$

where δ and $\delta + \Delta$ are the values of the parameter left and right of the discontinuity.

Replacing $\Delta C^*_j/\Delta c_j$ in the integral coherence condition with the value given in Eq. C20 results in

$$- X_T k_1 \frac{b_1}{a_1 \delta (\delta + \Delta)} = - X_T k_2 \frac{b_2}{a_2 \delta (\delta + \Delta)} = \ldots = - X_T k_{N_0} \frac{b_{N_0}}{a_{N_0} \delta (\delta + \Delta)} \qquad (C21)$$

The differential coherence condition (Eq. C2) with dC^*_j/dc_j replaced with Eq. C19 is

$$- X_T k_1 \frac{b_1}{a_1 \delta^2} = - X_T k_2 \frac{b_2}{a_2 \delta^2} = \ldots = - X_T k_{N_0} \frac{b_{N_0}}{a_{N_0} \delta^2} . \qquad (C22)$$

After having cancelled common factors in Eqs. C21 and C22, we get the same condition for the a_i, b_i and k_i:

$$k_1 \frac{b_1}{a_1} = k_2 \frac{b_2}{a_2} = \ldots = k_{N_0} \frac{b_{N_0}}{a_{N_0}} . \qquad (C23)$$

Thus the same line (Eq. C18) satisfies the differential as well as the integral coherence condition. In other words: k-shocks coincide with k-rarefaction waves.

Appendix D
Waves in a Two-Component System
with Langmuir Isotherm

Using c_j and C_j^* to denote soluble and adsorbed concentrations as functions of ξ, the equilibrium transport equation (Eq. 3) reads

$$(\frac{v_w}{\phi} - \xi) \frac{dc_j}{d\xi} - \frac{\xi}{\phi} \frac{dC_j^*}{d\xi} = 0 \qquad (j = 1, 2).$$ (D1)

The corresponding differential coherence condition (Eq. 15) is

$$\frac{dC_2^*}{dc_2} - \frac{dC_1^*}{dc_1} = 0.$$ (D2)

Writing the directional derivatives in Eq. D2 in terms of partial derivatives in composition space

$$\frac{dC_j^*}{dc_j} = \frac{\partial C_j^*}{\partial c_1} + \frac{\partial C_j^*}{\partial c_2} \frac{dc_2}{dc_1}$$

and multiplying with the derivative dc_1/dc_2 gives us an expression equivalent to (Eq. 19)

$$\frac{\partial C_2^*}{\partial c_1} (\frac{dc_1}{dc_2})^2 - (\frac{\partial C_1^*}{\partial c_1} - \frac{\partial C_2^*}{\partial c_2}) \frac{dc_1}{dc_2} - \frac{\partial C_1^*}{\partial c_2} = 0.$$ (D3)

At this point the specific form of the isotherm is introduced:

$$C_j^* = \frac{\Gamma_j c_j}{1 + B_1 c_1 + B_2 c_2} = \frac{\Gamma_j c_j}{\delta},$$ (D4)

i.e. for the 2-component Langmuir isotherm

$$\Gamma_j = X_T\, k_j,\; B_j = k_j \tag{D5}$$

For ion exchange in a 3-species system with constant total concentration $c = c_1 + c_2 + c_3$

$$C_j^*(c) = X_T \frac{\dfrac{\alpha_{j3}}{c}\, c_j}{\dfrac{\alpha_{13}-1}{c}\, c_1 + \dfrac{\alpha_{23}-1}{c}\, c_2 + 1}$$

we have

$$\Gamma_j = X_T \frac{\alpha_{j3}}{c},\; B_j = \frac{\alpha_{j3}-1}{c}\, . \tag{D6}$$

The partial derivative of the isotherm is then

$$\frac{\partial C_i^*}{\partial c_j} = \frac{\Gamma_i}{\delta}\left(\frac{B_j c_i}{-\delta} + \delta_{ij}\right), \tag{D7}$$

where the Kronecker symbol $\delta_{ij} = 0$ for $i \neq j$ and $\delta_{ii} = 1$.
 With Eq. D7, the Eq. D3 assumes the following form

$$\Gamma_2 B_1 c_2 \left(\frac{dc_1}{dc_2}\right)^2 + (\Gamma_1 B_2 c_2 + \Gamma_1 - \Gamma_2 B_1 c_1 - \Gamma_2)\frac{dc_1}{dc_2} - \Gamma_1 B_2 c_1 = 0\, . \tag{D8}$$

Dividing through by $\Gamma_2 B_1$ (note that $\Gamma_2 B_1 = \Gamma_1 B_2$) and abbreviating

$$k = \frac{1}{B_1}\left(1 - \frac{\Gamma_1}{\Gamma_2}\right); \; \bar{c}_j = \frac{c_j}{k} \tag{D9}$$

the differential Eq. D8 for $c_1(c_2)$ can be written as

$$\bar{c}_2\left(\frac{d\bar{c}_1}{d\bar{c}_2}\right)^2 - (1 + \bar{c}_1 - \bar{c}_2)\frac{d\bar{c}_1}{d\bar{c}_2} - \bar{c}_1 = 0\, . \tag{D10}$$

Glueckauf (1949) solved this equation by differentiating once more with respect to c_2

$$\frac{d^2\bar{c}_1}{d\bar{c}_2^2}\left\{2\bar{c}_2\frac{d\bar{c}_1}{d\bar{c}_2} - (1 + \bar{c}_1 - \bar{c}_2)\right\} = 0\, . \tag{D11}$$

This equation is satisfied by a set of lines

$$\frac{d^2c_2}{dc_1^2} = 0 \, ,$$

i.e.

$$c_2(c_1) = a \, c_1 + b \tag{D12}$$

that are tangential to the solution of the equation

$$2 \, \bar{c}_2 \frac{d\bar{c}_1}{d\bar{c}_2} = 1 + \bar{c}_1 - \bar{c}_2 \qquad . \tag{D13}$$

Using Eq. D13 in Eq. D10 gives

$$(1 + \bar{c}_1 - \bar{c}_2)^2 + 4\bar{c}_1\bar{c}_2 = 0 \, ,$$

which can be solved for c_2

$$\bar{c}_2 = 1 - \bar{c}_1 \pm 2\sqrt{-\bar{c}_1} \, . \tag{D14}$$

Equation D14 is a parabola that touches the axes at $\bar{c}_2 = 1$ and $\bar{c}_1 = -1$.

Appendix E
Solution of the Riemann Problem
with Riemann Invariants

The solution of the Riemann problem makes use of the fact that the waves are (straight) lines in composition space or – in other words – that the k-wave connecting constant states c_{k-1} and c_k adjacent to it is a line parallel to the right eigenvector r_k of the Jacobi matrix of the isotherms, \underline{Kd} (see Eqs. 33, C12, C13). With ~ denoting proportionality

$$c_k - c_{k-1} \sim r_k \tag{E1}$$

with

$$\underline{Kd}(c)\, r_k(c) = kd_k(c)\, r_k(c) \tag{E2}$$

and

$$r_k(c) \text{ being constant across the k-wave} . \tag{E3}$$

The Jacobian $\underline{Kd}(c)$ has been defined in Eq. 29.

In addition to Eqs. E1–E3 the solution uses only the biorthogonality of the left and right eigenvectors

$$r_k(c)\, l_j(c) = 0, \qquad (j \neq k) , \tag{E4}$$

where

$$l_k(c)\, \underline{Kd}(c) = kd_k(c)\, l_k(c). \tag{E5}$$

We will need the following theorem in this Appendix:

Theorem. The left eigenvector corresponding to the eigenvalue $kd_j(c_k)$ can be calculated from the Jacobians $\underline{Kd}(c_-)$ and $\underline{Kd}(c_+)$ as follows

$$l_j(c_k) = \begin{cases} l_j^+, & (j \le k) \\ l_j^-, & (j > k) \end{cases}. \tag{E6}$$

with $l_j^- = l_j(c_-)$ and $l_j^+ = l_j(c_+)$.

Proof. Because r_k is constant across the k-wave,

$$r_k(c_{k-1}) = r_k(c_k) \tag{E7}$$

so are (or can be chosen) the left eigenvectors that are orthogonal to r_k

$$l_j(c_{k-1}) = l_j(c_k), \quad (j \ne k). \tag{E8}$$

Then the left eigenvectors in the fan of constant states c_-, c_1, c_2,..., c_+ (see Fig. 11.1) are:

$$l_1(c_-) = l_1^-, \ l_2(c_-) = l_2^-, \ ..., \ l_{N_0}(c_-) = l_{N_0}^-$$

$$l_1(c_1) = l_1^+, \ l_2(c_1) = l_2^-, \ ..., \ l_{N_0}(c_1) = l_{N_0}^-$$

$$l_1(c_2) = l_1^+, \ l_2(c_2) = l_2^+, \ l_3(c_2) = l_3^-, \ ..., \ l_{N_0}(c_2) = l_{N_0}^-$$

$$l_1(c_+) = l_1^+, \ ..., \ l_{N_0-1}(c_+) = l_{N_0-1}^+, \ l_{N_0}(c_+) = l_{N_0}^+. \tag{E9}$$

This can be written in the compact form (Eq. E6), q.e.d.

Let $\underline{L}(c_k)$ be the matrix the rows of which are the left eigenvectors corresponding to the eigenvalues of the Jacobian $\underline{Kd}(c_k)$, $kd_1(c_k)$,..., $kd_{N_0}(c_k)$

$$\underline{L}(c_k) = \begin{bmatrix} l_1(c_k) \\ \\ l_{N_0}(c_k) \end{bmatrix}.$$

The j-th row in this matrix is either l_j^+ or l_j^- depending on c_k (see Eq. E9). So, $\underline{L}(c_k)$ is a known matrix.

The k-Riemann invariant is defined as the scalar function of the concentration vector that changes only across the k-wave and remains unchanged across all other waves.

Definition. Let ∇_c denote the gradient in concentration space, then the scalar function $w_j(\mathbf{c})$ is the j-Riemann invariant if

$$\mathbf{r}_k \nabla_c w_j = 0 \qquad (j = 1, 2, ..., k-1, k+1, ..., N_0) . \tag{E10}$$

Thus, both the left eigenvector \mathbf{l}_k and the k-Riemann invariant have the property that they change only across the k-wave. So, for the Riemann invariants there exists a relationship similar to Eq. E6

$$w_j(\mathbf{c}_k) = \begin{cases} w_j^+, & (j \le k) \\ w_j^-, & (j > k) \end{cases} \tag{E11}$$

with $w_j^+ = w_j(\mathbf{c}+)$ and $w_j- = w_j(\mathbf{c}-)$. The j-th component of the vector

$$\mathbf{w}(\mathbf{c}_k) = \begin{bmatrix} w_1(\mathbf{c}_k) \\ ... \\ w_{N_0}(\mathbf{c}_k) \end{bmatrix}$$

is thus either w_j^+ or w_j^-, depending on at what \mathbf{c}_k it is evaluated. Like the matrix $\mathbf{L}(\mathbf{c}_k)$, the vector $\mathbf{w}(\mathbf{c}_k)$ is known.

The Definition (Eq. E10) does not define a unique Riemann invariant. The following theorem gives us a convenient choice.

Theorem. The scalar product of the left eigenvector and the concentration vector \mathbf{c} evaluated at the k-th constant state

$$w_j(\mathbf{c}_k) = \mathbf{l}_j(\mathbf{c}_k)\,\mathbf{c}_k , \qquad (j, k = 1, 2, ..., N_0, j \ne k) \tag{E12}$$

is a k-Riemann invariant.

Proof.

$$\mathbf{r}_k \nabla_c w_j = \mathbf{r}_k (\nabla_c \mathbf{l}_j\, \mathbf{c} + \mathbf{l}_j\, \nabla_c \mathbf{c}) = \mathbf{r}_k\, \nabla_c \mathbf{l}_j\, \mathbf{c} + \mathbf{r}_k \mathbf{l}_j , \qquad (j \ne k) . \tag{E13}$$

Because of Eq. E8 $\mathbf{r}_k \nabla_c \mathbf{l}_j$ in the first term of Eq. E13, the derivative of \mathbf{l}_j in direction of \mathbf{r}_k, is zero:

$$\mathbf{r}_k \nabla_c \mathbf{l}_j = 0 , \qquad (j \ne k) . \tag{E14}$$

By Eq. E4 the last term in Eq. E13 is zero. Thus,

$$\mathbf{r}_k \nabla_c w_j = 0 \qquad (j \ne k) ,$$

which is the definition of the k- Riemann invariant, q.e.d.

The solution of the Riemann problem is now straightforward. Equation E12 can be written in matrix form

$$\mathbf{w}(\mathbf{c}_k) = \underline{L}(\mathbf{c}_k)\, \mathbf{c}_k \,, \tag{E15}$$

where $\mathbf{w}(\mathbf{c}_k)$ and $\underline{L}(\mathbf{c}_k)$ are known for all \mathbf{c}_k. Vector \mathbf{c}_k is the unknown. Inverting Eq. E15

$$\underline{L}^{-1}(\mathbf{c}_k)\, \mathbf{w}(\mathbf{c}_k) = \mathbf{c}_k \tag{E16}$$

Equation E16 gives the solution of the Riemann problem.

Glossary of Notation

The first parentheses give units. The sign (-) means dimensionless. The last parentheses indicate where in Chapter 11 the symbol is defined or first used.

α index specifying site type, (Eq. 64)

α_{iN} constant for A_j - A_N exchange, (-), (Eq. 37)

c vector of soluble concentration $\{M, H_t\}$, Eq. 6)

$c_j(x, t)$ soluble concentration of component A_j, (mol/l of liquid), (Eq. 1) Index j is omitted in 1-component system, (Eq. A1).

c_k vector of soluble concentrations in constant state between k- and (k+1)-wave, (Fig. 11.1 and Eq. E1).

$c_\infty(x, t)$ time asymptotic soluble concentration profile, (mol/l of liquid), (Eq. A3).

c_+ soluble concentration vector right of concentration jump at t = 0, (mol/l of liquid), (Eq. 4). Scalar c_+ is used for the 1-component system (Eq. A6).

c_- soluble concentration vector left of concentration jump at t=0, (mol/l of liquid), (Eq. 4). Scalar c_- is used for the 1-component system (Eq. A6).

$C_j(x, t)$ concentration of component A_j on the surface, (mol/l of system volume), (Eq. 1). Index j is omitted in 1-component system, (Eq. A1).

$C_j^*(c)$ equilibrium concentration of component A_j on the surface, (mol/l of system volume), (Eq. 2), except in Eqs. 37–99, where $C_j^*(c)$ means the concentration of a particular surface species. Concentration of component A_j on surface is sum of the concentrations of the surface species containing A_j. Index j is omitted in 1-component system, (Eq. A2).

D lumped hydrodynamic dispersion/molecular diffusion coefficient, (m²/a), (Eq. 1).

D^* coefficient of molecular diffusion in water (0.16 m²/a), (Eq. A10).

δ dispersivity, (m), $\delta \approx L/100$, (Eq. A10).

$\delta^\alpha(c)$ measure of sum of concentrations adsorbed on site type α, (-), (Eq. 64). α is omitted if there is only one site type, (Eqs. 101 and C3).

∇_c $(\partial/\partial c_1, \partial/\partial c_2, ..., \partial/\partial c_{N_0}) = \text{grad}$, (Eq. E10).

ε_j relaxation time for component A_j, (year), (Eq. 2). It is the time during which 37% of the equilibrium adsorbed concentration is reached. Index j is omitted in 1-component system, (Eq. A2).

$\mathbf{h}(\xi)$	vector of soluble concentrations $h_j(\xi)$, (Eqs. 5 and 8).
$h_j(\xi)$	component j of $\mathbf{h}(\xi)$, $h_j(\xi)$ is concentration $c_j(x, t)$ of A_j in soluble phase as a function of the speed $\xi = x/t$ of the concentration, (mol/l of liquid), (Eq. 5).
$H(\xi)$	vector of adsorbed concentrations $H_j(\xi)$, (Eqs. 7 and 8).
$\underline{H}(M, H)$	adsorbed proton concentration, (mol/l of system volume), (Eq. 73).
$H_j(\xi)$	component j of $H(\xi)$, $H_j(\xi)$ is adsorbed concentration $C^*_j(x, t)$ of component A_j as a function of the speed $\xi = x/t$ of the concentration, (mol/l of system volume), (Eqs. 7 and 8).
$H_t(\xi)$	total proton concentration in solution as a function of the speed $\xi = x/t$, (mol/l of liquid), (Eq. 87).
H_+	proton activity in pre-equilibrant (pore water before arrival of abrupt concentration change, see Eq. 4). $H_+ = H(x \geq 0, t = 0)$, (mol/l of liquid), (Eq. 85).
H_-	proton activity in feed (see Eq. 4), $H_- = H(x < 0, t = 0)$, (mol/l of liquid), (Eq. 88).
H_{t-}	$H_t(x < 0, t=0)$, (mol/l of liquid), (Eq. 86)
HFO	concentration of hydrous ferric oxide, Fe_2O_3, (mol/l of system volume), (Eq. B38).
j	index specifying chemical component A_j, (-), (Eq. 1).
I	ionic strength, (mol/l), paragraph after (Eq. 56).
\underline{I}	identity matrix, (Eq. 29).
κ, κ_j	Langmuir adsorption constants, (mol/l), (Eq. B21)
k	index specifying wave, the slowest wave has $k = 1$, (Fig. 11.1, Eq. 13).
k	threshold concentration of remobilizing component for accumulation of contaminant, (mol/l of liquid), (Eqs. 96 and 98).
k	1-component Langmuir adsorption constant, $(mol/l)^{-1}$, (Eq. B21).
k_1	equilibrium constant for formation of XOH_2 surface complex from XOH and H, (l/mol), (Eq. 49).
k_2	equilibrium constant for formation of XO surface complex from XOH and H, (mol/l), (Eq. 50).
$\underline{Kd}(c)$	Jacobi matrix of isotherms evaluated at location c in composition space, (-), (Eq. 29).
K_w	water dissociation constant, $(10^{-14} mol^2/l^2)$, (Eq. 87).
$l_k(c)$	left eigenvector of $\underline{R}(c)$ corresponding to eigenvalue $\rho_k(c)$, (Eq. 68).
L	length of column of porous medium, (m), (paragraph after Eq. 1, also Fig. A1).
$\underline{L}(c)$	matrix of left eigenvectors $l_k(c)$, (Eq. 69).
$M(\xi)$	soluble concentration of metal, (mol/l of liquid), (Eq. 77).
$\underline{M}(M, H)$	adsorbed metal concentration, (mol/l of system volume), (Eq. 74).

M_+	$M(x \geq 0, t = 0)$, (Eq. 83).
M_-	$M(x < 0, t = 0)$, (Eq. 86).
N_0	number of chemical components, (-), (Eq. 1).
P	electrostatic correction factor, (-), (Eqs. 42 and 57).
PZC	point of zero charge on hydrous ferric oxide in pristine water, i.e. at $I \leq 0.0001$ mol/l, (Eq. 57).
ϕ	porosity of medium, (-), (Eq. 1).
$\mathbf{r(h)}$	right eigenvector of \underline{R} evaluated at $\mathbf{c} = \mathbf{h}$, (Eq. 33).
$\mathbf{r_k(c)}$	right eigenvector of $\underline{R}(\mathbf{c})$ corresponding to eigenvalue $\rho_k(\mathbf{c})$, (Eq. 33).
$\underline{R}(\mathbf{c})$	retardation matrix evaluated at location \mathbf{c} in concentration space, (Eq. 30). The matrix degenerates to the retardation of the traveling wave, R(c), in the 1-component system, (Eq. A6).
$R_k(\mathbf{c})$	retardation of k-rarefaction wave, (-), (Eq. 23) and k-shock, (Eq. 24). R_k of k-rarefaction wave is equal to the kth eigenvalue ρ_k of \underline{r}, but \underline{R}_k of k-shock is kth root of Rankine-Hugoniot relation (Eq. 24) or (Eq. 26).
$\rho(\mathbf{c})$	eigenvalue of $\underline{R}(\mathbf{c})$, also called "retardation", (-), (Eq. 32).
$\rho_k(\mathbf{c})$	retardation of k-rarefaction wave, (-), (Eq. 34).
s	$x - \sigma t$, (m), (Eq. A4).
σ	speed of traveling wave, (m/a), (Eq. A5).
σ_k	speed of k-shock, (m/a), (Eq. 24).
t	time variable, (a), (Eq. 1).
τ	tortuosity, (-), (Eq. A10).
v_w	flux of water, (m/a), (Eq. 1).
v_0	speed of water, (m/a), (Eq. B38).
$w_j(\mathbf{c_k})$	k-Riemann invariant evaluated at $\mathbf{c_k}$, (-), (Eqs. 65 and 67).
$\mathbf{w(c_k)}$	vector of N_0 Riemann invariants evaluated at $\mathbf{c_k}$, (-), (Eq. 70).
x	spatial variable (m), (Eq. 1).
X_T^α	concentration of adsorption sites of type α, (mol/l of system volume), (Eq. 64). Index α is omitted when there is only one site type, (Eq. 39).
XO	concentration of empty sites, (mol/l of system), (Eq. 50).
XOH	concentration of sites covered with one proton, (mol/l of system), (Eq. 49).
XOH_2	concentration of sites covered with two protons, (mol/l of system), (Eq. 49).
XOHM	concentration of sites covered with M via a proton bridge, (mol/l of system), (Eqs. 44 and 59).
XOM	concentration of sites covered with metal, (mol/l of system), (Eq. 51).
ξ	speed of concentration \mathbf{c}, (m/a), (Eq. 5).
ξ_k	speed of k-rarefaction wave, (m/a), (Eq. 13).

References

Bear J (1979) Hydraulics of groundwater. McGraw Hill Book Co, New York

Berninger JA, Whitley RD, Zhang X, Wang N-H L (1991) A versatile model for simulation of reaction and nonequilibrium dynamics in multicomponent fixed-bed adsorption processes. Computers Chem Engng 15(11):749–768

Charbeneau RJ (1988) Multicomponent exchange and subsurface solute transport: characteristics, coherence, and the Riemann problem. Water Resour Res 24:57–64

Cederberg GA, Street RL, Leckie JO (1985) A groundwater mass transport and equilibrium chemistry model for multicomponent systems. Water Resour Res 21:1095–1104

Dzombak DA, Morel FMM (1990) Surface complexation modeling: hydrous ferric oxide. John Wiley, New York

Glueckauf E (1955) Ion exchange and its applications. Soc. of Chemical Industry 34, London

Harwell J (1992) Fate of organic pollutants in immiscible non-aqueous solvents, In: Petruzzelli D, Helfferich FG (eds) Migration and fate of pollutants in soils and subsoils. Proceedings of the NATO-Advanced Study Institute Summer School, Aquafredda di Maratea, Italy, May 24–June 6

Helfferich F (1967) Multicomponent ion-exchange in fixed beds. Ind Eng Chem Fundam 6:362–364

Helfferich FG (1992) Multicomponent wave propapation: the coherence principle, In: Petruzzelli D, Helfferich FG (eds) Migration and fate of pollutants in soils and subsoils. Proceedings of the NATO-Advanced Study Institute Summer School, Aquafredda di Maratea, Italy, May 24–June 6

Helfferich FG, Bennett BJ (1984) Weak electrolytes, polybasic acids, and buffers in anion exchange columns, I: Sodium acetate and sodium carbonate systems. Reactive polymers 3:51–66. II: Sodium acetate chloride system, solvent extraction and ion exchange 2:1151–1184

Helfferich F, Klein G (1970) Multicomponent chromatography – theory of interference. Marcel Dekker, New York

Hunt JR, Sitar N, Udell KS (1988) Nonaqueous phase liquid transport and cleanup, 1: Analysis of mechanisms. Water Resour Res 24:1247–1258

Hwang Y-L, Helfferich FG, Leu R-J (1988) Multicomponent equilibrium theory for ion-exchange columns involving reactions. AIChE J 34:1615–1626

International AVS Center, North Carolina Supercomputing Center, PO Box 12889, 3021 Cornwallis Road, Research Triangle Park, North Carolina, NC 27709, telnet: avsncscorg

Kinzelbach W, Schäfer W, Herzer J (1989) Numerical modeling of nitrate transport in a natural aquifer. In: Kobus H, Kinzelbach W (eds) Contaminant transport in groundwater. Balkema, Rotterdam, pp 191–198

Kirkner DJ, Reeves H (1988) Multicomponent mass transport with homogeneous and heterogeneous chemical reactions: the effect of the chemistry on the choice of numerical algorithm, Part 1: Theory. Water Resour Res 24:1719–1729

Lake LW, Helfferich F (1978) Cation exchange in chemical flooding, 2: The effect of dispersion, cation exchange, and polymer/surfactant adsorption on chemical flood environment. Soc Pet Eng J 18:435–444

Lax PD (1957) Hyperbolic systems of conservation laws II. Comm Pure Appl Math 10:537–566

Lax PD (1973) Hyperbolic systems of conservation laws and the mathematical theory of shock waves. Society of Industrial and Applied Mathematics, Philadelphia, PA 19103

Lichtner PC (1992) Time-space continuum description of fluid/rock interaction in permeable media. Water Resour Res 28:3135–3155

Liu T-P (1987) Hyperbolic conservation laws with relaxation. Commun Math Phys 108:153–175

Novak CF, Schechter RS, Lake LW (1988) Rule-based mineral sequences in geochemical flow processes. AIChE Journal 34:1607–1614

Ortoleva P, Merino E, Moore G, Chadam J (1987) Geochemical self-organization, I: Reaction-transport feedbacks and modeling approach. Am J Sci 287:979–1007

Pfann W (1958) Zone melting. John Wiley, New York

Reeves H, Kirkner DJ (1988) Multicomponent mass transport with homogeneous and heterogeneous chemical reactions: the effect of the chemistry on the choice of numerical algorithm, part II: Numerical results. Water Resour Res 24:1730–1739

Rhee H-K, Aris R, Amundson NR (1989) First-order partial differential equations, Vol. II: Theory and application of hyperbolic systems of quasilinear equations. Prentice Hall, Englewood Cliffs, NJ 07632

Schweich D, Sardin M, Jauzein M (1993) Properties of concentration waves in presence of nonlinear sorption, precipitation/dissolution, and homogeneous reactions, 1: Fundamentals, 2: Illustrative examples. Water Resour Res 29:723–741

Stumm W, Morgan JJ (1981) Aquatic chemistry (2nd ed). John Wiley

Temple B (1983) Systems of conservation laws with invariant submanifolds. Trans Amer Math Soc 2:781–795

Vogt M (1990) Ein vektorrechnerorientiertes Verfahren zur Berechnung von großräumigen Multikomponenten-Transport-Reaktions-Mechanismen im Grundwasserleiter. Dissertation, University of Karlsruhe, VDI Verlag

Walsh MP, Bryant SL, Schechter RS, Lake LW (1984) Precipitation and dissolution of solids attending flow through porous media. AIChE J 30:317–328

Wolfram Research Inc (1992) Mathematica®, Version 2.2, Wolfram Research Inc, 100 Trade Center Drive, Champaign, Illinois, IL 61820–7237, USA E-Mail: info@wricom

Zee SEATM van der (1990) Analytical traveling wave solution for transport with nonlinear nonequilibrium adsorption. Water Resour Res 26:2563–2577

Zielke C (1993) Transport von spezifisch adsorbierenden Schwermetall-Kationen: Anwendung eines chromatographischen Modells zur Beschreibung und Interpretation von Fronten in binären Systemen. Diplomarbeit, Technische Universität Hamburg-Harburg, Arbeitsbereich Umweltschutztechnik

Subject Index

Accumulation 219, 226
– intracellular, of metals 143
– intracellular, of organic pollutants 145
– of lipophilic substances 142
Acetylcholinesterase 184
– inhibition constant 185
– Paraoxon equivalents 185
Acid consuming capacity (ACC) 7–11
Acid producing capacity (APC) 7–12
Acid volatile sulphides (AVS)
– normalization 6, 7
S-adenosylmethionine 228
Adsorption
– constant 282
– equilibrium 61
– isotherms 64
– kinetic 57, 58
– matrix 61
– of colloidal particles 100–104
– of phenols on coated surfaces 96
– of phenols on mineral surfaces 94, 95
– of polymers 97–99
– rate 59
– sites 277
– theory 92, 93
Aluminium boats 249
Ammonia 28
Anions
– binding 135
– binding mechanisms 136
Anthropogenic source 250, 251, 254, 255, 258
Antifouling paints 245
Application visualization system (AVS) 282
As 198–203, 207, 211
Atomic absorption spectrometry (AAS) 39

Atrazine
– adsorption isotherm 64, 68, 70
– adsorption kinetic 58
– analysis 56
– desorption 63
Available sulphide capacity (ASC) 9

Bacillus licheniformis 139
Bacillus subtilis 138
Bacteria 222, 226, 231
– sulphate reducing 128
– undergoing starvation 163, 165, 168
– viable counts 163–165, 167–169
Benthos 15, 19, 25
Binding
– active 125
– metals on cell walls 138
– passive 125
Binding capacity
– gram negative cells 140
– gram positive cells 140
Binding proteins 1 60–161
Bioaccumulation 247, 251
– of pesticides 142
– of silver 144
Bioavailability
– metals 6
– organic chemicals 14, 15
Biocide 246, 250
Bioconcentration factor (BCF) 22
Biocorrosion 223
Biofilms 115, 164
– as a gel matrix 124, 125
– as nucleation sites 126
– sorption capacity 125, 127, 147
– sorption sites in 131
– "stickiness" 123
Bioindicator 247

Springer-Verlag
and the Environment

We at Springer-Verlag firmly believe that an international science publisher has a special obligation to the environment, and our corporate policies consistently reflect this conviction.

We also expect our business partners – paper mills, printers, packaging manufacturers, etc. – to commit themselves to using environmentally friendly materials and production processes.

The paper in this book is made from low- or no-chlorine pulp and is acid free, in conformance with international standards for paper permanency.